AF517890

Postharvest Management and Processing Technology

Postharvest Management and Processing Technology

(Cereals, Pulses, Oilseeds, Fruits and Vegetables)

U. D. Chavan
Senior Cereal Food Technologist
Department of Food Science and Technology
Mahatma Phule Krishi Vidyapeeth, Rahuri – 413 722
Dist: Ahmednagar, Maharashtra

V.D. Pawar
Principal
Rajiv Gandhi, College of Food Technology
Marathwada Agricultural University Parbhani
Dist: Parbhani, Maharashtra

2014
DAYA PUBLISHING HOUSE®
A Division of
Astral International Pvt. Ltd.
New Delhi - 110 002

ISBN 978-81-7035-787-2 (Hardbound)
ISBN 978-93-5124-157-7 (International Edition)
Reprinted: 2014
First Published: 2012

Publisher's Note:

Every possible effort has been made to ensure that the information contained in this book is accurate at the time of going to press, and the publisher and author cannot accept responsibility for any errors or omissions, however caused. No responsibility for loss or damage occasioned to any person acting, or refraining from action, as a result of the material in this publication can be accepted by the editor, the publisher or the author. The Publisher is not associated with any product or vendor mentioned in the book. The contents of this work are intended to further general scientific research, understanding and discussion only. Readers should consult with a specialist where appropriate.

Every effort has been made to trace the owners of copyright material used in this book, if any. The author and the publisher will be grateful for any omission brought to their notice for acknowledgement in the future editions of the book.

Published by : **Daya Publishing House®**
A Division of
Astral International Pvt. Ltd.
81, Darya Ganj, Near Hindi Park,
Delhi Medical Association Road,
New Delhi - 110 002
Phone: 011-4354 9197, 2327 8134
Fax: +91-11-2324 3060
E-mail: info@astralint.com
Website: www.astralint.com

Laser Typesetting : **Classic Computer Services**
Delhi - 110 035

Printed at : **Salasar Imaging Systems**
Delhi - 110 035

PRINTED IN INDIA

Acknowledgement

Authors are thankful to the Hon. Dr. T.A. More, Vice-Chancellor, Mahatma Phule Krishi Vidyapeeth, Rahuri for his critical and fruitful suggestions while the preparation of *"Postharvest Management and Processing Technology: Cereals, Pulses, Oilseeds, Fruit and Vegetables"* book. We are thankful to those scientists who worked on development of various processing technology, value-added products from cereals, pulses, oilseeds, fruits and vegetables and their contribution is acknowledged in this book as well their sources are cited in the reference section. Authors are also thankful to all scientists and friends those helped directly or indirectly while preparation of this manuscript. We would like to express over profound thanks to our family members for their constant support, patience and devotion, which inspired us to complete this manuscript.

U.D. Chavan

V.D. Pawar

Foreword

Latest advances are fast transforming postharvest management from a means of subsistence to an organized industry. This has stimulated a great deal of interest among widely diversified group of scientists. In this era of high yielding crop varieties and multiple cropping techniques, global competition in the quality of food products has put every country on the alert to improve upon postharvest techniques and value addition sector. In India, the processing of only 2 per cent of the total produce of fruits and vegetables is at present undertaken as compared to 70-80 per cent in the developed countries. However the Government of India has fixed a target of 10 per cent of such produce to be processed at the earliest. Also, the actual producer is not getting his due share from the prevailing market prices.

The Indian Council of Agricultural Research and the State Agricultural Universities have made concentrated efforts in increasing productivity of fruits and vegetables. In recent years, the Government of Maharashtra has launched an ambitious Horticulture Development Programme to increase area under fruit crops by providing grant-in-aid to the growers. It is envisaged that the area covered under fruit crops in Maharashtra would be 10 lakh hectares in the next few years. However, very little attention has been paid to reduce the postharvest losses that occur due to poor handling, storage and transportation. According to reports of Swaminathan Committee, postharvest losses of fruits and vegetables in India vary from 20-30 per cent of production with estimated value of Rs. 1500-2000 crores annually. If these losses are controlled by using improved postharvest technology, the supplies of fresh fruits and vegetables will be increased to the extent of their existing losses. This will help to increase the per capita availability of fruits and vegetables.

Recent developments in nutrition have created greater demands for fresh fruits and vegetables. The processing technology that increases the functionality without changing their fresh like properties has significant role in modern fruit and vegetable industry. This book addresses factors that are involved in maintaining the quality of fresh fruits and vegetables after harvest. It is vital that consumers in health-conscious society would be provided with food products which retain their quality characteristics leading to more healthy individuals.

Fruits and vegetables, being perishable, are processed into several products such as jams, jellies, marmalade, syrups, squashes and canned, dehydrated and fermented products by employing different methods of processing. This book incorporates information on principles involved in preparation of various products as well as methodology employed in home scale as well as industrial processing of fruits and vegetables. I hope, this information will be useful for students in the field of horticulture, food science and nutrition and professionals in food industries.

The ever-increasing scientific research on the subject has resulted in a large body of knowledge which can be found scattered in various scientific journals and technical reports and is not easily accessible. The available material needs to be assembled and briefed in order that the students/technologists/agriculturists/horticulturists/engineers and others actively engaged in quantitative studies can have it as a ready reference. This book provides a comprehensive but concise treatise on the subject matter.

J. K. Chavan

Ex. Head,

Food Science and Technology,

MPKV, Rahuri

Preface

Processing of agricultural produce and their postharvest management is an emerging area in the present scenario encompassing a widely diversified group of scientists. India has now attained the stage where apart from enhanced production of crops, emphasis is being laid on processing and value addition of cereals, vegetables and food crops. However, the utilization of fruits and vegetables for processing in the organized and un-organized sectors is around 2 per cent of the total production. Given the huge potential for preparation of processed food products and limitation of the domestic market in consuming these, it is imperative that concerted efforts are made to export processed food for which the private, public and co-operative sectors, all have to play their rightful role. Value addition is not only a science, it is an art too.

Fruits and vegetables are important constituents of our diet which provide dietary fibers, vitamins, and minerals in addition to proteins, carbohydrates and a small amount of lipids. These contribute more than 50 per cent requirement of vitamin A. The dietary fibers aid in digestion, bowel movement and utilization of more concentrated foods in human diet. Fruits and vegetables are highly perishable commodities. Losses to the extent of 20-30 per cent occur during postharvest handling of fruits and vegetables. If these losses are controlled by using improved postharvest technology, the supplies of fresh fruits and vegetables can be increased to the extent of their existing losses. This will certainly help to increase per capita availability of fruits and vegetables.

The postharvest technology of fruits and vegetables includes harvesting, sorting, grading, packaging, transportation, storage and processing. The consumption of fruits and vegetables in fresh, unprocessed form is economically cheaper and nutritionally more desirable. The minimal processing involves treatment to these commodities without changing wholesomeness and nutritional properties of fresh produce. It includes sorting, grading and packaging of fresh produce so that consumer can get processed product having physical, nutritional and aesthetic properties similar to that of fresh produce. In recent years, there are significant developments in canning, freezing and dehydration of fruits and vegetables. Novel product in tetra pack like Frooti has captured the Indian market as ready-to-serve beverage. Similarly, canned corn and canned mixed beans have become popular in Indian market. Ketchup and sauces are also produced and marketed under various trade names. The processed products like jams and jellies have become integral part of breakfast served in middle class families of our society. Considering the importance of fresh as well as processed products, an attempt has been made to present developments in processing technology of fruits and vegetables in this book. I hope, this will be useful book to under-graduate as well as post-graduate students in agriculture, food technology, horticulture and home science.

U.D. Chavan

V.D. Pawar

Contents

1

Postharvest Prospects and Opportunities

Introduction

The dynamic, non-liner system, easily affected by the factors beyond anybody's control, our agriculture has received many accolades but is also increasingly criticized for its sluggish growth and declining contribution to the national Gross Domestic Production (GDP), at present 27 per cent but going down. However, no sector of economy has been able to dislodge it from its exalted status of being its backbone-the declining clout notwithstanding. Having said this, it is also to be recognized that agriculture is fast losing its shine and urgent re-look at our priorities is over-due. Indian agriculture, which has hitherto been a way of life has to transform itself into a new *avatar* of commercial agriculture and do what it takes to become such. Many models have been suggested from brilliant to hare-brained for achieving this breakthrough. However, any model, which talks about complete break from the past, may not succeed in our milieu because in a country of our size it takes more than a single factor, how so ever important, to change the scenario. There will have to be many paradigm shifts. To start with, we must come out from the blame-game and go ahead with whichever model of change we are comfortable with.

It has been observed that our agriculture and allied sector continue to be geared up for raw material supply and not much beyond. Selling raw material is a sign of primitive economy and agriculture is the biggest follower of this practice. The return on investment from sale of raw material can never reach the level of return from sale of value-added produce for which any number of examples can be cited. One example would nevertheless, be highly illustrative. In the glut season, potato sells for less than Rs. 2-3/kg, but the branded potato chips sell for Rs. 250-300 per kg, both the farmers and consumers getting a raw deal. This happens because of many natural and man-made reasons, but the main reason is that there are so many intermediaries between the producer and the consumer which need to be brought down to more manageable level (as it happens in the developed countries). The farm gate price in India as a percentage of the consumer retail price is quite low and needs to be raised. The less remunerative agriculture is a cause of great worry and efforts need to be made to convert it into highly remunerative one as it will ensure greater infusion of money (in the form of inputs etc.) and technology forming a virtuous circle. The postharvest management involves reduction of huge losses, safe handling, transport, storage and most importantly processing and value-addition. This will ensure more money into the pocket of farmers and also availability of high quality raw and processed product at competitive price to the consumers (domestic as well as foreign).

Share of Indian Agriculture

Indian agriculture has made great strides, which has been often praised but seldom replicated. The country with a production of 750 million tonnes of food commodities is only second to China with 830 million tonnes and there is no doubt that we shall overtake it in the next a few years. While that sounds good and heartening, it is also a fact that out of this 750 million tonnes, the country may actually be getting around 600 million tonnes only, thus roughly losing around 764 billion rupees every year as shown in Table 1.1 and 1.2.

Present Status of Postharvest Processing

India with 16 per cent of world's population, 4.5 per cent water and 2 per cent land area is the second largest producer of food (next to China), which includes 200 million tonnes of food grains and about 132 million tonnes of fruits and vegetables. Due to the multitude of postharvest factors, the postharvest loss in fruits and vegetables is four to five times higher than in food

Table 1.1: India's share of raw-materials in the world

Raw Material	*Share in per cent*
Fruits and vegetables	
Mangoes	52
Bananas	18
Lemons	19
Dry onions	11
Cauliflowers	38
Cereals and pulses	
Wheat	11
Rice	21
Pulses	27
Dry beans	22
Chick peas	67
Lentils	28
Oilseeds	
Groundnut	27
Castor seeds	70
Rapeseed	20
Nuts	
Coconuts	20
Cashew nuts	21
Milk and its products	
Buffalo milk	62
Goat milk	20
Butter and *ghee*	21
Evaporated milk	10
Plantation crops	
Tea	27
Sugarcane	22

Source: Ministry of FPI, Govt, of India, New Delhi.

grains. The wastage-cost of fruits and vegetables alone is estimated to be 300 billion rupees each year. While the production scenario is heartwarming, hardly 2 per cent of the horticultural produce goes for commercial processing, whereas this figure is more than 75-80 per cent in developed countries. Due to improper postharvest management and lack of adequate processing facilities, a huge quantity of nutritive food materials is lost in our country. In the horticultural sector, the situation is worse since it is often found that there is glut of a particular horticultural commodity in one part of the country when it is scarce in other parts. Prices of seasonal fruits and vegetables fluctuate greatly and during the period of maximum availability, the prices become un-remunerative to the farmer. At other times, these commodities are so highly priced that the ordinary consumer finds it beyond his purchasing power. It is unfortunate that unscrupulous traders and middlemen (5-7 in India as compared to 2-3 in developed countries) get away with most of the profit leaving the producer high and dry. It is reckoned that the farmer/producer get only 25-30 per cent of the price paid by the consumer.

Table 1.2: Level of production of food commodities and estimated postharvest losses

Type of Food Commodity	*Present Level of Production*			*Postharvest Losses*	
	Quantity, Million Tons	*Average Price, Rs./Ton*	*Value Billion Rs.*	*Quantity Million Tons*	*Monetary Value, Billion Rs.*
Durable (cereals, pulses and oil seeds)	215	1000	2150	21.5	21500
Semi-perishables (potato, onion, sweet potato)	40	3000	120	8.0	2400
Perishables (fruits, vegetables, milk, fish, eggs)	140	15000	2100	35.0	52500
Total	395	–	4370	64.5	76400

Source: Pandya and Ali, 2000.

Scope

The avoidable wastage of valuable commodities can be checked if these are processed into value-added products or adequately distributed in different parts of the country. There is a great scope of both domestic and export trade in our country by improving the postharvest distribution and processing facility through an organized yet decentralized system. If the fresh and processed fruits and vegetables are evenly marketed from the places of abundance to the place of scarcity, not only will the consumers get the produce at a reasonable price but also the producer will not be forced to sell at throw-away prices. Some of the techniques are elaborated which are generally not followed in our country and that can be of some practical use to achieve our national objectives.

To attract and retain youth in agriculture, it is essential that agriculture becomes both intellectually stimulating and economically rewarding. For this, especially in the context of resource poor farmers and women, it is important that the economic benefits of agro-processing and agri-business are taken to the rural areas. Most agriculture and food processing industries around the world operate in a manner that would ensure economies of the scale. This also leads to economization of farming and processing. Incremental job opportunities in the agri-food sector are therefore, limited, especially in advanced countries. Developing countries such as India are somewhat differently placed. It is reckoned that this sector will provide maximum opportunities for income and employment generation.

National Food Processing Policy

The Government has come out with a draft national food processing policy with a vision to motivate farmers and food processors and to provide interactive coupling between technology, economy, environment and society for steady development of food processing activities to build up a substantial base for production of value-added agro-food products for domestic and export markets with a strong emphasis on food safety and quality enabling the farmers especially to realize direct benefits of new technology and marketing network and to ensure adequate availability of quality food products for the consumers at economic prices.

The policy seeks to create an appropriate environment for the entrepreneurs to set up food processing industries through:

1. Creating enabling environment
2. Infrastructural development
3. Backward linkage
4. Forward linkage

Food Processing Industry

Food Processing Industry is of enormous significance for India's development because of the vital linkages and synergies that it promotes between the two pillars of the economy, namely industry and agriculture. It will give India the potential to become number one in food production with sustained efforts. The growth potential of this sector is enormous and it is expected that the food production will double in the next 10 years, marking improvement in consumption of value-added foods.

Growth of the food processing industry will also help in bringing immense benefits to the economy, raising agricultural yields, meeting productivity, creating employment and raising the standard of very large number of people through the country, especially in the rural areas. Economic liberalization and rising consumer prosperity is opening up new opportunities for diversification in food processing sector. Liberalization of world trade will further open up new vistas for growth. Food processing has the largest employment generation potential. It generates employment for 54,000 persons per 10 billion rupees investment, whereas in textiles and paper industries it is 45,000 and 25,000 persons respectively. Altogether 1.27 million job opportunities are likely to arise in the food-processing sector with the implementation of the approvals given and industrial entrepreneurial memoranda filed so far.

Although we are one of the largest producers of raw materials for the food processing industry in the world, the industry itself is extremely underdeveloped in

India. Less than 2 per cent of fruit and vegetable production is processed with 30 per cent in Thailand, 70 in Brazil, 78 in Philippines and 80 in Malaysia. The value addition in the food sector is still very low at 7 per cent.

Agro-processing

Primary Processing of Cereals, Pulses and Oilseeds

Cereals

Milling of cereal used to be done by women mostly using manual and foot-operated chakkies for their household consumption and occasionally to help the relatives. Now the milling has been replaced by *chakkies* for making wheat flour and dehulling/shelling of rice. The pulse processing has entirely gone into hands of *dhal* mills.

The flour *chakkis* are of 5-10 hp and produce wheat flour of different fineness. Mini burr mill developed at Central Institute of Agricultural Engineering (CIAE), Bhopal (capacity 10-75 kg/h, 0.5 hp electric motor) have been found as versatile size-reduction equipment for the production of flour, *dhal, besan* and powdered spices (chili, coriander, turmeric etc.). Central Food Technological Research Institute, Mysore has developed grain mill of the capacity of 100 kg/h and it produces flour *(atta)*, refined flour *(maida)* and grits *(suji, dalia)*.

Rice milling which was earlier done by hand/foot pounding has been replaced by paddy hullers/shellers. There are a large number of rice mills of varying capacity in the country. Parboiled rice is prepared in certain parts of the country. Low cost methods, in addition to commercial large-scale methods, have been developed for parboiling. Improved versions of machines for use at small scale or as cottage industry have been developed and are commercially available. Women can easily operate these equipments.

The primary grain processing, particularly processing and milling of rice and wheat has substantial capacity in the country. In the last five years, there has been a marginal increase in the number of flour mills adding to the capacity. Table 1.3 gives an idea about the large number of milling equipment in India. India is the largest exporter of Basmati rice and due to high competitiveness, the rice mills have installed most modern milling equipment like colour sorters etc. Solvent extraction milling is also quite promising.

Table 1.3: Rice, dhal and wheat milling equipment

Type	*Number*
Hullers	90,000
Shellers	4,500
Huller-cum-sheller	8.300
Modern rice mills	33,000
Dal mills	12,000
Roller mills	750
Flour mills	2,66,000

Pulses

About 75 per cent of the pulses produced in India are milled into *dhal.* Traditionally *dal* was produced after simple pre-treatment with the help of lightweight stones (grinder/*chakki)* which leave some husk (5-10 per cent) having low market acceptability. Now simple *dhal* mills have been developed at many research and development institutions and can be easily operated for *dhal* production at cottage level (Table 1.3).

The main objective of pulse processing is to remove hulls as cleanly as possible. This is accomplished by giving treatment (mostly hydro-thermal) for weakening of bond. In the traditional mills, water is sprayed by hand on the pulses and then sun-dried or in some cases dried in mechanical dryers. In further improved mills, a flight mixer does water application. After water application, tempering is required to get uniform effect as well as curing action. In the modem method, pitting (cracking of hull) is quite popular for facilitating water penetration. Pitting also helps in crushing shriveled grains and mud particles of the size of grain which could not be removed during cleaning and grading.

Traditional practice of *dhal* milling is time-consuming, more labour intensive and there is much loss of material with average milling yield of 65-74 per cent only as against 80-84 per cent which can be achieved. The milling of pulses on a commercial scale is generally based on dry-processing technique, where the major unit operations of de-husking and splitting are mechanized. Dehusking and splitting are either done in a single operation or more advantageously independent operations. There are presently about 700 *dhal* mills processing different pulses in the country. Significantly higher yields may be obtained when legumes are milled by improved method as compared to traditional methods as can be seen from Table 1.4.

Table 1.4: Average yield of Dal

Sl.No.	*Legume*	*Kernel Content After Removing Husk, per cent*	*Head Dal Yield by Improved Method, per cent*	*Yield by Traditional Method, per cent*
1.	Chick pea			
	(*a*) Big grains	88.5	84	74
	(*b*) Small grains	87.5	83	72
2.	Pigeon pea			
	(*a*) Big grains	88.5	85	75
	(*b*) Small grains	85	82	68
3.	Mung gram	89.5	84	65
4.	Black gram	87.5	83	71
5.	Lentil	88.5	78	76
6.	Lathyrus bean	88	83	76
7.	Cow pea	89.5	84	–
8.	Kidney bean	90.3	86	–
9.	Horse gram	87.5	81	–
10.	Soy-bean	89.5	86	–

Source: Kurien, 1977.

Dhal milling units are notorious for their noise and dust pollution, which is required to be reduced/removed. Further, energy use is also poor and needs to be avoided and better power transmission mechanism required. For pre-treatment, work on the use of enzymes is encouraging but much more is required.

Importance of Legumes in Human Nutrition

The vast population growth of developing nations has created global food shortages. This problem has been further compounded by periodic droughts in Africa and Asia. It is estimated that the world population in 2025 will be over 12 billion. Increases in population are expected to occur mostly in developing countries. Hence, the resources to produce more food to fulfill the demand of their growing population are limited. Scientific developments in crop production have increased the yield of principal food crops. However, the impact of such increases is not apparent due to increases in population in these countries. In recent years, there has been a rapid rise in the prices of agricultural inputs. This has restricted food produces to intensive farming in areas where the density of population is high and the major proportion of the population subjects below the poverty level.

The world food supply depends upon success in agriculture. The availability of cheep and abundant energy that facilitates the use of fertilizers, pesticides and irrigation influences the production of food. Energy is no longer cheap nor abundant. As the pressure of the population increases, the shortage of energy and fertilizers becomes acute. Techniques that can increase food production without demanding large quantities of energy are beginning to assure importance. The most important task facing mankind today is a solution to problems of world hunger and malnutrition using methods that lessen the exploitation of nonrenewable energy sources. The huge increase in the cost of nitrogen fertilizer in developing countries is perhaps the most insidious effect the recent energy crisis has had. Legumes and biological nitrogen fixation assume a new importance under this condition.

Legumes are crucial to the balance of nature. They are able to convert nitrogen gas from the air into ammonia which is readily utilized by plants. The nitrogen that legumes contribute can be vital in maintaining soil productivity over a long period. By using nitrogen-fixing bacteria, legumes may meet their nitrogen needs without requiring a significant quantity of fertilizer. Cereals lacking this system contain less seed protein and require more nitrogen fertilizers for satisfactory growth and yield.

Food legumes have the exceptional, immediate potential for alleviating human malnutrition in tropical countries by virtue of nutritional and agronomic advantages. Legumes contain two to three times more protein than do cereals. The excellent nutritional value of most food legumes in terms of protein, calories, vitamins, minerals and fibers is highly complementary to a cereal-based diet in developing countries. Most legumes have a distinct advantage over other food crops because of their simplicity of preparation and multiplicity of edible forms, e.g., tender green seeds and leaves, unripe whole pods, green peas, and dry seeds. Protein occupies an important and unique place in the human diet. Although various classes of compounds can satisfy our energy needs, the protein requirement is basic and irreplaceable. Food legumes are an important dietary source of both energy and protein, while cereal seeds are regarded in developed countries as a source of energy in the diet. On a global scale, food legumes and cereals are the main sources of protein. In fact, cereals and legumes supply about two thirds of human protein intake. Thus, seed proteins must inevitably increase in importance if the greater number of future generations is to be adequately fed.

The potential for seed proteins to cope with the population explosion indicating that the present world grain production could come close to satisfying world population protein requirements in the year 2025 if it is utilized directly for human consumption. The need for the increased direct consumption of seed products has been realized and is actively being encouraged. However, sociological and logistic barriers should be overcome. A more attractive food product (made by processing technology) will make seed proteins more acceptable for consumption. It is, therefore, necessary to improve nutritional quality and the balance of essential amino acids and to remove toxins and antinutritional factors. The amounts of protein available for consumption (but not actually consumed) in various regions of the world vary from 50 to 100 g/day per inhabitant for populations living under different socioeconomic conditions. For half the world population, only about 50g of the proteins per day per inhabitant are available, of which 80 per cent are plant proteins. In relation to the total quantity of available energy proteins represent from 10 to 15 per cent of available calories, but this value may reach 20 per cent if estimated relative to the energy actually consumed in developed countries.

The lack of dietary protein was once considered to be the primary cause of malnutrition in the world. This view has changed most people now agree that if caloric needs are satisfied, protein intake will be adequate. Although true, this statement does not negate the importance of providing sufficient protein in the diet. The development of an adequate world supply of dietary proteins requires increased utilization of direct plant proteins by humans.

Adequate supplies of dietary protein do exist, assuming that all the plant proteins which are produced each year and are edible for humans are actually consumed in an equitable manner by the human population. However, from a practical point of view, this is unlikely to occur. A significant protein of available edible plant proteins will undoubtedly continue to be fed to livestock because of the demand for animal proteins for human consumption. For major shifts to occur in the utilization of plant proteins, suitable vegetable protein products must be developed. Some are similar to and complete with animal protein products as desirable components of the diet. However, it is important that new products be developed which will be accepted as preferred food items on their own merit and not merely as imitations of animal protein products.

The protein content of food legumes varies widely, depending upon environmental factors and genotypes. Several studies have indicated a negative correlation between protein content and sulfur-containing amino acid in the proteins of legumes. As increase in protein content beyond certain levels by genetic manipulation may yield proteins with low levels of sulphur-containing amino acids, creating an imbalance of amino acids. A change in seed protein results in the modification of amino acid composition. Seed amino acid composition is controlled by a molecular mechanism that only a very few mutants have been shown to be able to escape. Therefore, research needs to be carried out on the molecular mechanism which controls seed storage proteins. This would help increase the level of sulphur-containing amino acids. The successful *in vitro* molecular hybridization between subunits resulting from storage proteins of dicotyledonous seed belonging to a distinct botanical family led to the assumption that such substitution will be viable and accordingly usable from a practical point of view.

Food legumes contain several antinutritional compounds. These antinutritional compounds have a protecting effect on the see embryo and subsequently on whole seeds. They act as a natural pesticide, but also play a very important nutritional role. Estimation of the ability of new food types (*e.g.*, protein plants) to satisfy the protein requirements of an organism depends not only on their amino acid content but also on the repercussions of anti-enzymes and other compounds on the digestibility of these amino acids. Chemical analysis is not enough. It must be accompanied by further testing in particular, biological tests such as protein efficiency ratio (PER) and biological value (BV) of proteins. The action of heating on the destruction of anti-enzymes and hemagglutinins has been well established. However, these treatments may destroy some of the essential amino acids and vitamins. In order to maintain the nutritional value of food subjected to such treatments, it is necessary to make certain that heating temperature and processing time reach, and do not exceed, the optimum temperature required to eliminate the effects of inhibitors and hemagglutinins without altering the basic nutrients. The products should be evaluated using both biological and analytical techniques.

In developing countries, legumes are processed using traditional methods. These include milling, soaking, germination, fermentation, and cooking. It has been shown that such processing of legumes has several nutritional advantages and disadvantages. It is therefore necessary to optimize processing conditions in order to yield a product which has improved nutritional functionality. This involves chemical and biological evaluation of product different stages of processing.

Proteins are known to interact with lipids, tannins, phytates, flvour compounds, and pigments. These interactions occur when legumes are processed and converted into products. This decreases the bioavailability of proteins. Factors favoring these interactions should be investigated in detail and possible modifications in processing conditions should be worked out to minimize or avoid such interactions. When proteins are extracted from food legumes, physical and chemical changes may occur which modify the nutritive value of proteins or induce certain physical effects. These interactions must be controlled using the parameters involved in these modifications. Some processes should be abandoned (alkali treatment, chlorinated solvent); oxidizing agents must be carefully manipulated, and the conditions which develop the maillard reaction must be avoided.

Legume-based fermented foods are popular in Southeast Asia, Near East, and part of Africa. The prospect for increased consumption of legume based fermented foods are promising for both developed and developing countries. Most legume based fermented foods are introduced to the Western World by immigrants from the orient. Some fermented foods such as miso, soy sauce, and tempeh are becoming popular in the Western Hemisphere. With the increasing awareness of the nutritional value of legume-based fermented foods, the popularity and consumption of these foods is on the rise in the Western world. This will continue to increase in underdeveloped and developing countries in the future. Improvements in the organoleptic characteristics of legume-based fermented foods may increase their production and consumption. Thus, the use of legume-based fermented foods undoubtedly will continue to increase in the orient, in other parts of Asia and in Africa since these are traditional foods in these parts of the world. Similarly, there is a considerable potential for the development of new markets for traditional legume-based fermented foods in the Western world.

Food legumes are relatively high in fiber (non-digestible food components) compared to other foods and these are physiologically beneficial. Fermentation of non-digestible food components by anaerobic bacteria in the intestine gives rise to gas formation and formation of lactic acid and volatile fatty acids. These acids are reported to promote the rapid intestinal transit of feces of a more bulky and softer stool. Lack of fiber in the Western diet is believed to result in constipation and to be a main factor in the appearance of directicuar and colon related diseases. It is suggested that ingestion of appreciable amounts of beans along with other foods eases or relieves constipation and other colon-related diseases.

In the Western world, consumption has recently shifted from animal fat to vegetable oil products because of their ability to lower the blood cholesterol level. Many food legumes have been shown to lower serum cholesterol levels in the liver and aorta of rats. This has been attributed in part to a high content of polyunsaturated fatty acids such as linoleic and linolenic acid in legumes. In recent years, hypocholesterolemic effects have been obtained by using texturized and extruded soybean isolates which contain, in addition to proteins, carbohydrates, dietary fiber and saponins with possible independent hypocholesterolemic effects. The reported positive response obtained in hyperlipidemic patients was due to decrease of LDL cholesterol. This is nutritionally advantageous to those having high cholesterol levels. Cluster bean gum has been used for the treatment of diabetes. In addition, it has been shown to have a hypocholesterolemic effect. These studies indicate that food legumes have an important role to play in the nutrition of people in both developing and developed countries.

During legume storage, several chemical and biochemical changes occur in seeds. The result is a change in the cooking behavior of seeds which leads to the hard-to-cook phenomenon. It is well established that prolonged cooking results in a decrease in protein digestibility. Investigations into histochemical and biochemical changes which take place during storage and the conditions which favour the seeds and make them hard to cook will help in understanding the mechanism involved. Such studies can preserve the nutritional quality of legumes during storage. Process can be made in the near future on the protein malnutrition problem only if basic food grains such as legumes are upgraded in nutritive quality. If this happens, the nutrition of the economically handicapped section of the population can be improved without special educational efforts or added expense.

Oilseeds

India has a multi-tier extraction involving traditional *kolhus/ghanis* located in 'illages, which leave 10-15 per cent oil with the cake followed by mechanical oil xpellers located in semi-urban and urban areas which may leave 5-7 per cent oil vith cake. In oilseed processing, dehulling/decortication are an important unit peration. A number of such units have been developed and are available in the narket for various oilseeds. Women can easily operate groundnut decorticators, unflower decorticators, castor shellers, soybean dehullers etc. For oil extraction, hey can use power-driven portable power *ghanies* and Table oil expellers, which are ommercially available. There are about 3, 00, 000 mechanical oil expellers (2, 50, 00 *hanis* and 65,000 expellers) and 600 solvent extraction plants producing oil worth 0 billion rupees. Considerable work has been done on single and double stage crew press, steam cookers, disintegrators, filtering gadgets/processes, solvent xtraction equipment and refining for processing of oilseeds. Work has been done on hysical, chemical and enzymatic pre-treatments of oilseeds and results have been ncouraging.

Solvent extraction of oil from oilseeds by using solvents (generally hexane) is ecoming very popular worldwide as it leaves less than 1 per cent oil in the oilcakes. here has been wide concern against the removal of traces of solvent from the cake nd as a result safer solvents like isopropyl alcohol (IPA) are being preferred. ontinuous counter current percolation system is however more popular because of ts better efficiency. Refining follows oil extraction, which is achieved through legumming, neutralizing, bleaching, winterizing and deodorization. Use of oilcake ias been mainly as cattle feed. Even if half of the estimated cake/meal in India (about -6 million tonnes/year) is upgraded and utilized for human consumption, it would dd 2-3 million tonnes food and even a million tonnes of protein for diet. Non-raditional oilseeds (1.1 million tonnes annually) are quite important for Indian conomy especially for the livelihood of tribals. If these oil sources are exploited in a ystematic way, sizeable quantity of edible/non-edible oils could be added to the oil rid and it can provide employment for 500 million man-days or 1.6 million workers er annum.

Recent advances in oil expelling is supercritical CO_2 extraction with or without co-solvent (ethanol or IP A) at very high pressure and high temperature, which has hown promise. Supercritical fluid (SCF) has versatile extraction capability, which omes from the innate solvent power that originates from the molecular association f the fluid in its supercritical domain. Once the liquid molecules are placed at a emperature and pressure higher than their critical point, the inter-molecular distances nd forces can be closely controlled by independently controlling either pressure or emperature. This control is exploited in supercritical fluid extraction.

Processing and Preservation of Fruits and Vegetables

Most of processed fruits and vegetables (juice and pulp, jams and jellies, beverages, *urees*, ketchups, squashes, concentrates, pickles, resins etc.) are prepared in cottage r small scale sector on account of round the year availability for a wide variety of ruits and vegetables as well as due to large scale domestic level processing done by women. Defective, small size products left after grading for Table uses can be easily processed to a great economic advantage. New techniques for extending shelf-life of fruits have been developed and commercialized like waxing, packaging in perforated polybags, cold modified/controlled atmosphere storage etc. Presently postharvest operations for commercial organized marketing are undertaken at the urban or consumer level. If undertaken at production catchments level rural women can be encouraged for sorting, cleaning, washing, and waxing, packaging and also minimal primary processing.

Pre-harvest spraying of fungicides on kinnow plants at 45, 30 and 15 days before harvest with 0.2 per cent Biltox-50 is better in reducing fruit-drop followed by 0.3 per cent Dithane M-45 and 0.2 per cent Bavistin. It has also been reported that Blitox-50+ 1 per cent Calcium nitrate or 0.3 per cent of Bavistin was sprayed at 15 days before harvest to minimize fruit-rot in un-waxed fruits during storage under cold-store condition. Although, it is a higher dose hence 0.2 per cent concentrated Bavistin either with calcium-nitrate or alone may be applied before harvest. Wax application reduced fruit-rot as well as weight-loss during storage. Total 10 fungi were identified from the surface of kinnow fruit and out of those *Penicillium digitatum, P. italicum* and *A. citri* were found most prominent fungi to cause rotting. Postharvest treatment of fruits with 1000 ppm concentration of Bavistin was significantly effective against *P. italicum* in un-punctured fruits which could be stored safely upto 60 days under cool chamber. Fruits treated with fungicide in combination with wax-emulsion (Zividar-Bavistin 0.1 per cent) and stay-fresh (shine) + Bavistin 0.1 per cent were found better than other treatments. Fruits treated with hot water immersion at 50°C for 2 minutes did not rot even up to 90 days of storage at 5 ± 1°C. It was also observed that if fruit got injury then no treatment except *D. hansenii* is able to check the growth of *P. italicum* on the its surface. *D. hansenii* + Mustard oil in 50:50 ratio reduced fruit-rot as well as further growth of *Penicillium* spp. on the surface of the fruit. Corrugated fiber board boxes (7-ply) were found suitable for packaging of kinnow fruit and it reduced the transportation and subsequent storage losses as compared to other packaging materials like wooden boxes, bamboo baskets and gunny bags.

Tomatoes harvested at breaker-stage and carried in plastic crates with grass as the cushioning material account for the least losses (4 per cent for 400 km distance of transportation). In other packaging and cushioning materials, losses are 0.2 to 0.5 per cent higher than that for breaker-stage and 5.0 to 6.0 per cent higher at pink-stage. Tomatoes harvested at breaker-stage using straw as the cushion material during transportation cause 1.1 per cent losses for 100 km distance of transportation in plastic crates. In other materials, losses are 0.6 to 0.7 per cent higher for breaker-stage and 2.3 to 4.8 per cent for pink-stage. The postharvest losses of tomatoes at different levels are given in Table 1.5.

With proper postharvest management practices, which include right stage of harvest with proper harvesting technique followed with sorting, packaging and transporting as per the recommendations described above, better quality of tomatoes with the least postharvest losses can be ensured.

Table 1.5: Different types of losses at various levels

Types of Losses	*Range of Losses*
Losses at farm level	2-10 per cent
Losses at wholesale level	Monetary loss of 30-40 per cent price (5·8 per cent tomatoes are sold at 30-40 per cent reduced price).
Losses at retailer level	Physical loss 2-3 per cent (Monetary loss of 70-80 per cent of price in about 10 per cent of tomatoes)
Losses at cold store level	6–8 per cent
Losses at processing industry	Loss of 2 kg of tomato pulp for preparing each 1 kg of ketchup of TSS 26-28 °Brlx

Value-addition in Agriculture Produces

The value-addition of farm-produce could be undertaken due to different factors including increase in the price of the produce, sale before or after the glut and higher market demand etc. The farmer must have means and capacity to hold the farm produce for longer periods for which he requires an appropriate storage structure and the desired technology to reduce storage losses and transform the produce into a product which fetches higher price, is able to hold little longer and is available out of season etc. Thus processing is a major value-addition activity to the farm produce. The processing technology depends upon the type of raw-produce and is location and crop-specific. Various postharvest operations include cleaning, grading, drying, dehydration, dehulling, decortications, shelling, peeling, grinding/milling, polishing, extraction, curing, treatment, storage, packaging, transportation etc. The postharvest losses of cereals, pulses and oilseeds are about 10-12 per cent which is quite large. Even if these losses can be reduced by half, it will make a difference to the economy of the country Table 1.6.

Table 1.6: Estimated value addition to agricultural produce (1995-96)

Level of Processing	*Value Addition, Billion Rs.*
Primary processing	1080
Secondary and tertiary processing	220
Products produced/processed in organized sector	100
Products/processed in un-organized sector	1200

Source: Singh, G. 1997.

Causes for Low Level Processing in Agriculture Produce

1. Most Indians prefer to consume fresh fruits and vegetables for a variety of reasons mainly due to prevalence of warm sub-tropical climate where the chances of spoilage are high.
2. High cost of processed foods which is beyond the reach of the poor people? Enhanced prices may be due to low turnover, low productivity, and seasonality of produce and high extent of spoilage.

3. Poor quality of raw material which affects the quality of processed goods
4. A plethora of rules and regulations resulting in long processing time, whicl inhibit a budding entrepreneur. Many of the laws were framed some 5(years ago and need revision.
5. General lack of knowledge about the processes, equipment and technology due to lack of availability of information from reliable sources. There is a need for developing a strong data base.
6. Fewer opportunities for human resource development resulting in shortage of trained manpower.
7. Lack of financial support on soft terms and absence of incentives like tax exemption, relaxation/rebate etc. for starting a new unit. It may be noted that in India, tax on processed food is amongst the highest in the world. A tax holiday for a period of 10 years would be a good idea.
8. Lack of infrastructure including electricity, good quality water, communication facilities etc. makes it very difficult to sustain processing units. The power blackouts for hours/days are not unknown in many parts of India.
9. Inadequate market intelligence due to lack of infrastructure creates problems for marketing of goods to fetch better price. Market intelligence network needs to be established.

Factors Attracting Farmers for Agro-processing Enterprise

1. Sensitization about procedures, agencies involved, institutional help, single window clearance.
2. Skill development including specialized short term training
3. Assured supply of raw materials
4. Easy availability of credit on soft terms
5. Special incentives including subsidy in purchase of inputs, tax relief etc.
6. Market support including market intelligence
7. Easy access to suitable technology
8. Exposure to latest technologies through visits/study tours/consultancy
9. Awareness about standard Hygienic Food Manufacturing Practice (GMP) and Hazard Analysis and Critical Control Points (HACCP)
10. Positive approach by the government including commitment to promote food processing.
11. Establishment of a corpus fund for development of agro-processing by woman.

Agro-processing Centre (APC)

The model concept of establishing agro-processing centers in production catchments has been evolved in the All India Co-ordinate Research Projects on

Postharvest Technology. Agro-processing centre is an enterprise where the required facilities for primary and secondary processing, storage, handling and drying of cereals, pulses, oilseeds, prices, fruits and vegetables are made available on rental basis to the rural people. Value-added agro-based products and processed food items are also prepared and marketed by the centre. An individual/Cooperative/community organization/voluntary organization may manage this type of centre. The centre meets the processing, preservation, handling and marketing needs of the excess produce available in a village or a cluster of villages. Thus, the centre is a means of providing income and employment to rural people through agro-based processing activities of the various products.

Generalized agro-industrial models cannot be suggested for a country as these are location specific. The agro-industrial model for the area could be developed in a manner, which provides sufficient job opportunities to an entrepreneur-unit's adequate profit margin. The APCs also provide good opportunity for women. The research and development institutions as well as commercial houses have developed the agro-processing machines suiting the various requirements.

For developing an agro-industrial model for a particular area, following points should be given due consideration:

1. Existing potential demand of processing produce
2. Availability of raw material *i.e.,* produce
3. Technology/process to be used for processing
4. Volume of production (based on assessment of demand through market survey)
5. Identification of suitable technologies, plants and machinery for the desired volume of production
6. Training facility for operation, repair and maintenance
7. Credit availability at soft rates/subsidies given by the government
8. Facility for storage and marketing.

In order to derive the maximum benefit, the APCs should be properly laid out to optimize available space and utilize the energy efficiently. The selection of equipment (capacity, power requirement etc.) is of utmost importance. Since most of the postharvest operations are seasonal in nature, 2-3 technologies/products may be taken-up to develop an agro-industrial complex to provide regular employment throughout the year. It is estimated that an investment of rupees 300 thousand will create jobs for two persons and generate decent income. With an investment of rupees 500 thousand, it will create employment opportunities for 4 persons and generate income around 90 to 100 thousand rupees.

2

Processing Technology and Value Addition in Cereals

Physico-Chemical Properties of Grains

A grain is a living biological product which germinates and respires also. The respiration process in the grain is externally manifested by the decrease in dry weight, utilization of oxygen, evolution of carbon dioxide and release of heat. The rate of respiration is dependent upon moisture content and temperature of the grain. 'The rate of respiration of paddy increases sharply (at 25°C) at 14 to 15 per cent moisture content which is called the critical point. On the other hand the rate of respiration increases with the increase of temperature to 40°C. Above this temperature the viability of the grain as well as the rate of respiration decreases significantly.

Structure

Wheat and rye consist mainly of pericarp, seed coat, aleurone layer, germ and endosperm whereas oats, bailey, paddy, pulses and some other crops consist not only of the above five parts but an outer husk cover also. The husk consists of strongly lignified floral integuments. The husk reduces the rate of drying significantly. The embryo or germ is the principal part of the seed. All tissues of the germ consist of living cells which are very sensitive to heat. The endosperm, which fills the whole inner part of the seed, consists of thin-walled cells, filled with proto-plasm and starch granules and serves as a kind of receptacle for reserve foodstuff for the developing embryo.

Chemical Composition

The grain is composed of both organic and inorganic substances, such as carbohydrates, proteins, vitamins, fats, ash, water, mineral salts and enzymes. Paddy, corn, wheat, buck wheat seeds are especially rich in carbohydrates whereas legumes are rich in proteins and oil seeds rich in oils. Generally, percarp (and floral integuments also) contains cellulose, pentosan and ash, the aleurone layer contains mainly albumin and fat. The endosperm contains the highest amount of carbohydrate in the form of starch, small amount of reserve protein and a very little amount of ash and cellulose whereas the germ contains the highest amount of fat, protein and a small amount of carbohydrate in the form of sugars and a large amount of enzyme.

Effects of Temperature on the Quality of Grain

Proteins

The proteins present in cereal grains and in flour are hydro-phillic colloids. The capacity of flour proteins to swell plays an important role in the preparation of dough. At temperatures above 50°C denaturation and even coagulation of proteins take place. As a result, the water absorbing capacity of the proteins and their capacity for swelling decrease.

Starch

Starch is insoluble in cold water. It swells in hot water. Up to a temperature of 60°C, the quality of starch does not change appreciably. With a further increase in temperature, particularly above 70°C, and especially in the presence of high moisture in the grain, gelatinization and- partial conversion of starch to dextrin take place. In addition, a partial caramelisation of sugars with the formation of caramel may take place which causes deterioration in colour of the product. These effects have been discussed in detail in Section II on Parboiling.

Fats

Fats are insoluble in water. Compared to alburmns and starch, fats are more heat resistant. But at temperatures above 70°C, fats may also undergo a partial decomposition resulting in an increase of acid numbers. In the range of temperatures from 40 to 45°C, the rate of enzymatic activity on fats increases with the increase of moisture and temperature. With a further rise of temperature the enzymatic activity begins to decrease, and at temperatures between 80 and 100°C the enzymes are completely inactivated.

Vitamins

The heats sensitive B-vitamins present in the germ and aleurone layer are destroyed at high temperature.

Physical Properties

The knowledge of important physical properties such as shape, size, volume, surface area, density, porosity, colour, etc., of different grains is necessary for the design of various separating and handling, storing and drying systems. The density and specific gravity values are also used for the calculation of thermal diffusivity and Reynolds number. A few important physical properties have been discussed here.

Sphericity

Sphericity is defined as the ratio of surface area of sphere having same volume as that of the particle to the surface area of the particle.

Porosity

It is defined as the percentage of volume of inter-grain space *to* the total volume of grain bulk. The per cent void of different grains in bulk is often needed in drying, air flow, and heat flow studies of grains. Porosity depends on (a) shape, (b) dimensions, and (c) roughness of the grain surface. Porosity of some crops is tabulated as follows:

Grain	*Porosity per cent*
Com	40-45
Wheat	50-55
Paddy	48-50
Oats	65-70

Coefficient of Friction and Angle of Repose

Angle of repose and frictional properties of grains play an important role in selection of design features of hoppers, chutes, dryers, storage bins and other equipment for grain flow.

Coefficient of Friction

The coefficient of friction between granular materials is equal to the tangent of the angle of internal friction for the material. The frictional coefficient depends on (*a*) grain shape, (*b*) surface characteristics, and (*c*) moisture content.

Angle of Repose

The flowing capacities of different grains are different. It is characterized by the angle of natural slope (angle of repose). The angle of repose is the angle between the basic and the slope of the cone formed on a free vertical fall of the grain mass to a horizontal plane. The angle of repose for a few important grains is tabulated as follows:

Grain	*Angle of Repose (degree)*
Wheat	23-28
Corn	30-40
Millets	20-25
Rye	23-28
Oats	31-44
Barley	28-40
Paddy	30-45

Thermal Properties

The raw foods are subjected to various types of thermal treatment namely, heating, ooling, drying, freezing, etc., for processing. The change of temperature depends on he thermal properties of the product. Therefore knowledge of thermal properties, amely, specific heat, thermal conductivity, thermal diffusivity, is essential for the lesign of different thermal equipments and for solving various problems on heat ransfer operation.

Specific Heat

The specific heat of a substance is defined as the amount of heat required to raise he temperature of unit mass through 1°C. The specific heat of wet grain may be onsidered as the sum of specific heat of bone dry grain and of its moisture content.

General Grain Milling Operations

Food grains are naturally endowed with outer protective husk/bran layers omposed of rough, fibrous, pigmented and waxy substances which are undesirable or edible purposes. It also consists of oily germs which are undesirable for storage purposes. Removal of these parts constitutes the most fundamental prerequisite in grain milling or flour milling technology of cereals, In grain milling the outer husk/ bran layers are removed from the grain with its shape retained whereas in flour nilling, flour without or with negligible bran content is prepared without the grain shape. In general, milling refers to the size reduction and separation operations used or processing of food grains into edible form by removing and separating the inedible and undesirable portions from them. Milling may involve cleaning/separation, husking, sorting, whitening, polishing, grinding, sifting, etc. To increase the milling quality of the food grains or to improve the quality and quantity of their end products or to facilitate milling operations for the desired products, food grains are sometimes subjected to hydrothermal treatment prior to milling called conditioning. Basic milling operations and hydrothermal treatment involved in cereal milling technology have been discussed.

Classification of Separation Methods

Any mixture of solid materials can be separated into different fractions according to their difference in length, width, thickness, density, roughness, and drag in moving air, electrical conductivity, colour and other physical properties.

Each of the various types of separators employed in flour and grain milling is designed on the basis of the difference in the following physical characteristics of grain: (1) Width and thickness of the grain for sieves, screen cleaners, sifters, thickness graders, grading reels, inclined sifters, etc.; (2) Length of the grain for indented type or disc type pocket separators; (3) Aerodynamic properties for husk aspirators, cyclone separators. (4) Form and state of the surface for separators for coarse grain, spiral separators, belt type separators; (5) Specific gravity and coefficient of friction for separating tables, stone separator; (6) Ferromagnetic properties for magnetic and electromagnetic separators; (7) Electrical properties for electrostatic separators and (8) Colour for electronic separators.

Separation According to Aerodynamic Properties

The pneumatic separation is based on the difference in aerodynamic properties of the different 'components. The aerodynamic properties of a particle depend on the shape, dimensions and weight of the particle, the state and position of the particle with respect to the air current. The aerodynamic properties have been discussed in detail in other section

Separation According to Specific Gravity

If the components of a mixture are different in densities and subjected to a reciprocating movement of an inclined table or screen then the 'Components of the mixture are readily separated into different fractions according to their densities. This type of separator works on the principle of self sorting or stratification. The heavier particles of the mixture sink to the bottom by the to and fro movement of the inclined table and are then separated by any suitable method. In a mixture of components of same density, the finer particles sink while in a mixture of components of equal dimensions but of different densities, the heavier particles sink to the bottom. This is the principle of operation of stone separators. Composite stone separators of pneumatic grading-table type are used to enhance settling of stone. The stone separator has been discussed in detail in Rice Milling Chapter. The effectiveness of operation of a stone separator depends on many factors. Of them kinematic parameters are most important. Continuous and uniform feeding of mixture is also important.

Separation According to Magnetic Properties

Metallic impurities in the grain accelerate wear and tear of different parts of the milling machinery. Moreover, even minute quantity of meta-Ilk impurities present in the milling products can make them unfit for human consumption. The magnetic impurities like steel, pig iron, nickel and cobalt particles present in a grain mixture can be separated on the basis of their differences in magnetic properties. Since effectiveness of removal of metallic impurities depends on the force of; attraction of the magnet, an electromagnet is preferred to an ordinary permanent magnet as the force of attraction of the former can be increased by increasing the strength of electric current only. Magnetic separation consists of three steps: (*a*) distribution of feed over the magnet, (*b*) collection and retention of the ferromagnetic impurities by the magnet and (*c*) cleaning of the magnet from the impurities.

Separation According to Electric Properties

When different particles are charged with statically electricity and are passed through another electric field, then the action of the outer electric held on the electric field of the charged 'particles- produces some mechanical work which is used for separation. The electrical separation consists of two stages: Preliminary charging of the particles with electricity and separation of the charged particles by electrostatic forces in accordance with the magnitude and nature of the charges on the particles. Magnitude of the preliminary charge is determined by the following factors: electrical conductivity, dielectric constant, and other properties of the particles such as particle size and shape, specific gravity and design of the separators also.

Separation According to Colour

Difference in colour can be used for separation. With the help of an electronic separator some fruits, vegetables and cereal grains can be sorted from the discoloured or defective ones in. accordance with their differences in colour. In an electronic separator the seeds are uniformly fed to the optical chamber. Two photo cells are set at a 'certain angle in order to direct both beams to one point of the parabolic trajectory of the seeds. A needle connected to a high voltage source is placed on the other side. When a beam through photoelectric cells falls on a dark object, a current is generated on the needle. The end of the needle receives a charge and imparts it to the dark seeds. The grains are then allowed to pass between two electrodes with a high potential difference between them. As a result two fractions of the mixture are separated according to difference in colours. Electronic separators have been used in various industries for a long time. Recently electronic sorters are being used for the separation of discoloured grains in some advanced countries. But the sorting capacity of these units is limited resulting in very high cost of separation.

Frictional Separations

Separation according to surface properties: The frictional properties can be utilized for the separation of a mixture of grains of almost same size. Sizes of oats and hulled oats, millets and bind weed are almost the same. These mixtures can be separated by frictional separators consisting of an inclined plane surface; the operation of the separators is based on the differences between the friction angles of two types of grain, when grains are allowed to move along an inclined plane, frictional forces of different magnitudes act upon these particles. Therefore, different particles move on inclined surface at different velocities. In this case also heavier particles will sink at the lower layer and move at a lower velocity, while lighter particles will float at the top and move downward at a higher velocity.

Husking/Scouring/Hulling of Grain

In general, husking refers to the removal of outer seed coat from the grain kernel. The terms hulling and scouring are also used in cereal milling. In grain milling, husk is removed from the grain retaining its original shape whereas in flour milling bran is removed from the grain to produce flour without any emphasis on its shape. Husking and scouring are the most important operations in grain milling or flour milling technology of cereals. Different types of machines are employed for husking and scouring operations because of the differences in anatomical structure, type of bonds and strength properties among husk, bran and kernel of different grains.

Methods of Husking

The operation of husking and scouring machines can be divided according to the following three basic principles: (1) Compression and shear: Compression and shear can compress, split and strip off the husk from grain. Concave type of husking machine, rubber roll husker, etc., are designed on the basis of this principle. (2) Abrasion and friction: Hollanders are based on the friction of grain on an abrasive surface (emery). (3) Impact and friction: Husk can be stripped off by the action of impact and frictional force. Centrifugal type paddy Sheller comes under this group.

Concave Type Husker

This machine consists of a horizontal rotating cylinder, called roll, and a stationary cylindrical surface known as concave. The husk/bran layer 'can be removed from buck wheat and millets keeping the original shape of the kernel by applying mild shear and compression. These types of machines are fairly efficient and require low power. On feeding into the mill the grain is first caught up by the rolls and passed through the husking zone between the roll and the concave where it is subjected to shear and compression simultaneously. One part of the husk is shared by the rotating roll while the other part is pressed against the stationary Concave and subjected to breaking forces. The minimum clearance between the roll and the concave must be greater than the dimensions, of the grain kernel; otherwise the kernel will be crushed. The radius of curvature of the concave is usually the same as the radius of the roll. Composition of roll and concave varies with the type of grain to be husked. As for example roll made of abrasive material and concave made of commercial rubber are used for millets. Usually the following specifications of rolls and Concaves are used:

Diameter of concave	50 to 60 cm
Length of concave	19 to 30 cm
Angle of contact	40 to 70°
Peripheral speed	10 to 15 m/sec

Husking by the Action of Rubber Rolls

In case of paddy, deformation caused by shear and compression of the two rotating rubber surfaces are sufficient to split and separate the husk from the grains. The paddy is passed through the clearance between two rubber rolls, rotating in opposite directions at different speed. The clearance between them is smaller than the mean thickness of the paddy. One part of the husk is subjected to shearing forces whereas the other part in contact with the slower roll IS under compression and is thus subjected to breaking forces. Husking is done by the action of these forces.

Grinding

Cereal grinding system can be divided into two groups: plain grinding and selective grinding. In plain grinding hard bodies are grounds to a free flowing material consisting of particles of sufficiently uniform size. This material is either the final product or a product ready for further processing. In selective grinding, the grinding operation is carried out in a number of stages successively using differences in structural and mechanical properties of the components of the body. It should be noted that the power consumption for grinding is about 50 to 80 per cent of the total power required for all operations. Therefore, the following important points are to be considered for power design of the grinding system: rational utilization of the raw material, quality of the products, size and colour of the products, efficiency of the grinders, specific power consumption and costs of production.

The Characteristics of grinding operation are affected by the following grain parameters:

(*i*) Type of cereal grain, (*ii*) Variety, (*iii*) moisture content, (*iv*) extent of hydrothermal treatment given to the grain, (*v*) mechanical properties. Each type of grain has optimum moisture content for highest efficiency of grinding. The maximum formation of new surface for each type of wheat is related to the optimum moisture content. Moreover, power consumption is also dependent on moisture content. The starch granules are separated from the proteins due to deformation of hard wheat. Hardness of rice and barley endosperms is 11.5 to 14.5 kg/mm^2, and 6.0 to 9.0 Kg/mm^2 respectively.

Effectiveness of Grinding

The main criteria for evaluation of effectiveness of grinding of any solid body, including grain are: degree of grinding, specific power consumption and the specific load of the initial product on the working tool of the grinder.

Machinery Used in Cereal Grinding

Depending on the objective of grinding and mechanical properties of cereal grain, effects of the following are utilized for grinding: Compression and simultaneous shear of the material, impact followed by crushing of the material etc.

Grinding of Grain in Roller Mills

The roller mill consists of two cylindrical steel rolls revolving in opposite directions at different speeds. In the roller mill the grain or its parts are ruptured in a space, which is narrowed towards the bottom. Certain degree of grain rupturing starts at above the line connecting the centers of the rolls. The slow roll holds the grain during the action of the fast Toll. In the grinding zone, the grain or its parts are simultaneously subjected to compression and shear resulting in deformation of grain. Rupturing of endosperm without much grinding the bran of wheat under grinding conditions is the characteristic of the first breaking system (Figure 2.1 and 2.2).

Factors Affecting the Effectiveness of Roller Mills

- ✰ *Clearance between rolls:* Even a small variation in rollers' clearance leads to considerable variations in the products.
- ✰ *Geometrical parameters of rolls:* Conditions of rupture of the particles to be ground depend upon the roll diameter, clearance and initial size of the particle. Shape, number, slope, mutual position and shape of the cross-section of corrugations of the rolls have significant effects on the quality and yield of flour, total output and specific power consumption of the roller mills. Position of the corrugation edges has also the same effects on the product.
- ✰ *Kinematic parameters of the rolls:* The important kinematic parameters of the rolls are speed of the fast *roll* and ratio of the speeds between fast and slow rolls. The efficiency of grinding depends upon the kinematic parameters.

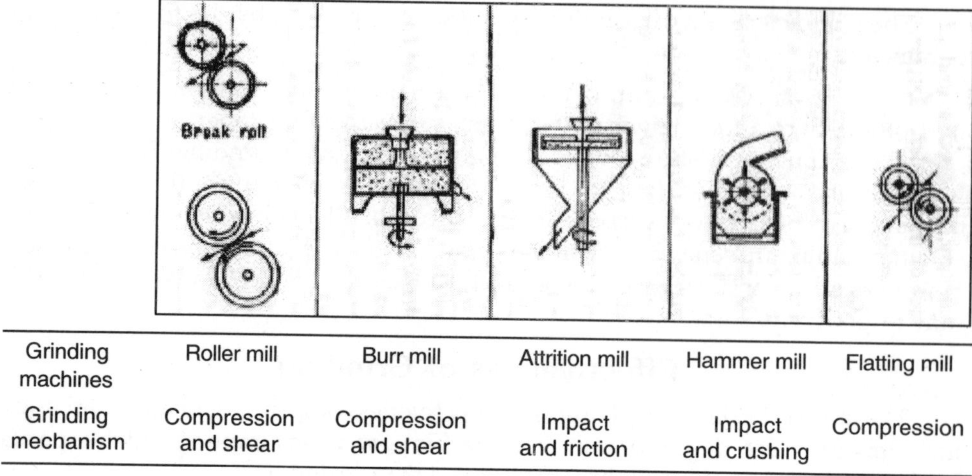

Grinding machines	Roller mill	Burr mill	Attrition mill	Hammer mill	Flatting mill
Grinding mechanism	Compression and shear	Compression and shear	Impact and friction	Impact and crushing	Compression

Figure 2.1: Classification of grinding machinery

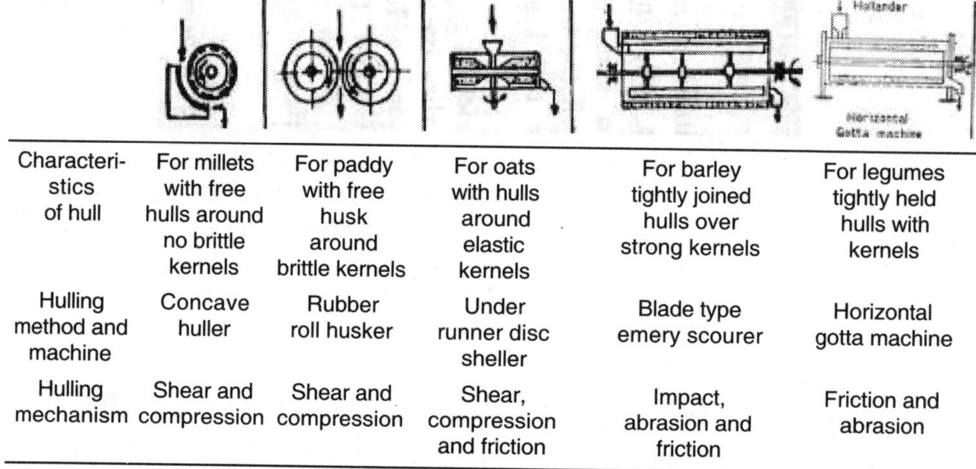

Characteristics of hull	For millets with free hulls around no brittle kernels	For paddy with free husk around brittle kernels	For oats with hulls around elastic kernels	For barley tightly joined hulls over strong kernels	For legumes tightly held hulls with kernels
Hulling method and machine	Concave huller	Rubber roll husker	Under runner disc sheller	Blade type emery scourer	Horizontal gotta machine
Hulling mechanism	Shear and compression	Shear and compression	Shear, compression and friction	Impact, abrasion and friction	Friction and abrasion

Figure 2.2: Classification of husking/scouring/hulling methods

☆ *Capacity of the roller mills and power consumption for their operation:* The capacity of a roller mill is the amount of product in kilograms, ground per unit time.

Grinding Grain in Hammer Mills

Grinding in a hammer mill involves impact on the material followed by crushing. The main working tools of this type of mill are hammers made of high quality steel, screens, and metal lines. The output of hammer mill depends on the peripheral speeds of the hammer, the clearance between the screen and hammers, the area of the screen, the size of the screen openings, and the structural-mechanical properties of the material.

Corn Milling

Introduction

Corn is one of the world's most versatile seed crops. Its botanical name is *Zea mays.* Corn is used as food and feed. Corn can be processed into various foods and feed ingredients, industrial products and alcoholic beverages. But the modem corn milling technology developed for the above products is mostly confined to some of the developed countries only. However, modem corn milling technology is to be suitably adopted for producing the types of products required for other countries. At present, there are two modern methods of milling of corns, dry milling and wet milling. Besides germ for corn oil extraction and husk and deoiled germ, etc., for feed, grits (mainly used for the breakfast cereals) are the main products of corn dry milling whereas pure starch, germ and feed are the major products of wet milling.

Composition and Structure

The mature corn kernel is composed of four major parts: (a) endosperm (82 per cent), (b) germ (12 per cent), (c) pericarp (5 per cent) and (d) tip cap (1 per cent).

Pericorp: The pericarp is mainly composed of four successive layers, namely, outermost thick layer of tough cells, spongy layer of cells, seed coat or testa layer and aleurone layer. The spongy layer is continuous with spongy cells of tip cap.

Germ: The germ is mainly composed of scutellum and embryonic axis. The major parts of lipids and proteins are reserved in the scutellum.

Endosperm: The endosperm of corn is composed of floury and horny parts. The proportion of horny to floury parts varies widely from dent corn to floury corn variety. The ratio of horny to floury endosperm is about 2: 1 in dent whereas the floury corn contains little horny part. The horny parts being hard and floury parts being soft, the latter is milled into flour easily during rolling. The horny regions have 1.5 to 2 per cent higher protein content than the floury regions.

Corn Dry Milling

Corn dry milling system can be divided into two groups: the traditional non degerming system and modem degerming system. In the non degerming system, the whole corn is ground into meal of high fiber as well as high protein contents by a stone grinder without removing germ. After grinding certain amount of germ and hull can be removed from the meal by sifting. In the degerming system the corn is moistened with a little amount of water and tempered for moisture equilibration. After degerming the stock is dried, milled and classified into different products. The purpose of all dry degerming corn milling methods is to remove hull, germ and tip cap from the corn kernel as far as practicable and primarily produce corn grits with some meals and flours. The germ is then used for oil extraction and deoiled germ, hull, etc., are used as feed which is known as hominy feed. The yield of endosperm products and hominy feed are about 70 per cent and 30 per cent respectively.

Tempering-Degerming (T.D.) Method of Dry Milling

The major objectives of this method are: (*a*) to remove essentially all germ and hull so that endosperm contains as low fat and fiber as possible, (*b*) to recover a maximum amount of the endosperm as large clean grits without any dark speck, and (*c*) to recover a maximum amount of germ as large and pure articles.

Description of the T.D. System

The basic operations/processes involved in the T.D. methods are as follows:

1. Cleaning of the corn.
2. Conditioning of the corn by addition of control amount of moisture either at ordinary temperature or at an elevated temperature to toughen the germ and husk and facilitate their removal from the endosperm.
3. Releasing hull, germ, and tip cap from the endosperm in a degermer.
4. Drying and cooling the degermer products obtained from the degermer.
5. Fractionating degermer stock by multistep milling through a series of machines namely roller mills, sifters, aspirators, gravity table separators, and purifiers to separate and recover the various products.
6. Further drying of the products is done as and when necessary.
7. Blending and packaging of products.

Cleaning of Corn

Thorough cleaning of corn is essential for the subsequent milling operations. Pieces of iron, etc., are removed by magnetic separators. Dry cleaners consisting of sieves and aspirators and sometimes a wet cleaner consisting of a washing destoning unit and a mechanical type dewatering unit, known as whizzer, are used for cleaning of corn.

Hydrothermal Treatment/Conditioning

Predetermined amount of moisture is added to the corn in the form of cold or hot water or steam in one, two or three stages with appropriate tempering times after each stage; the tempering times (rest periods) vary according to the hydration methods. So also tempering temperatures vary from room temperature to about 50°C accordingly. The optimum moisture content for degerming in the Beall. Degermer is 21-25 per cent. Either cold or hot water is used for the addition of moisture. A little heat in the form of open steam is added as and when necessary.

Degerming

The purpose of degerming is to remove hull, tip cap, and germ as far as practicable and leave the endosperm into large grits. However, the products from degermer consists of a mixture of kernel components, freed from each other to varying degrees, with the endosperm particles varying in sizes from grits to flour.

The Beall Degermer consists of a rotating cast iron conical roller mounted on a horizontal shaft in a conical cage. Part of the cage is fitted with perforated screens and the remainder with plates having conical projections on its inner surface the rotating cone has similar projections over most of its surface. The feed end of the cone has spiral corrugations to move the corn forward whereas the large end has corrugations in an opposite direction to retard the flow. The product leaves the unit in two streams. The major portions of the released germ, husk and fines as well as some of the grits are discharged through the perforated screens. Tail stock containing large amount of grits, escapes through an opening fitted with the large end of the cone. A hinged gate with an adjustable weight adjusts pressure inside the chamber and controls the flow of the stock.

Drying and Cooling of Degermer Stock

The degermer products are to be dried to 15 to 18 per cent moisture content for proper grinding and sifting. Generally rotary steam tube dryers are used for drying the product. Rotary Louver type dryer 'Can also be employed. The stock is heated to about 50°C. Counter-flow or cross-flow rotary, vertical gravity or fluidized bed types of cooler can be used for cooling the dried products.

Rolling and Grading

Recovery of various primary products is the next step. Further release of germ and husk from the endosperm products occurs during their gradual size reduction roller mills. The germ, husk and endosperm fragments are then separated by means of sifters, aspirators, specific gravity table separators or purifiers. Sifting is an important operation and is variously referred to as scalping, grading, classifying, or bolting depending upon the means used and purpose. Sifting is actually a size separation operation on sieves. Scalping is the coarse separation made on the product leaving a roller mill or degermer. Grading or classifying is the separation of a single stock (usually endosperm particles) into· two or more groups according to particle size. Bolting is the removal of hull fragments from a corn meal or flour.

Another modem corn dry milling method, known as Miag process. The flow diagram of the Miag process is given in the Figure 2.3 but detailed process has not been described here.

Corn Wet Milling

It has been discussed earlier that pure starch, pure germ and feed are the basic products of corn wet milling. But a few hundreds of byproducts can be produced from these three main products. A list of these byproducts with their uses is given in Table 2. 1. The raw corn for wet milling should contain 15-16 per cent moisture and it should be physically sound. Insect and pest infested, cracked and heat damaged corns (treated at temperature around 75°C during drying) are unsuitable for wet milling. The heat damaged corn affects the quality of oil extracted from its germ. Sufficient amount of moisture is added to the corn during steeping in the wet milling process in order to prepare the corn for subsequent degerming, grinding and separation operations. The wet milling process consists of the following steps: (a) cleaning, (b) soaking, (c) germ separation and recovery, (d) grinding and hull recovery, and (e) separation of starch and gluten.

Table 2.1: Corn wet milling products

Product	Feed/Food Uses	Industrial Uses
Germ oil and meal foots	Livestock	Soap, glycerine, leather dressing, etc.
Refined oil	Salad and table oils, cooking oils, margarine	Pharmaceutical
Steep water	Yeast-food	Phytic acid, inositol
Gluten and hulls	Livestock and Poultry feed	-
Starch	Corn starch, chewing gum, bakeries, baking powder, brewing confectionery	Textiles, laundry, paper and paper boxes, explosives, cosmetics, adhesives.
Syrup	Bakery products, canned fruits, ice cream, confectionery, soft drinks, chewing gums, mixed syrups, and jellies.	Textiles, leather tanning, pharmaceuticals, tobacco.
Sugar	Bakery products, pharmaceuticals, jams and jellies, ice cream, canned foods, confectionery.	Rayon, tanning, fermentation, brewing, vinegar, caramel colour, fermentation products, tobacco.

Cleaning

Impurities such as dust, chaff, cobs, stones, insect-infested grain and broken grain, and other foreign materials are removed from corn by screening and aspirating. The clean grains are conveyed to the storage bins.

Steeping

The major objectives of steeping are (1) to soften the kernel for grinding, (2) to facilitate separation of germ, (3) to facilitate separation of gluten from the starch granules, and (4) to remove solubles, mainly from the germ. Water impregnated with SO_2 (*i.e.*, acidulated: water with H_2SO_4) is used for steeping. It helps in arresting certain fermentation during long steeping process. The steeping is carried out at about 50°C for a period varying from 28 to 48 hours in different plants. The steeped corn attains a moisture content of about 45 per cent. The flow diagram of corn wet milling process is shown in Figure 2.3.

Germ Recovery

The wet and softened corn kernels containing about 45 per cent moisture are conveyed to the degerminating unit. This machine consisting of a metallic stationary plate and a rotating plate with projecting teeth is employed only for tearing the soft kernels apart and freeing the germs without grinding them. The pulpy mixture containing germs, husk, starch and gluten is passed through hydroclones, where the germ being lighter is separated from other heavier ingredients, by centrifugal force. Only modern starch plants employ hydro clones for germ separation. Otherwise the floatation method of germ separation is still in use in old types of mills.

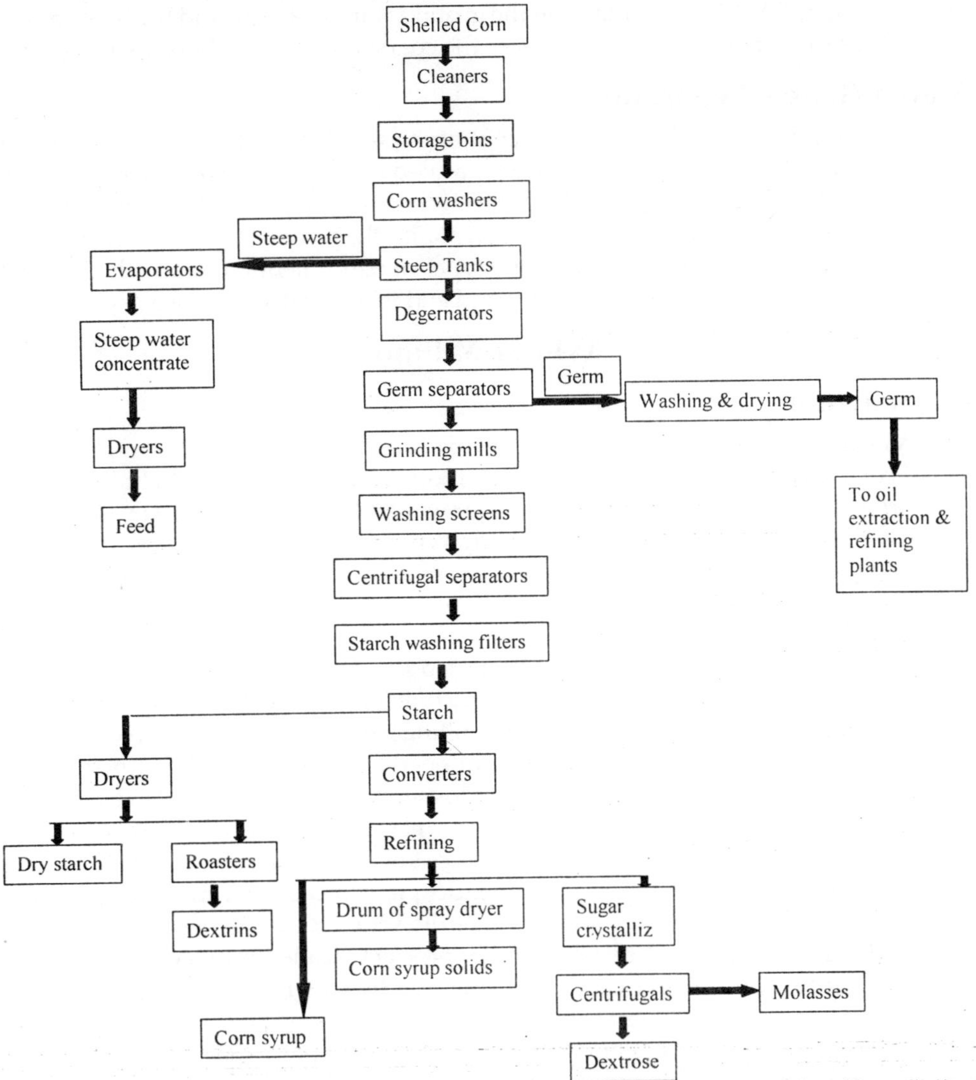

Figure 2.3: Flow diagram of corn wet milling and refining processes

Milling and Fiber Recovery

After separation of germ and screening of the coarse particles, the mixture contains starch, gluten and hulls. Mainly horny endosperm and hull are then generally ground by either traditional Burstone mill or modern entoletor impact mills to release the rest of the starch. Material to be ground enters the machine through a spinning rotor and is thrown out with great force against the impactors at the periphery of the rotor and also against a stationary impactor resulting in considerable reduction in particle size. Here only the starch is readily released, with a very little size reduction of hulls. The milled slurry, containing the ground starch, gluten, and hulls, is passed through a series of hexagonal reels where the coarser hulls and fibers are removed.

Starch-Gluten Separation

In the modern process, the slurry containing starch and gluten is concentrated and then the lighter gluten particles are separated from the relatively heavier starch particles by the centrifugal force in high speed centrifuges. The centrifuging of starch is carried out in two stages. In many modern plants, the second stage of centrifugation is performed by a number of hydroclone types of equipments. The starch obtained from the second stage of separation is filtered and then dried to produce dry starches.

Wheat Milling

Introduction

Wheat is the principal food grain in many countries of the world. It is one of the most important cereals and is used as staple food in the form of flour. In India, a large proportion of wheat is 'Used as the familiar *atta* and *maida* (white flour). The hard wheat's are also ground into *suji* (semolina). Whole wheat is ground into atta by the traditional stone grinder without prior separation of bran and germ from it.

Flour Milling

The objective of modern flour milling is to obtain the maximum amount of white flour from the wheat endosperm without any bran or germ content. Conditioning of wheat by hydrothermal treatment prior to milling helps in the separation of barn and germ from the endosperm. If wheat is conditioned by hydrothermal treatment, bran and germ become rubber-like while the endosperm becomes soft. It also eliminates the difference in grinding characteristics between soft and hard wheat. When the conditioned wheat is sheared by the corrugations of first break roll during the milling operation, it splits open releasing small endosperm pieces and thus exposing the remaining endosperm which could be carefully scraped off the bran in successive break rolls. The yields of white flour and byproducts (called mm feed) from white flour milling are about 70 per cent and 30 per cent by weight, respectively. The mill feed is composed of 12 per cent bran, 3 per cent germ and 15 per cent shorts. Wheat consists of bran (12 per cent), germ (3 per cent) and endosperm (85 per cent). Modern flour milling consists of six steps: (1) receiving, drying and storage of wheat, (2) cleaning, (3) conditioning; (4) milling into flour and byproducts, (5) packaging and storage of finished products, and (6) blending. Of them the most important operations namely, cleaning, conditioning and milling have been discussed here.

Cleaning

Wheat is thoroughly cleaned to remove all fine impurities and the dirt sticking to the surface of the grain. To remove loose fine impurities a set of cleaners is employed. Small pieces of sticks, stones, sand, etc., are removed by sieving and the light impurities like chaff, etc., are removed by aspirations. Then the wheat is allowed to pass over powerful magnetic separators to remove pieces of ferromagnetic materials. The seeds of other food grains, defective grains and weed are removed by disc evaporators. The next step in the cleaning operation is the removal or dirt sticking to the surface by scouring. Usually, wheat is moved by paddles against stationary emery-coated surface. Then the dirt and loose outer coating are aspirated off. The scratches and cracks formed in wheat during scouring help in increasing the rate of moisture absorption at the time of washing and conditioning. The final cleaning step is washing by water which allows the dirt and bits of metal to sink. The moisture content of wheat is increased by about one per cent during washing.

Conditioning/Hydrothermal Treatment

The conditioning of wheat can be done either at room temperature, elevated temperature or at high temperature. But the temperature of wheat grain should not be raised above 47°C otherwise the gluten quality will be affected which deteriorates the baking quality of the flour. Generally the moisture contents of soft and hard Wheat's are increased to 15 to 17 per cent and 16 to 19 per cent respectively by soaking and then the moisture of the grain is equilibrated by tempering for 18 to 72 hours in the tempering bin. In a modern system conditioning of wheat is performed in four stages. The conditioner mainly consists of three sections namely, preheating section, moistening section, and cooling section. In the first section wheat is preheated to the proper temperature, in the second section wheat is moistened to the desired moisture level and in the third section soaked wheat is cooled to the room temperature. Finally the treated wheat is kept in a separate tempering bin for 18 to 72 hours.

Hydrothermal treatment of grain by direct steaming has been popular for the last few years. It has many advantages over heating by air because both moistening and heating are carried out simultaneously in a single operation. Moreover, the grain is heated within 20 to 30 seconds to about 47°C. But the grain temperature above 47°C may adversely affect the quality of the flour. The rapid rate of heating weakens the intermolecular bonds in various parts of the grain to a considerable extent resulting in easier separation of bran, more effective grinding of endosperm and stronger action on proteins and enzymes.

Grinding (Milling)

Milling of wheat is carried out by roller mills. The roller milling system is mainly divided into the break roll and reduction roll systems. In addition most of the flour mills keep a stand-by system known as scratch system. The scratch system is nothing but an extension of the break roll system. The break rolls and the reduction rolls are differentiated with the variation in their surface conditions. The surface of the reduction roll is smooth whereas the surface of break roll is corrugated. In the break rolls, the bran is cracked, the kernel is broken open. The endosperm adhering to bran is milled away successively in a few steps. Generally a series of four sets of break rolls are used. Each set of rolls takes stock from the preceding one. After each break, the mixture of free bran, free endosperm, free germ and endosperm still adhering to the bran is sifted and separated. The endosperm adhering to bran is passed through the next break roll while the middle size endosperms called middling's are sent to the reduction rolls for proper size reduction to flours. Therefore, the break rolls are mainly used for the production of middlings and the reduction rolls are used for grinding of free middlings into proper flour size. After each reduction of endosperm (middlings) the flour is sifted away from the bigger size middlings and the remaining middlings are passed to the next reduction rolls. The above operations are continued until the desired products are obtained. As many as 12 to 14 reduction rolls are used in most flour mills. But all reduction rolls are not used for all break products.

Storage of Finished Products

The flour and the mill feed (bran, germ and shorts) are bagged in waterproof bags, stitched and stored in cold dry condition in flat godowns. The flow sheet of all process is shown in Figure 2.4.

Components of a Wheat Mill

Break Roll

Break roll consists of, twin pairs of corrugated steel rolls. One roll of a pair revolves faster than the other, differential speed being in the proportion of 2.5 to 1.

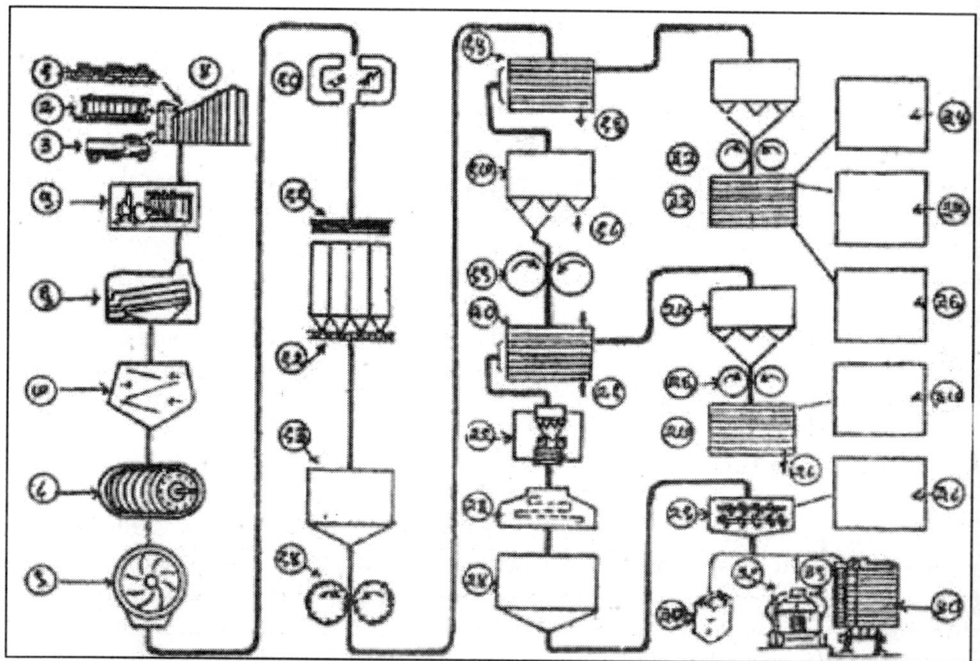

Figure 2.4: Flow sheet of wheat processing

1. Ship 2. Railway 3. Truck 4. Storage (godown) 5. Examination and combining 6. Sieving 7. Air cleaning 8. Cleaning by disc 9. Scorers 10. Magnetic cleaning 11. Condition of wheat 12. Combing 13. Milling tank 14. First breaking by roller 15. Sieving 16. Maida 17. Purification 18. Brain + shorts 19. Reduction rolls 20. Sieving 21. Maida 22. Reduction rolls 23. Bleaching 24. Large storage 25. Purification 26. Germ rolls 27. Sieving 28. Maida 29. Addition of minerals and vitamins 30. Bagging 31. Maida transport 32. Reduction rolls 33. Sieving 34. Brain 35. Shorts 36. Maida 37. Germs 38. Patient Maida 39. Transport by truck 40. Transport by train.

Break Sifting System

This can be divided into two parts-plan sifters and purifiers. (a) *Plan sifter*: Plan sifter is a scalping system removing large bran pieces adhering with endosperm at the top. The next series, which are finer, remove the bran and germ. The next layer of still fine sieve removes the endosperm middling and the bottom rough flow. (b) *Purifier*: The middling containing finer bran particles are removed by purifier before they move to reduction roll.

Reduction Rolls

The reduction roll comprises of two smooth rolls. The rolls in the reduction system are further divided into coarse rolls and fine rolls depending on the clearance between the rollers. It is possible to grind flour into very fine particles by gradual grinding. But under high grinding pressure the starch is ruptured and this should be avoided.

Reduction Sifting System

The same plan sifting system is used here. After each reduction the product is separated by plan sifters where the finished flour is sifted by 120 mesh sieve (silk) and removed and oversized material is sent back to the reduction rolls for further processing.

Scratch System

If the mill is functioning properly, *i.e.*, good release of endosperm is obtained on the break rolls, the scratch system can be bypassed, if not, the scratch system is employed to maintain proper release of endosperm from bran. The scratch system is an extension of the break system and thus used as stand-by system only.

Rice Processing

Parboiling of Paddy

Rice parboiling process consists of five major steps-soaking, steaming, drying, conditioning and cooling. During soaking the water should penetrate the centre of the grain. The water content of paddy after soaking increased to about 50-55 per cent on dry basis. During the steaming process the rice is gelatinized. Finally the paddy is dried to desirable moisture content so that it toughens enough and does not break easily during milling. A few Traditional Methods of the traditional premilling treatments to improve the milling, nutritional, cooking and keeping qualities are:

A Tapa

This treatment originated in Bengal. The paddy is soaked in water at room temperature for 24 hours and then dried in the sun (hence the name *atapa* or sun-dried). The dried paddy is then milled by traditional methods. The relative breakage of rice is more in this process.

Balam

This treatment also originated in Bengal. It is slightly better than *atapa*. Here the paddy is sprinkled with water, which inflates the grain. When the paddy is dried in the sun and milled, the hull is easier to remove.

Josh

This treatment was developed in Larkhana, Pakistan. Large earthen pots filled with paddy and water is placed on 15'cm layer of hull which is used as fuel. The pots remain on the fire overnight. The next day the water is drained and the paddy is placed on shallow iron pans and heated over fire for one hour. The paddy is then dried in sun.

Sela

This treatment was originated in Saharanpur, India. Paddy is soaked in water at room temperature for 24-48 hours and then gently roasted in hot sand (80-90°C). The roasted paddy is then further dried in sun and milled.

Siddha

This treatment originated in Bengal. Paddy is soaked in water at room temperature for 20 hours, and then boiled for few minutes. It is finally dried in sun, In this method, rice from over-soaked paddy becomes coarse in appearance, and rice from over-dried paddy shows poor milling quality.

Parboiling

Parboiling is the latest premilling treatment which improves the quality of rice. The traditional parboiling process in India is carried out in different ways. The principal methods are single-boiling and double-boiling process.

The Double Boiling Method (Dobhapa)

This involves double steaming in its sequence of operations. The vertical steaming kettles, generally two in number, are made of mild steel plates. The size of each kettle is generally 700-900 mm in diameter and 1.2-2 meters in height with about 600 mm deep conical bottom having a 300 mm diameter flanged outlet, fitted with a sliding valve. The flat top cover is fitted with a 300 mm diameter central opening for feeding the raw paddy. The kettles are provided with steam pipe extending vertically through the centre about half way down through the top cover. When the direct steam from the boiler at 6 to 7 kg/cm^2 is used, a 20 to 25 mm diameter steam pipe is supplied. The soaking is done in large masonry water tanks constructed on the ground floor. These are generally two in number working each in line with a steaming kettle. Each tank usually of 23 meters in depth hold's 20 to 35 tons of paddy. The water level in the tank is maintained such as to cover the paddy completely during the soaking period.

The process involves filling the dry paddy in the steaming kettle and opening the steam valve. During steaming the top opening of the kettle is 'covered with a gunny bag. Depending on the size of the batch the steam starts blowing out through the bag on the top in three to eight minutes. Steam is turned off and the paddy is discharged from the bottom and then quickly dumped into the water in the soaking tank. The temperature in the tank rises, as more and more paddy is dumped. Finally at the end the temperature of soak water is around 70 to 80°C. But because of large exposed surface, the temperature of soak water drops to 50 to 60°C within two hours. The paddy is allowed to soak for 24 to 72 hours after which the water is drained off. The colour of the soak water turns brown. In most cases signs of fermentation have been observed within 18 hours. These 'result in a bad odour and brown colour in the milled rice. The soaked paddy is then lifted back into the steaming kettles for the second steaming or actual parboiling operation. The procedure followed is the same as is done in case of the first steaming. The parboiled paddy is discharged from the kettles and allowed to dry in the drying yard or in a mechanical drier.

The Single Boiling Method

The paddy is soaked in cold water in the cement tanks for a few days and then steamed in the usual manner. The soaking time is generally more in the case of the single boiling process. In both these methods, during prolonged soaking of the paddy, the rice prepared out of it produces a bad odour, as a result of fermentation during soaking. These difficulties are eliminated by the improved methods of parboiling developed in India at the Central Food Technological Research Institute, Mysore, Jadavpur University, and Kolkata and in certain other parts of the world. In the modern methods of parboiling, the long steeping and steaming cycle using law temperatures is replaced by those with short cycles using high temperatures and pressures and the process is carried out either in a batch plant or a semi-continuous or continuous plant. The treatment will depend on the variety and quality of paddy and the characteristics of the final product desired. The choice of any parboiling technique will also depend on the initial investment, running costs, local conditions and amount of automation desired.

Parboiling Method of CFTRI, Mysore

The system developed by CFTRI (Central Food Technological Research Institute) was primarily aimed at improving the yield and quality of rice with a lower capital investment. The soaking and steaming is Jane in the same mild steel cylindrical tank. Steam enters through the perforated pipe at the centre and there are more perforated pipes arranged radically at the bottom of the tank. The base of the tank is cone-shaped and is closed at the bottom by a water tight hatch. At the side of the hatch there is a valve for draining of the steeping water. During parboiling the tank is filled with water heated by steam injection to 85°C. The paddy is then poured inside the tank. The temperature of water drops to 70-75°C. After two to three and a half hours steeping, the water is drained off. Pressurized steam is then passed through the perforated pipes until the husks just begin to crack open. After steaming the hot paddy is unloaded through the bottom hatch and then spread over the drying floor.

Rice Milling

Rice milling machinery used in different countries range from crude hand pounding equipment and small scale hullers to highly sophisticated and capital intensive units. However, the rice milling machinery can be broadly classified into two groups: traditional and modern rice milling machinery (Figure 2.5).

Traditional Rice Milling Machinery

Traditional rice mills include hand pounding equipments, single huller and battery of hullers, sheller-cum-huller and sheller mills.

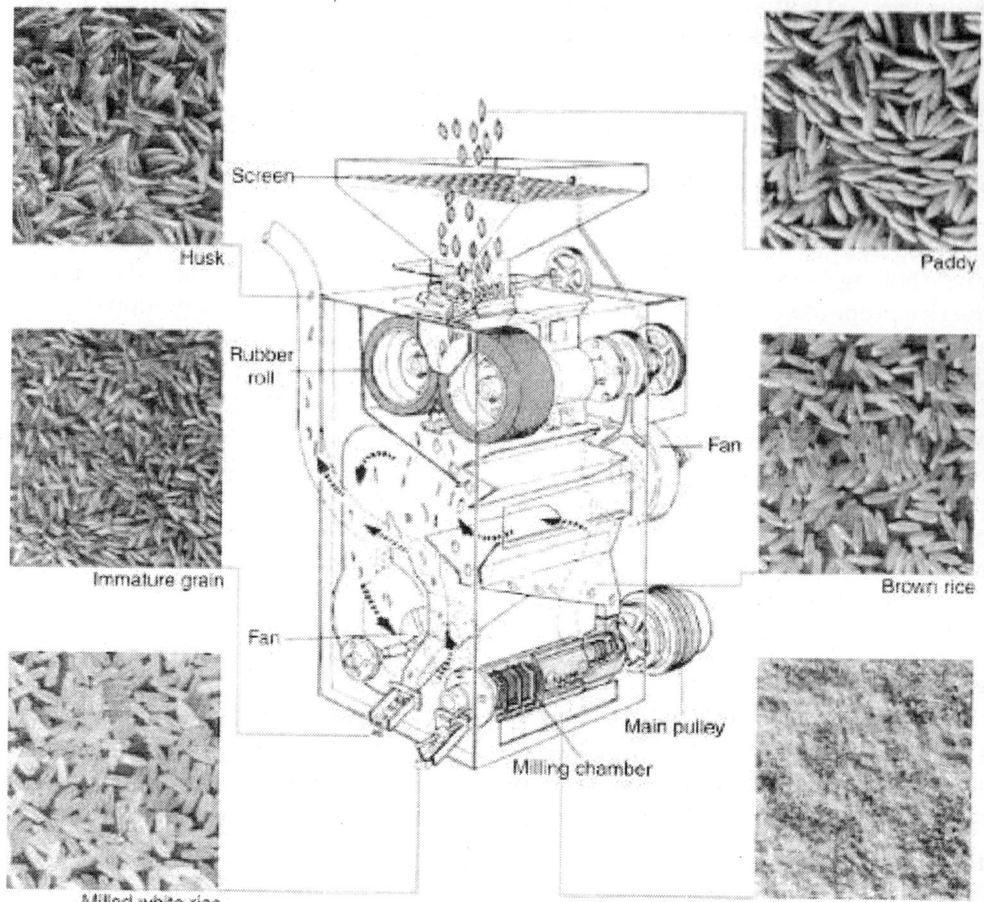

Figure 2.5: Rice pearler combined with husker

Hand Pounding

A variety of implements are used for the purpose of hand pounding, the more 'Common being; (a) mortar and pestle, (b) Dhenki and (c) hand stone *(chakki)*.

Single Huller

An iron bar called knife or blade is protruded to resist the movement of grain in the milling chamber and to control the degree of whitening. The blade can be moved forward and backward manually to adjust the clearance. Paddy is fed at one end of the upper half and discharged from the other end of the half of the housing. Paddy handling capacities of the huller from 0.25 to 0.75 tons/hr are 'common. Generally in a single huller paddy is husked and whitened in a single operation by applying high pressure resulting in a high degree of whitening and high percentage of broken in the processed rice. The total and head yields can be slightly increased by milling paddy in a number of passes. Single hullers of small capacity are located in the village for custom milling. But rice mills consisting of two to six hullers are located in tons market areas and cities. The total rice out turns is about 56 and 64 per cent for raw and parboiled paddy respectively.

Huller Mill

The huller mill consists of battery of hullers, sieves for cleaning paddy, reciprocating sieves for removing brokens, etc. and bucket elevators. Sometime, aspirator is also employed for removing husk. The machines are arranged in proper sequence. Sun dried raw or parboiled paddy is first cleaned to remove foreign materials. The clean paddy is fed to the firs, huller, used mainly for husking. Then the mixture of husked rice, paddy and husk IS fed in equal amount to the second and third hullers who are operated with narrower clearance between the blade and rotating ribbed roll. In the second and third hullers complete husking and whitening takes place. The mixture of whitened rice, husk and some bran obtained from these two hullers are separated in the sieve sifter.

Sheller Mill

As far as basic milling operations are concerned, the sheller mill is almost identical to a modern mill except that the husking operation is done by the under runner disc husker (sheller) in place of modern rubber roll husker. This type of mill consists of: (a) a cleaner, (b) one or more disc shellers, (c) aspirator (to remove husk), (d) one or more paddy separators, and (e) one or more cone type rice whitener. Commonly, the capacities and total power consumption vary from 1 to 2 tons/hr and 35 to 50 BHP respectively. Among all traditional rice mills, total and head yields are highest in sheller mill. However, in huller and sheller-cum-huller and sheller mills certain amount of husked rice, smaller and lighter paddy escape along with the husk due to defective design of the aspiration system and small brokens are also lost with the bran. Moreover, immature paddy cannot be separated in these mills and is mostly lost in the pile of husk.

Modem Rice Milling Machinery

The modern rice milling machinery can be divided into two major groups: (1) rice milling machinery developed in Japan, and (2) rice milling machinery developed in Europe. Both European and Japanese modern rice milling machines have been described in this chapter. The major operations performed by modern rice mills are as follows: (1) storage, (2) cleaning, (3) husking, (4) separation, (5) whitening and (6) grading.

Cleaning

The paddy procured from the farmer is cleaned with the help of paddy cleaners. The removal of impurities from the grains is essential to protect the subsequent milling machinery from unusual wear and tear and to improve the quality of the final product. If the procured paddy contains excessive amounts of foreign materials or if paddy is parboiled prior to milling then the procured paddy is sometimes pre-cleaned in an open double sieve type pre-cleaner or enscalper installed outside the mill room.

Husking

The purpose of a modern husking machine is to remove husk from paddy without damage to the bran layer and rice kernel. Husking machines are known by different names such as huskers, dehuskers, shellers, and sometime hullers also.

Impact Type Paddy Husker

The working principle of the impact or centrifugal type of husker is based on the utilization of impact and frictional force for husking of paddy. In the impact type husker, paddy is thrown against a rubber wall by a rotating disc. The impact on the rubber wall due to the centrifugal force of the rotating disc causes cracking of hulls with a minimum damage to the kernel. Relations between husking percentage and rpm of the disc, husking percentage and angle of inclination of the husking plate with japonica varieties of paddy are shown s that the husking percentage increased sharply from 2,000 rpm and reached its maximum at 4,300 rpm. But the head yield decreased as rpm increased. Therefore, optimum rpm of the disc for different varieties of paddy is very important. It was also observed that the most important factor affecting husking ratio was not the diameter of the rotating disc but the circumferential speed.

Rubber Roll Husker

When paddy is passed between two resilient surfaces rotating at different speeds in different directions, its husk will be split and stripped off. If paddy is allowed to pass through the gap (smaller than the thickness of paddy) between two rubber rolls rotating at different speeds then it makes contact 'with the two rolls for different periods of time. The contact of the faster revolving roll is longer than the slower revolving roll. As a result the paddy is sheared and compressed and its husk is stripped off and removed.

Both rolls have the same diameter. Diameter, of the rubber rolls varies from 150 of 250 mm depending on the capacity of the husker. The wear of the rubber is considerable and with the reduction of the roll diameter, capacity is also reduced. The main reason for the capacity reduction is the decrease in the relative speed of the, two rolls. In general only 85 to 90 per cent of the paddy fed is husked due to variation in moisture content, size, and degree of maturity of grain, eccentricity of the revolving rubber, rolls and uneven wearing of rubber roll surfaces. The clearance between the two rolls can be adjusted manually. A sophisticated pneumatic device for automatic adjustment of the clearance has been introduced in place of manual adjustment. Nowadays rubber roll huskers are equipped with a blower to blow air on the rubber rolls' surface and bring down their temperature rise due to friction between husk and roller during husking.

When the husker is in operation for a certain period, the faster revolving roll wears more than the slower revolving roll. As a result the diameter of the former becomes smaller than the latter and the difference in peripheral velocity between the two rolls becomes less. Though the speed ratio is kept constant, yet it causes lowering of husking efficiency. To encounter this difficulty application of high pressure is not the proper means as it would break the grains. However, the rubber rolls are to the interchanged at a regular interval of time.

Separation

The husked rice separated from the mixture of paddy and husked rice by the paddy separator. The paddy separator consists of a number of identical inclined trays with dimples over the surface. These dimpled trays are in reciprocating motion. When a mixture of husked rice and paddy is delivered at the upper comer of the tray, the husked rice being heavier occupies the bottom layer and comes in contact with the dimpled tray, while the paddy being lighter floats on the husked rice. The size of the dimples is slightly bigger than the brown rice but smaller than the paddy. The downward movement of brown rice is thus partially restricted by the dimples while the paddy is free to move downward. With the reciprocating movement of the tray, the husked rice, in contact with the tray picks up the movement and moves to the side, while the paddy flow downward fast by gravity. At the same time, paddy and husked rice gradually move to the right side of the tray separating the husked rice at the upper part and the paddy at lower part and leaving the mixture at the central part. The tray overflow is, therefore, received by the three compartments divided by flaps. The two flaps divide the whole receiving chamber into three compartments. The position of the flaps is adjustable so that pure brown rice can be received in the upper compartment, paddy in the lower compartment, and the mixture of paddy and brown rice in the middle compartment of the chamber.

Whitening

The term whitening refers to the operation of removal of germ, pericarp, tegmen and aleurone layers from husked rice kernels. It is also called polishing or pearling or scouring. There are three major kinds of whitening machines used in the modern rice processing industry. They are: (1) vertical abrasive whitening cone, (2) the horizontal abrasive whitening roll, and (3) the horizontal metallic friction type roll.

Vertical Whitening or Pearling Cone

Basically, the machine consists of an inverted cast iron frusteconical rotor covered with abrasive material mounted on a vertical spindle, revolving inside a crib. The crib is lined with steel wire cloth or perforated metal sheets and it is equipped with rubber brakes which are placed vertically and spaced equally and protruded into the gap between cone and crib. The pressure inside the machine can be adjusted by pushing in or pulling out the rubber brakes. The peripheral speed of the cone should be 13 m/sec. The husked rice enters the gap and is dragged along by the rough surface of the rotating cone. The rubber brakes tend to stop it and cause it to pile up against their side. While pressed up against the brakes, the grains undergo a strong swirling and revolving movement because of their oval shape and smooth surface. Each grain is scoured by the abrasive surface of the cone. It also rubs against the surrounding grains and the rough lining of the crib as well. The grains meet almost same conditions all the way round the cone until they sink lower and lower by gravity and are finally discharged at the bottom of the cone. The bran is finally ground due to scouring and rubbing and escapes through the lining of the crib and regularly removed. As the vertical abrasive roll is an inverted truncated cone, the peripheral speed at the upper part is higher than the lower part. But the density and pressure at the lower part are higher than at the upper part. As a result, the grinding action is predominating at the upper part while frictional force is stronger at the lower part. In the strict sense this machine is a combination of abrasive and friction types of machines, generally, a series of two to four whitening cones are used to whiten rice successively in two to four passes.

Grading

After rice milling and processing the grading is done on the basis of grain size and shape as well as colour using various types of machines available in the market.

3

Processing and Value-Addition of Coarse Cereals and Millets

Sorghum/Jowar Processing

Introduction

Sorghum (*Sorghum bicolour* L. Moench) is one of the major cereal crops among the all millets grown mostly in dryland. It is grown as rainfed crop in both season *Kharif* (mansoon) and *Rabi* (winter). It is mainly grown in the Deccan plateau, Central and Western India apart from a few patches in northern India. Almost entire grain produce is used for human consumption in India. It is nutritionally superior to other fine cereals such as rice and wheat and hence it is known as nutritious cereal. Nutritionally sorghum grain contains 4.4 to 21.1 per cent protein, 2.1 to 7.6 per cent fat, 1.0 to 3.4 per cent crude fiber, 57.0 to 80.6 per cent total carbohydrates, 55.6 to 75.2 per cent starch and 1.3 to 3.5 per cent total minerals (ash). Sorghum also provides 350 Kcal energy, calcium, phosphorus, potassium, carotene and thiamin as well as antioxidants through phenolics and various types of tannins. It is mainly consumed as *Bhakari* (*roti*) *i.e.* un-leavened pancake in various states of India. Apart from the traditional products like *bhakari, bhatwadi, papad, popped grains, kurda, high fiber cookies, biscuits, flakes, thalipeeth, upama, rawa idali, dosa, uttappa, chiwada, chakali, papadi, ambil, shankarpale, cookies and cakes* are also prepared from sorghum and consumed as snack food items. Sorghum grains are not only good source of nutrients but it also contains special constituents such as phyto-chemicals, dietary fiber as well as resistant starch, which are more essential to the human nutrition. However, the grain sorghum consumption has remained restricted to the weaker sections of the society due to poor nutritional quality of grain and inferior quality of the products such as Bhakari as well as very low price of this food item as compared to the other cereal grains such as wheat, rice and other millets.

Sorghum is of African origin. A large variety of wild and cultivated sorghum are grown in the tropics and subtropics of the world. In India, sorghum constitutes an important article of food, after rice and wheat. The sorghum grain is small and rounded, varying in colour from off-white to white to varying shades of red, yellow or brown. The grain size varies, the weight ranging from 7.0 to 61g/1000 grains, with most sorghums weighing 20 – 30 g/1000 grains. The chemical composition of grain sorghum is similar to that of maize. Generally, sorghum has more protein than maize, a lower fat content and about the same amount and proportions of carbohydrate components. The proximate analysis of Indian sorghum grain indicates moisture, 11.9; protein 10.4; fat 1.9; fiber 1.6; carbohydrates 72.6 and minerals 1.6; minerals present in the grain are calcium, magnesium, potassium and iron.

In comparison with maize, sorghum grain contains approximately the same quantities of riboflavin and pyridoxine but more pantothenic acid, nicotinic acid and biotin. Nicotinic acid occurs in the grain in available form.

Starch is the major carbohydrate of the grain. The other carbohydrates present are simple sugars, cellulose and hemicelluloses. The amylose content of starch varies from 2 1 to 28 per cent. Starch from waxy varieties contains little amylose. Both waxy and regular starches contain free sugars upto 1.2 per cent. Sucrose being the major constituent (0.85 per cent) followed by glucose (0.09 per cent), fructose (0.09 per cent), maltose and stachyose. Sorghum grain contains no detectable amount of glucoside, but on germination within 3 days old seedling gives 3.5 per cent Dhurin which leads to the poisoning of animals consuming such sorghum seedlings.

The percentage of different protein fractions to the total protein of sorghum grown in India is albumin 5; globulin 6.3; prolamin 46.4 and glutelin 30.4. Prolamin and glutelin are principally present in the endosperm. Amino acid analysis of various protein fractions shows that there is better distribution of all essential amino acids in globulins than in prolamins. Sorghum protein is superior to wheat protein in biological value and digestibility. A vegetarian diet based on some varieties of sorghum is somewhat better than a rice-based diet. Sorghum lipids mostly consist of triglycerides, which are rich in the unsaturated fatty acids, oleic and linoleic, their percentage being 33 and 47 respectively.

The nutritional quality of grain sorghum can be improved by malting, fermentation and by mixing of flour from other cereal grains or pulses. Malting or germination of sorghum grain have been reported to improve the nutritional quality in terms of increased soluble proteins, free amino acids, reducing sugars, digestibility of proteins and starch. Such type of grain processing treatments increases dietary fiber content in the final processed products. The higher proportion of dietary fiber of grain sorghum confer many advantages such as nutritional, digestive and physiological benefits such as hypocholesterolemic and hypoglycemic effects, incidence of colon cancer, constipation and gastro-intestinal complications etc. A large number of diabetic patients are regularly consuming sorghum bhakari with fenugreek or bitter gourd *subji* through every day meals. However, the bhakari prepared from malted sorghum flour has been found to contain higher proportion of dietary fiber with superior nutritive and digestive values with additional physiological benefits. Natural sorghum roti consumption has been restricted to only weaker section of the society due to its coarse nature and palatability. This type of problems can be overcome by using malting and fermentation technology for sorghum.

Sorghum is undoubtedly the staple cereal of Botswana and is grown widely in the country. Statistical surveys indicated that 96 per cent population of southern and Southeastern districts of Botswana mainly consuming sorghum as a major food. Several varieties of sorghum and millet, both indigenous and exotic are grown in the country. All these varieties have proved to be of good food value and may be use for different industrial applications. The development of these varieties and their subsequent use has been steered by a number of institutions in the country and abroad, notably the ministry of Agriculture, SADC/ICRISAT. The Botswana Technology Center (and its international collaborators-CIRAD, NRI and IICT) and others. An identified constraint that continues to higher the national objective of food security is postproduction losses. These losses are believed to be high in magnitude although there are no loss assessment studies carried out in the country. These losses are caused by a number of agents, mainly insects and postproduction practices.

In many parts of the world sorghum has traditionally been used in several food products and various food items, like porridge, unleavened bread, cookies, cakes, couscous and malted beverages are made from this versatile grain. The traditional food preparation of sorghum is quite varied according to the region. Boiled sorghum is one of the simplest uses and small, corneous grains are normally desired for this type of food product. The whole grain may be ground in to flour or decorticated before grinding to produce a fine particle product or flour, which is then used in various traditional foods. Sorghum is a good source of lactobacilli, which is used in souring of foods used in traditional fermented foods and drinks. Nigerians produce "*Ogi*" a traditional fermented thin porridge. Fermented breads called "*masa*" and unleavened bread called "*waina*" are also produced in Nigeria. "*Tuwo*" is a stiff porridge produced in Nigeria from adding boiling water to sorghum flour and continuously stirring. Couscous can be made from sorghum. The grain is also used in making starch, dextrose, syrup, gur, vodka, malted flour for weaning foods, beers, and other alcoholic beverages.

Sorghum plays very important role in animal feed production in various countries like the USA, Mexico, South America and Australia. The nutritional value of good quality sorghum has an average feeding value of 96 to 98 per cent of corn. Sorghum can be processed to further improve its food as well as feed value and techniques such as grinding, crushing, steaming, steam flaking, popping and extruding have all been used to enhance the grain for food and feeding purpose. The products are then fed to beef or dairy cattle; egg laying hens, poultry, and pigs are also used in pet foods.

Industrial uses for sorghum include wallboard and biodegradable packaging materials. In addition to the wallboard grain sorghum can be processed interchangeably with corn for the production of ethanol-many ethanol producing factories within the United States use grain sorghum as their principal grain due to their favourable geographic location. A very valuable co-product created from the processing of grain sorghum for ethanol is dry distillers grain. Grain sorghum having high protein animal feed supplement generates additional revenues for the ethanol processing plant.

Benefits of Sorghum Consumption

1. Sorghum food provides non-glutinous flours which is useful for avoiding damage to the lining of the intestine and make easy for nutrients to be absorbed
2. It avoid celiac disease problems
3. It also avoid dermatitis herpetiforms disease
4. Sorghum is non-acid forming food
5. Sorghum foods are a soothing and easy to digest
6. Sorghum foods are least allergenic and mostly digestible
7. It is good source of dietary fibers
8. It also good source of vitamins and minerals
9. It also provides antioxidants for controlling cancers
10. Malted sorghum flour can be used for anti-constipation drinks
11. It provides sufficient amount of dietary fiber in human diet
12. Selected sorghum genotypes can be provide enough amount of lysine to meet human requirement
13. Sorghum diet is very good to the people suffering from the Jaundice
14. Sorghum has very low glycimic index so it helps to human health
15. Sorghum helps in loose and control of body weight
16. It increases body's sensitivity to insulin
17. It also helps in diabetes control
18. It reduces risk of heart diseases
19. It also reduces blood cholesterol levels
20. It also reduces hunger and keeps fuller for longer time
21. It also helps to manage polycystic ovary syndrome in women

Anti Nutritional Factors and Mycotoxins in Sorghum

Sorghum contains polyphenols, which are generally associated with grain pigmentation. Polyphenols commonly referred to as tannins, interfere with the bioavailability of nutrients. It has been reported that sorghum grains contain substantial quantities of condensed tannins and trace amounts of hyrdrolyzale tannins. Brown sorghums with high tannins are known to be resistant to bird depredation and have reduced preharvest germination and grain moulding.

Sorghum grain is known for its hardness compared to other food grains. The hardener of the grain is due to higher content of protein prolamin. Prolamin content varies from 36.0 to 51.0 per cent. The antinutritional factors present in sorghum grain are polyphenols, and phytic acid. Polyphenols are the secondary metabolites produced and the presence of polyphenols/tannins in diet result in poor digestibility of proteins as they bind the proteins present in grain and make them unavailable for the intestinal absorption. Mostly polyphenols present in colored sorghums. In white sorghum, flavan 3-ols or flavan 4-ols (monomers of polyphenols) are present in very low quantity. Apart from flavan 4-ols, phytic acid is also present in sorghum in the form of 6-inositol phosphate. Phytic acid usually forms insoluble compounds with minerals like calcium, iron, magnesium and zinc thus making them unavailable.

Presently the new research investigations on polyphenols and phytic acid consider these compounds as health factors and consumption of these factors increase immunity in animal and human systems against several diseases. Due to grain mould the kharif grain gets severely affected by *Fusarium* and *Aspergillus* fungi, which produce harmful toxins. These mycotoxins namely aflatoxins and fumonisins cause deleterious effects to human and animal health. Hence the grain that is used for food preparation and consumption should be free of these toxins. However these toxins have a safety limit beyond which they cause ill health. The safety limit for aflatoxins is 20 ppb and for fumonisins it is 200 ppb as per the CODEX committee. Aflatoxin contamination was relatively less compared to Fumonisin contamination.

Nutritional Enrichment of Sorghum with Other Cereal/Pulse Grain

Since the need for nutritive food at low cost is in demand, and majority of the population cannot afford to buy variety of food grains for their bread and *roti* preparation, there is a great necessity to make a readymade *atta* with all the possible ingredients in a desired combination. The concept of multigrain flour is more ideal not only for nutritional and cost benefits but to improve the texture and shelf life of the flour. Efforts have been made to make composite flour with sorghum, wheat, maize, ragi, bajra and soybean keeping sorghum up to 50-60 per cent. The kneading quality of the composite flour was very good. Also the roti made of the composite flours was very soft and good in taste. The dough has good viscosity and the mean diameter of the roti made from composite flours was 24.16 cm. as compared to sorghum grain flour, which was 22.39cm. The composite flour is also ideal for making different foods listed in this book. The composite flour is very useful for children with malnutrition and under nutrition. The composite flour is also ideal to include in the mid day meal programme as the school children need complete balanced diet.

The substitution of sorghum flour to poor vegetarian diets based on sorghum and finger millet increased the protein content of the diet and markedly improved the rate of weight gain in experimental trials. Consumer acceptability tests indicated that groundnut flour could be incorporated up to a level of 20 percent with sorghum in familiar food preparations. Soybean flour when incorporated in sorghum flour, it was found that the blend of 70 to 30 per cent was found to be the best. This blend had 19 per cent protein and 2.26 PER, which is very close to that of casein (2.5). The blend was used in popular jowar recipes such as *roti* and *laddu*, which were found to be highly acceptable by the people. This blend also satisfied the I.S.I. specifications for infant foods. Therefore, preparations of roti and laddu from sorghum-soy bean blend seem to be quite promising in bridging the protein-calorie gap.

Comparison with Other Cereals

Grain sorghum is mostly used for human food as well as in small quantity for animal feeds. The utilization of sorghum grains in the world is shown in Table 3.1. From this table it is conformed that most of the sorghum grains are used as food in the Asia and Africa. Developed countries are using very small quantity of sorghum grains as a food and remaining mostly using for animal feeds.

Table 3.1: Sorghum utilization (average million tons)

Region	*Food*	*Feed*	*Other Uses*	*Total*
Africa	8.0	0.4	2.3	10.7
Asia	15.1	6.3	2.1	23.5
Central America	0.3	8.4	0.2	8.9
South America	-	4.6	0.3	4.9
North America	-	12.6	0.1	12.7
Europe	-	1.4	-	1.4
USSR	-	2.3	0.3	2.6
Oceania	-	0.4	-	0.4
World	23.4	36.4	5.3	65.1
Developing countries	23.2	15.6	4.8	43.6
Developed countries	0.2	20.8	0.5	21.5

Sorghum grains are rich in fiber and minerals apart from having a sufficient quantity of carbohydrates, proteins, and fat. The comparison of sorghum grain with other cereals is given in following Tables 3.2–3.7.

Table 3.2: Average composition of cereal grains

Grains	*Moisture (per cent)*	*Crude Protein (per cent)*	*True Protein (per cent)*	*Oil/ Ether Extract (per cent)*	*Crude Fiber (per cent)*	*CHO (NFE) (per cent)*	*Ash (per cent)*
Barley	15.0	9.0	8.5	1.5	4.5	67.4	2.6
Sorghum	11.0	9.6	8.6	3.8	1.9	71.3	2.4
Maize	13.0	9.9	9.4	4.4	2.2	69.2	1.3
Millet	13.0	10.5	9.9	3.9	8.1	60.7	3.8
Oats	13.0	10.4	9.5	4.8	10.3	58.4	3.1
Rice	11.4	8.3	7.2	1.8	8.8	64.7	5.0
Rice polished	13.0	6.7	6.4	0.4	1.5	77.6	0.8
Rye	13.0	11.6	10.7	1.7	1.9	69.8	2.0
Wheat	13.0	12.2	11.0	1.9	1.9	69.3	1.7

Table 3.3: Amino acid composition of sorghum, wheat and rice grains (g/16 g Nitrogen)

Amino Acid	*Sorghum*	*Wheat*	*Rice*
Lysine	2.60	2.72	3.68
Histidine	2.06	2.08	2.08
Arginine	4.20	4.64	7.68
Aspartic acid	7.21	3.40	4.85
Threonine	3.00	2.88	3.68
Serine	3.66	4.30	4.56
Glutamic acid	20.30	32.50	11.69
Proline	6.25	11.60	5.27
Glycine	3.45	3.20	6.41
Alanine	8.64	2.00	3.23
Cystine	0.97	2.24	1.44
Valine	4.39	4.48	6.08
Methionine	1.39	1.44	2.40
Isoleucine	3.83	3.52	4.80
Leucine	12.27	6.56	8.00
Tyrosine	3.25	2.88	4.64
Phenylalanine	4.55	4.48	4.48
Trptophan	1.12	1.12	1.28
Total	93.14	96.04	86.25
Energy (Kcal)	349	341	345

Table 3.4: Mineral composition of sorghum, wheat and rice flour (mg/100 g)

Mineral	*Sorghum*	*Wheat*	*Rice*
Sodium	21.00	19.3	2.2
Potassium	537.00	315.0	7.4
Calcium	25.00	48.0	10.0
Phosphorus	526.00	183.0	160.0
Magnesium	212.00	132.0	90.0
Iron	8.48	4.90	0.70
Zinc	3.91	2.20	1.40
Copper	0.86	0.51	0.14
Manganese	3.50	2.29	0.59

Table 3.5: Vitamins content in sorghum, wheat and rice grains

Vitamin	*Sorghum*	*Wheat*	*Rice*
β-Carotene (mg/100g)	47.00	64.00	-
Thiamine (mg/100g)	0.37	0.45	0.06
Riboflavin (mg/100g)	0.13	0.17	0.06
Niacin (mg/100g)	3.10	4.30	1.90
Total B6 (mg/100g)	0.21	0.57	-
Folic acid (mg/100g)	20.00	35.80	8.00

Table 3.6: Oxalic acid, phytic phosphorus and dietary fiber in sorghum, wheat and rice grains

Constituent	*Sorghum*	*Wheat*	*Rice*
Oxalic acid (mg/100g)	10.00	8.00	3.00
Phytic Phosphorus (mg/100g)	172.00	238.00	83.00
Phytic P as percent of total P	77.00	80.00	52.00
Total dietary fiber (per cent)	12.69	11.40	-

Table 3.7: Synergistic effects of legume (green gram) on PER of sorghum

Source of Protein	*PER*
Casein	2.87
100 per cent Sorghum	0.93
90 per cent Sorghum + 10 per cent green gram	1.11
80 per cent Sorghum + 20 per cent green gram	1.36
70 per cent Sorghum + 30 per cent green gram	1.75
50 per cent Sorghum + 50 per cent green gram	1.56

Grain Structure and Quality

Many scientists have described the kernel structure of sorghum. The kernel characteristics of sorghum described by using light, fluorescence and electron microscopy. The sorghum kernel can be divided into the following main and sub-components:

Seed Coat

1. Pericarp
   - (*a*) Epicarp
     1. Epidermis
     2. Hypodermis
   - (*b*) Mesocarp

- (*c*) Cross-cell layer
- (*d*) Tube-cell layer

2. Testa (not present in all grain types)

Embryo or Germ

1. Embryonic axis
2. Scutellum

Endosperm

1. Based on hardness (structure)
   - (*a*) Peripheral
   - (*b*) Corneous, vitreous, horny or flinty
   - (*c*) Intermediate
   - (*d*) Floury or soft
2. Based on chemical composition
   - (*a*) Waxy
   - (*b*) Non-waxy
   - (*c*) Sugary

Sorghum grain processing involves converting shelled grain into consumable or suitable for further processing. The primary processing events involves cleaning, dehulling (decorticating), pounding and milling into fine powder. Then comes secondary processing, which involves turning raw material into food that includes blending, roasting, fermentation, frying or cooking etc.

Sorghum grain requires the removal of the outer layer before any further processing because the milling and sensory qualities of sorghum for *roti/bhakari* making are influenced by the kernel characteristics. More recently, Chavan and Salunkhe (1984) have described the kernel characteristics and their implications in the nutritional and processing properties of sorghum. The sorghum kernel can be divided into seed coat, embryo or germ and endosperm (Plate 3.1). The seed coat contributes 3 to 10 per cent of dry weight of the kernel. It is made up of four sub-layers: epicarp, mesocarp, cross-cell layers and tube cell layers. The epicarp is subdivided into epidermis and hypodermis. Many times, epidermis contains pigments depending upon the genetic background of the seed. If the kernel is enclosed in pigmented glumes, these pigments migrate to the endosperm, especially, during humid weather or if crop receives rains at late maturity.

The presence of pigments (red, brown, lemon yellow or black) in seed coat or in endosperm poses a serious problem of their separation by pearling or milling which affects the quality of final product. The layer beneath the pericarp is called testa. The testa layer is complete in some sorghum grains, partial, sporadically around the pericarp in some while totally absent in others. These layers contain the polyphenolic pigments called tannins. Due to its fibrous irritating nature and presence of pigments,

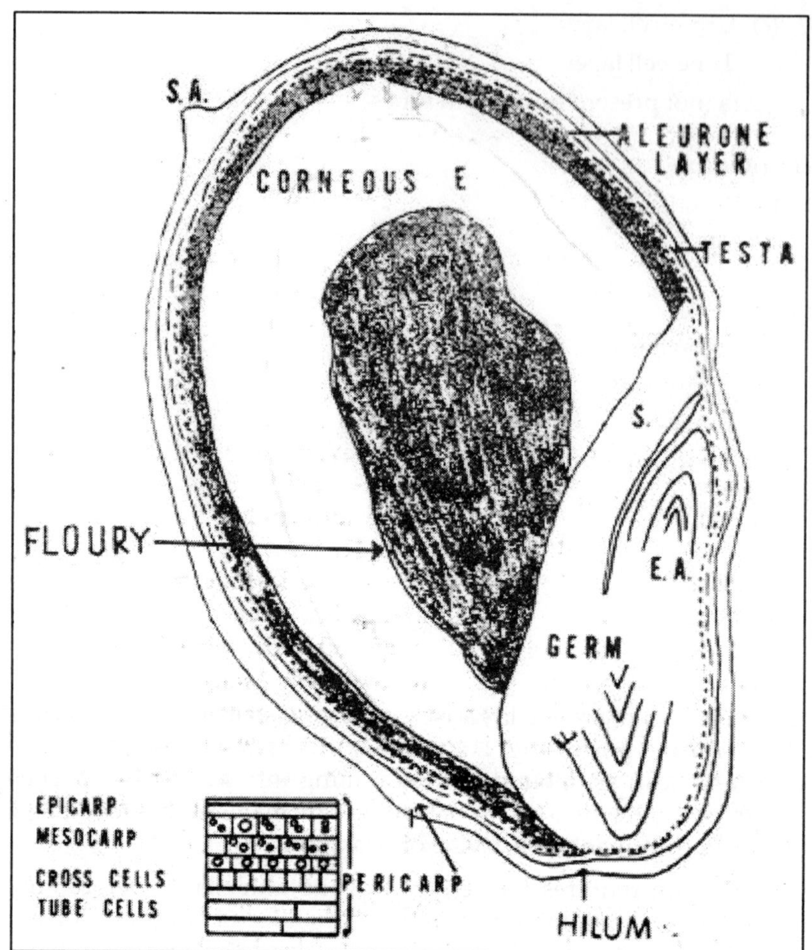

Plate 3.1: Sorghum grain structure

the seed coat and testa are considered as undesirable parts of the grain from processing and utilization point of view. In the traditional milling process, part of the seed coat and testa are ground to powder, which is not fully separated by shifting the flour through a common sieve leaving considerable amount of bran with tannins in the resultant flour. Therefore, these components need to be effectively removed during milling to improve the quality and acceptability of sorghum products. The embryo or germ contributes to about 8 to 12 per cent of the dry weight of the kernel. It contains oil globules, protein bodies and few starch granules. The germ of some sorghum cultivars is more deeply embedded inside the endosperm and is difficult to remove. The germ contains more proteins that rich in lysine. Hence, a deep placement of embryo is desirable to avoid its loss during pearling or decortications.

The endosperm is the largest component contributing about 81 to 85 per cent to the dry weight of the kernel. It consists of aleurone layer, peripheral (hardest), corneous (hard), intermediate (semi hard) and floury (soft) portions. The aleurone is a single layer of block like rectangular cells and contains a large amount of minerals, water-soluble vitamins, autolytic enzymes, oil, and protein bodies. The remaining portions of the endosperm contain mostly of nutrients such as proteins, starch, oil and minerals of the sorghum grain. The highest protein concentration is in the peripheral region and decreases towards floury area. The major proteins in the outer portions are prolamine, a lysine deficit protein while the proteins in floury region are relatively rich in lysine. The relative proportions of the corneous to floury endosperm within the sorghum kernel (endosperm texture) are known to influence processing properties of sorghum. Usually corneous endosperm is associated with more hardness, less breakage during pearling, easy removal of bran, resistance to pest infestation during storage and higher yields during milling. The physical properties of the grain influence *bhakari* making quality of sorghum. In general, pale yellow grain with intermediate corneous endosperm and with thin pericarp without testa produced *bhakari/roti* with acceptable quality. Waxy and floury grains produce poor quality *bhakari/roti*. The detailed physical properties and chemical composition of various types of grains with typical endosperm character are given in Table 3.8 and 3.9. The protein concentration is highest in peripherial endosperm and decrease towards floury. The major protein in the outer portions is prolamin, a lysine deficit protein. This results in the lower concentration of lysine in the peripherial and corneous while the floury endosperm is higher in the lysine content.

Table 3.8: Kernel characteristics, physical properties and proximate analysis of grain sorghum with different endosperm structures

Parameter	*Corneous*		*Intermediate*		*Floury*	
	Ref: 1	*2*	*1*	*2*	*1*	*2*
1000 kernel weight (g)	20.0	19.0–28.5	24.0	24.5–48.7	22.0	11.6–28.8
Hardness, per cent	33.5	17.9–63.8	15.7	1.2–44.1	7.2	0.1–30.8
Density (g/cm^2)	1.4	1.31–1.4	1.37	1.32–1.38	1.2	1.22–1.36
Protein, per cent	12.9	13.8–17.8	11.8	12.4–15.6	16.6	12.6–14.6
Lipid, per cent	2.8	2.8–4.1	2.6	3.2–4.2	3.4	3.3–4.2
Ash, per cent	1.6	–	1.3	–	1.9	–
Starch, per cent	73.4	61.5–70.5	75.2	68.1–71.4	66.7	62.3–67.5
Amylose (per cent in total starch)	29.0	–	29.0	–	28.5	–

Source: 1. Sullins and Rooney, 1974; 2. Karim and Rooney, 1972.

Various in the structural features of the sorghum grain are large. Such differences in relative proportions of the seed coat, germ and endosperm structure and types affect the nutritional and processing quality of the grain. Identification of the cultivars early in the breeding programme for specific and uses can be possible by selecting the lines with certain physical characters such as pericarp thickness, presence or absence of testa, endosperm texture, placement and proportion of germ, hardness, and shape of the grain.

Table 3.9: Kernel characteristics, physical properties and chemical composition of sorghum grains with different endosperm types

Parameter	*Waxy*	*Normal*	*Sugary*	*Yellow*	*High–Lysine*
1000 kernel weight (g)	27.6–34.6	24.5–48.7	22.4–33.1	26.6–59.2	19.3
Hardness, per cent	4.3–43.5	1.2–44.1	5.0–29.1	16.4–29.6	–
Density (g/cm²)	1.32–1.33	1.32–1.38	1.29–1.30	1.32–1.36	–
Protein, per cent	13.3–15.7	12.4–15.6	14.7–17.3	11.4–14.0	11.7
Lipid, per cent	3.4–3.7	3.2–4.2	5.2–5.4	2.7–3.7	–
Starch, per cent	68.8–72.6	68.1–71.4	60.9–64.1	69.7–75.9	–
Amylose per cent (in total starch)	0.79–4.8	27.72	26.80	29.82	–
Lysine (g/100g protein)	–	2.34	–	–	3.51

Source: Karim and Rooney, 1972; Figures from Miller and Burns, 1970; Sullins *et.al.*, 1975.

The outer layers of certain sorghum varieties seed contain tannins, which are slightly toxic or have a bitter taste. Traditionally the processing of sorghum and millet has been carried out by grinding the whole grain between two plates of stones or by pounding grain using a pestle and mortar. The latter method is the commonest. The grain is then winnowed using levels to remove the bran. Pounding and winnowing are repeated several times for getting good quality flour. The objective of hand pounding is thus twofold, the first is to remove the bran and the second is to produce the good quality flour, but this is time consuming and backbreaking system.

The improved methods of processing sorghum and millets improve two stages, which use a dehuller and a hammer mill. The dehuller is used to remove the bran from the sorghum before it is milled. A dehuller has an abrasive disc or stones set on a horizontal shaft rotating at a high speed inside a casing. The grinding action of the spinning stones and the friction of the other grains rub off the bran. After primary processing, cereal products, flour or whole grain are further processed in the home and by small cottage industries into final products. Commonly final products from sorghum and millets include foods with porridge, fermented drinks and weaning foods. There is still a great potential for sorghum to be processed into snack foods, breakfast cereals, whole grain foods and baked products.

Milling

Dry milling separates the grain in to three components, germ, endosperm and seed coat. Milling techniques practiced mostly depend on the end use of the products (Plate 3.2). Pounding in a mortar and pestle to remove outer bran, with dry or slightly wet grain, does traditional milling (Plate 3.3). After pounding, the outer bran is removed by winnowing and the endosperm is pulverized in the same pestle and mortar or ground as small *chakkai*. Pounding is a very laborious and time-consuming process. Hence the usage of hand pounding is not advantageous to practice in any food enterprise. The dry milling process starts with the cleaning of grains. The cleaned grain is conditioned by addition of water, to soften the endosperm, and milled by the

Plate 3.2: Flourmills; 1. Vertical mill; 2. Horizontal mill and 3. Pin mill

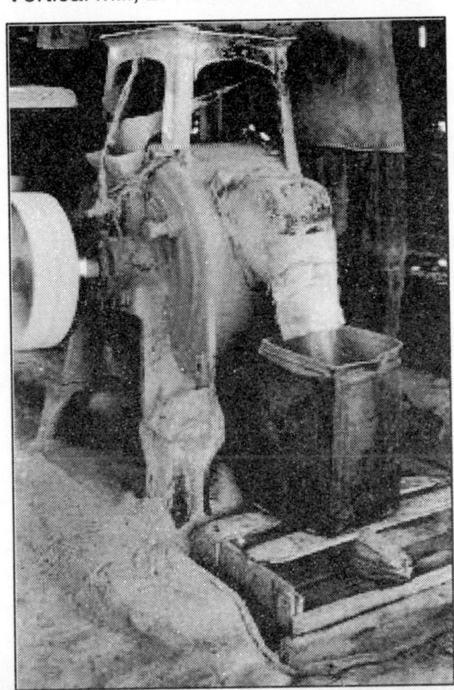

Contd...

Plate 3.2–*Contd...*

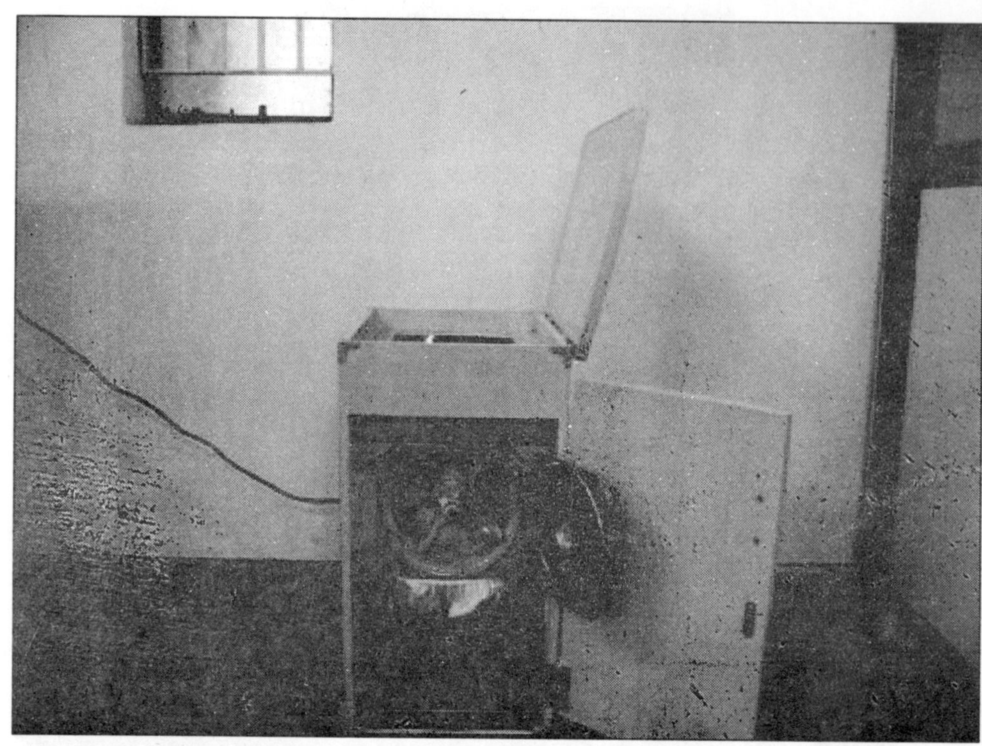

conventional roller mills to separate the endosperm, germ and bran from each other. The endosperm is recovered in the form of grits, with the minimum production of flour. Yields of various fractions from the dry milling process are grit, 76.7; bran 1.2; germ 11 and fiber 10 per cent. Bran and germ are further processed as in the case of maize by dry milling for the preparation of oils and feeds.

Milling of Sorghum

Milling process of sorghum is called pearling and decortications. In this case cleaned grains are wetted by spraying water for 2 – 3 min and immediately milled in a rice huller to remove a major part of the coarse fiber, pigment, with minimum degree of cracking of the grain. A maximum of 12 per cent polishing can be carried out. This type of milling can give products rich in protein (up to 27 per cent) and which are also high in fat and give high yield of ash, but are low in fiber. These products are used in the preparation of food products of high protein content.

Wet Milling of Sorghum

Wet milling of sorghum is carried out by methods similar to that of wet milling of maize. Clean sorghum seeds are steeped for 48 hrs in warm water (50 °C). Steeping of water softens the kernel and assists separation of the hull, germ and fiber from each

Plate 3.3: The stone mill and the pestle-mortar system used for milling cereals in India

other. After steeping, the steep water is drained off, and the sorghum is coarsely ground in degerminating mills to free the germ from the grain. Then the ground material flows down separating through in which hulls and grits settle, while the germ overflows. The degerminated material ground in to fine flour.

However, the milling of sorghum is more difficult than that of maize because of the small size and spherical shape of the sorghum kernel and the dense high-protein peripheral endosperm layer. Manufacture of starch is the main purpose of wet milling. However, some of the pigments of the pericarp and sub coat of the grain leach out and stain the starch. Thus grains with dark coloured outer layers are not satisfactory for wet grinding. Sorghum starch is indistinguishable from cornstarch and can be used interchangeably with cornstarch in most industrial applications. Sorghum oil obtained from germ fraction, after refining is used for salads and general cooking. Kernel residues containing bran and gluten are processed as cattle feed.

The starch of waxy sorghums consists almost entirely of amylopectins. These types were developed for starch in the wet milling process. There is some evidence that waxy sorghums may be more digestible. Sticky millet grown in Burma and Hawaii (var. Splendidum) had a more glutinous endosperm and was used to make cakes and sweet meats.

Sorghum can be milled for starch and oil etc. using very similar (wet milling) technology to that of maize. Products were starch (regular or waxy), germ oil and livestock feed. The starch was used in the same ways as maize starch, *e.g.*, paper and board, in the food industry etc. Some was converted to sugar, syrup, and dextrin. There were two livestock feeds, gluten feed (21 per cent protein) and gluten meal (41 per cent protein) product.

Improved Milling Methods

Improved milling methods include roller, milling, pearling, impacting, grinding, and air classification. The roller milled products are reported to have a higher production costs partly due to lower extraction rate. This may not therefore be suitable for adoption of millets commercially. The pearling technique has been applied successfully to sorghum and other millets.

For pearl millet, extraction rates varied between 80 and 90 per cent for similar dehulling times. For elusions only 3 min dehulling is required with extraction rate of 90 per cent. Mechanical dehulling improved physical appearance and functional properties. Dehulling effectively removed the colour of all varieties of sorghum. The detail procedure of mechanical dehulling process is shown in following flow chart (Figure 3.1, Plate 3.6). Mechanical dehulling reduces the drudgery. It converts millets into convenient ready-to-cook grain and improves the quality of flour.

Roller Milling

Basically it involves a series of rolls. First, the grooved or corrugated set of rollers break opens the grains. Subsequently, the exposed endosperm is crushed between a series of smooth reduction rollers and freed from the toughened bran. Screening separates the fine grains and the coarse fraction is further fed to the next set of rollers after every pass. The bran is generally, removed by screening and aspiration. The utility of dry milling largely depends upon the maximum yield of endosperm and the use of germ and bran as valuable byproducts. The break flour obtained from the first set of rollers is mostly floury endosperm. It is low in protein, oil, fiber and ash; amounts to 10-15 per cent of the grain; behaves like starch and appears specky. Five fractions of grain sorghum obtained by specific gravity separators (Table 3.10) and found that the separation of germ more difficult from waxy sorghum than that of non-waxy.

Attrition Milling

The attrition mill fitted with saw-tooth blade plates, which provide the dehulling/ pearling action. The grains were further abrased by a drum rotating in a cylindrical screen. Finally, the hulls were separated from the kernels by an air separator. Most attrition type dehullers comprise two stone or metal discs, either or both of them rotate around a vertical or horizontal axis. The attrition is provided by introducing metal pins or blades into the surface of either or both of the rotors or the rotor and stator.

Table 3.10: Yield and proximate composition of dry-milled non-waxy and waxy sorghum

Fraction	*Yield*	*Protein*	*Oil*	*Fiber*	*Ash*
		Non-Waxy sorghum			
Whole grain	100	9.6	3.4	2.2	1.5
Grits	67	9.6	0.6	0.8	0.5
Bran	12	8.9	5.5	8.6	2.4
Germ	11	15.1	20.0	2.6	8.2
Fines	10	7.1	2.4	1.3	1.0
		Waxy sorghum			
Whole grain	100	9.8	3.2	2.2	1.6
Grits	65	9.8	1.0	0.8	0.6
Bran	15	7.1	4.8	11.3	2.1
Germ	14	15.3	18.7	3.0	8.5
Fines	6	6.5	2.3	1.5	1.0

Abrasive Milling

The abrasive mill consists of 13 carborundum stones (12-inch diameter), which can be driven at a speed up to 200 rpm (Plate 3.5). In continuous operation, grains are fed through a hopper at one end and released after stone action through an overflow

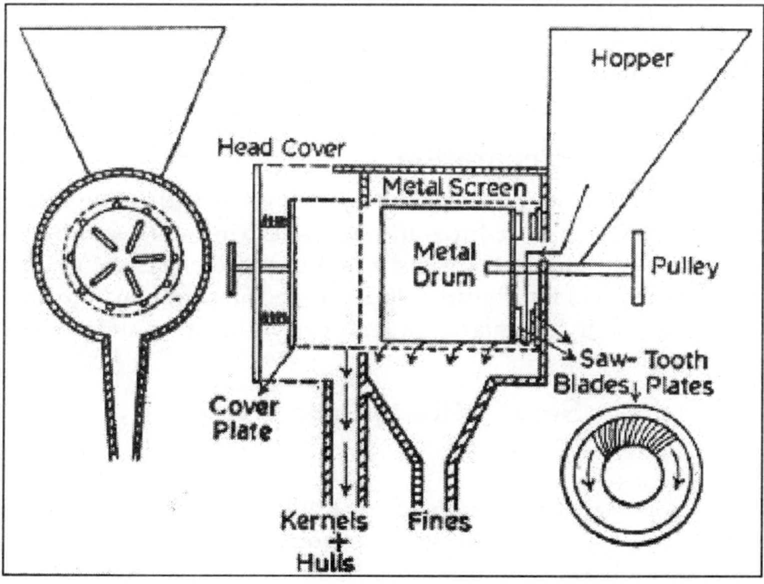

Plate 3.4: Schematic diagram of the attrition mill (*Source*: Reichert and Young, 1976)

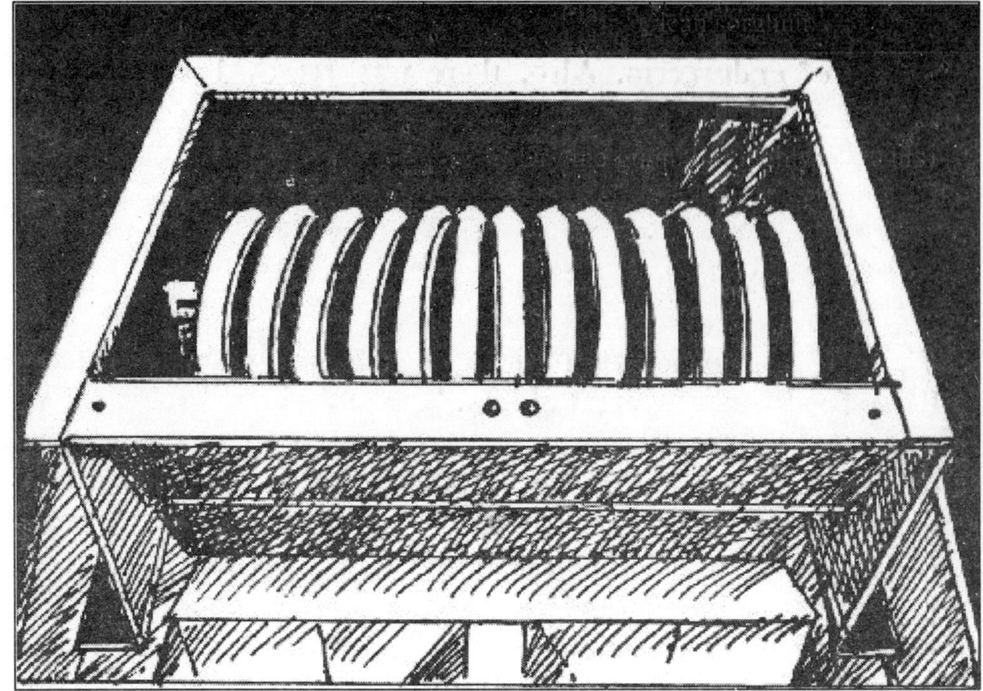

Plate 3.5: Dry milling of grain sorghum (*Source*: Reichert and Young, 1976)

outlet at the other end. The amount of kernel removed as fine is determined by the retention time in the mill, which in turn, depends upon the grain-feeding rate. After discharge from the mill the grain was passed through the air separator on the attrition mill to remove the fines. In addition to efficiency, other factors such as size, maintenance requirements and convenience favour the use of abrasive over attrition type mill for a village scale milling operation.

Sorghum Semolina Preparation

Improvement in primary process itself has helped in getting fine sorghum flour. Use of appropriate milling procedures and sieves help in getting sorghum semolina. Fine quality sorghum flour can be substituted in place of Bengal gram flour and wheat flour or wheat semolina. The process developed for sorghum flour and semolina is shown in Figure 3.2. In a similar manner, pearl millet, ragi and maize flour and semolina can be obtained.

Small Millets Processing

Introduction

Small millets provide essential food for millions of people around the world. Most of the millets are grown as subsistence crops and therefore do not receive the international research and political support these deserve. Millets have certain specialties, which make these yield products of superior nutritional and technological

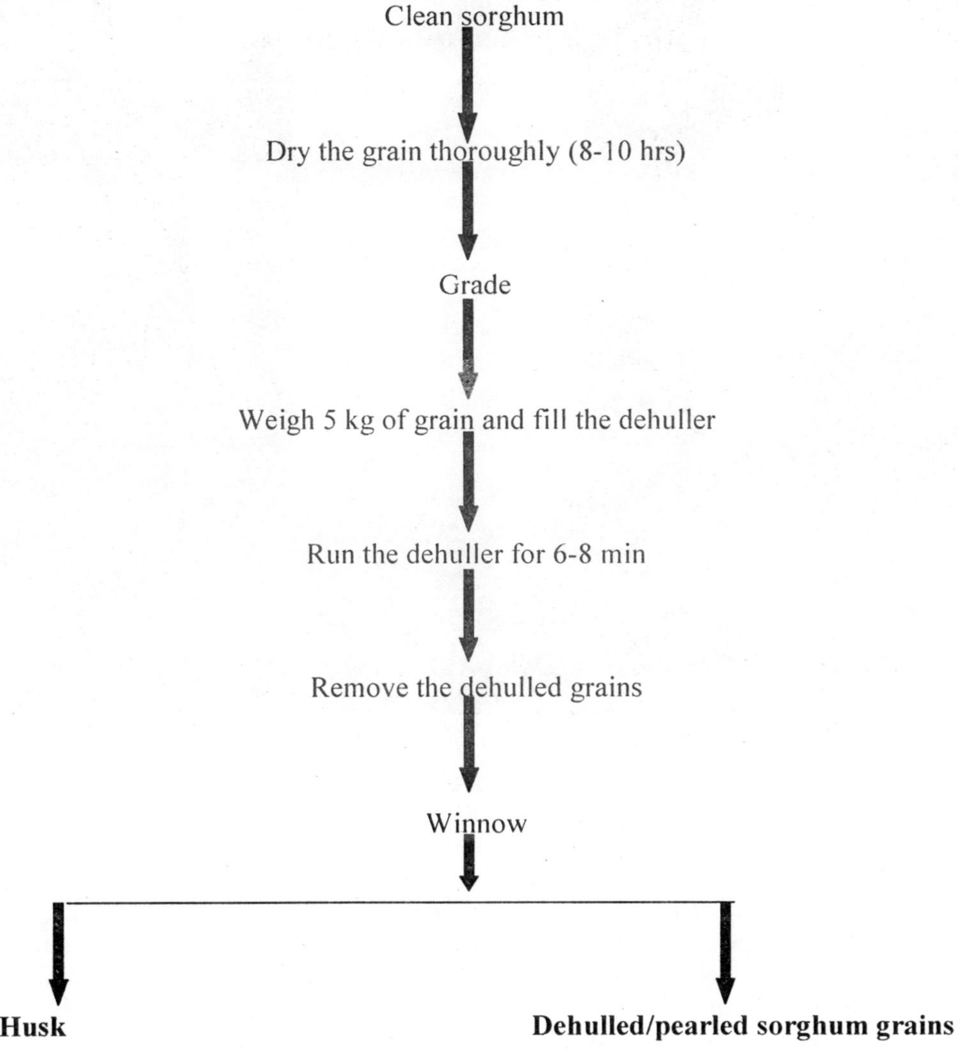

Figure 3.1: Mechanical dehulling of sorghum grains

characteristics than the major cereals. Millets have remained as the food for the people of lower socio-economic strata and traditional consumers because of their coarse texture, characteristic flavour, intense coloured seed-coat and cultural attachments. With constantly increasing awareness of good nutrition for healthy living, the consumption of millets is increasing among the affluent people also. However, the non-availability of processed products similar to rice or wheat is the primary reason for their consumption being confined to traditional consumers. There is a great need, especially in developing countries, to improve the productivity of the millets. India is the largest producer of coarse cereals. Conventionally coarse cereals are considered

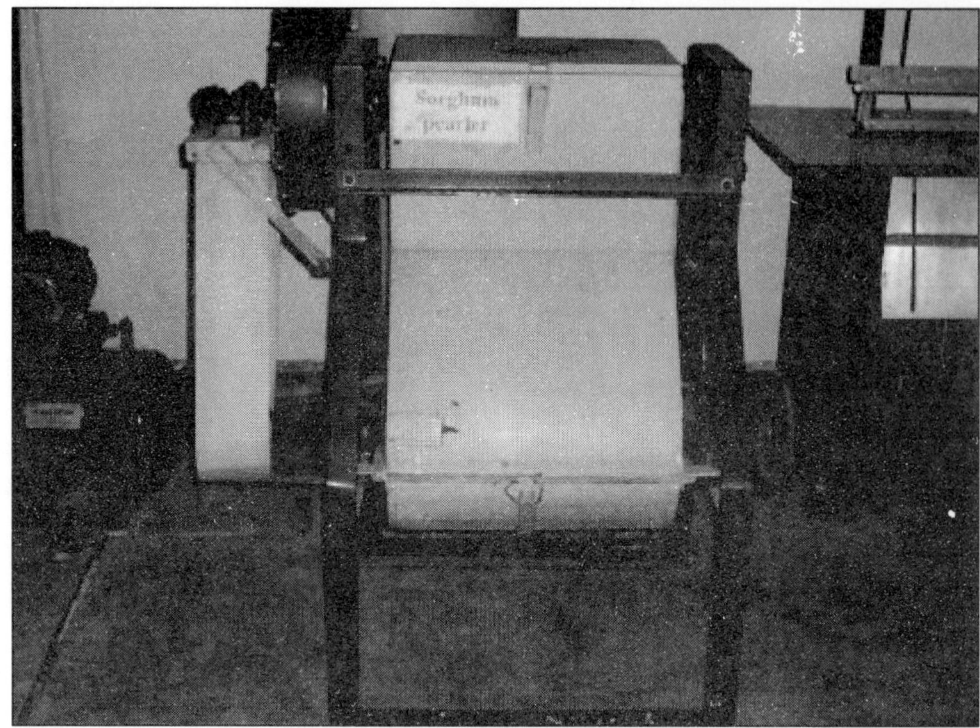

Plate 3.6: Dehulling/pearling machine

to be all cereals except rice and wheat. The important coarse cereals generally referred are sorghum *(jowar)*, barley, pearl millet *(bajra)* and maize. Finger *(R.agi)*, *kodo (kodo)*, foxtail *(kallgmli)*, proso *(cheena)*, barnyard *(sawa)* and little millets *(kutki)* are a few other common types of millets. The nutrient composition of all these grains is comparable to rice and wheat. Some of the coarse cereals are even nutritionally superior. Mineral matter, especially calcium and phosphorus is high in most of the coarse cereals. Realizing their nutrient superiority, these grains are considered as *nutri-cereals* (Nutritious grains).

Being highly resistant to extreme environmental conditions, the millets are valued as potential crops which stretch the availability of food during seasons of mansoon failures or other agronomic stresses. Yet another important property most of the millets have is that due to their tough outer layers, insects find it difficult to bore through or cause damage and thus these can be stored without getting infested for years together.

The millets belong to the grass family. The grain size of the millets is very small ranging mostly between 2 to 6 mm. About 10 to 12 types of millets are grown in our country and the major among these are the pearl millet and finger millet (Table 3.11).

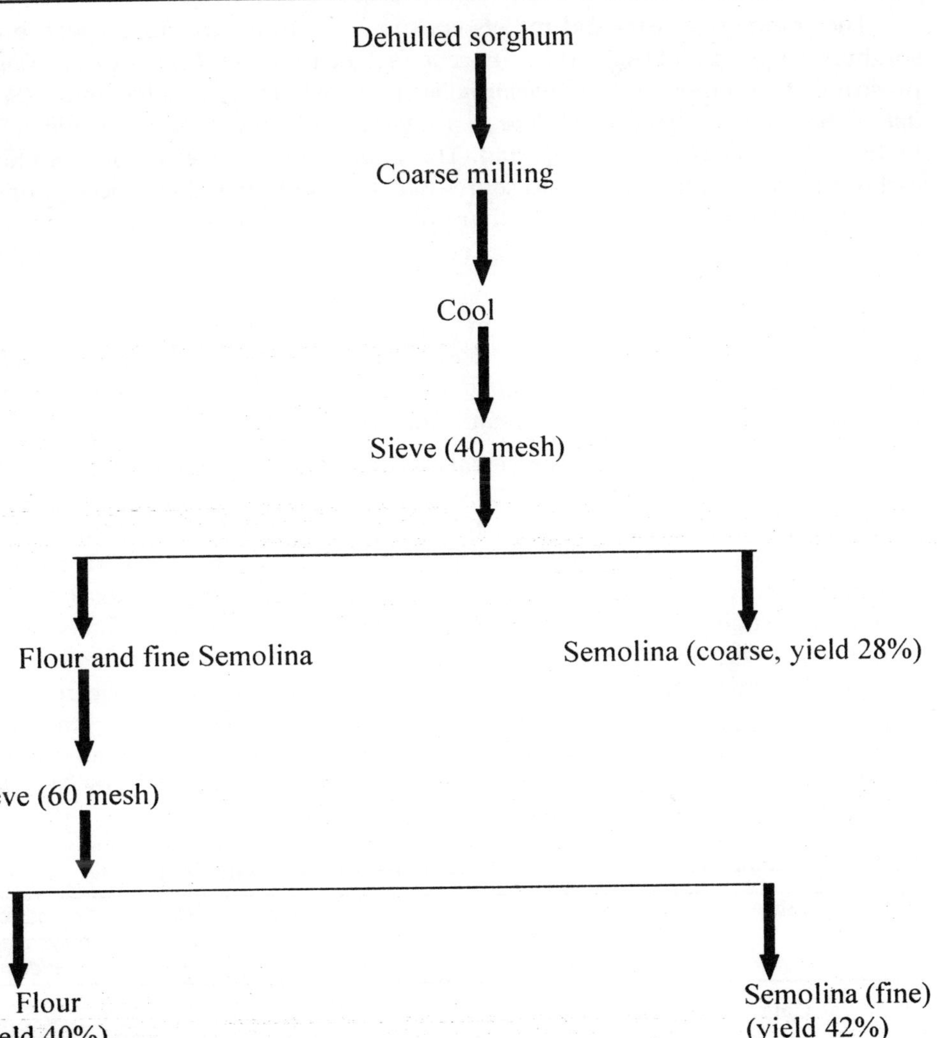

Figure 3.2: Processing of dehulled sorghum in to semolina

Table 3.11: Common, colloquial and botanical names of important millets grown in India

Common Name	*Colloquial Name*	*Botanical Name*
Sorghum	Jowar	*Sorghum bicolor*
Pearl millet	Bajra	*Pennisetum americanum*
Foxtail millet	Kangni	*Setaria italica*
Proso millet	Cheena	*Panicum miliaceum*
Barnyard rnillet	Sawan, Madira	*Echinochloa frumentacea*
Little millet	Kutki, Samai	*Panicum sumatrense*
Finger millet	Ragi	*Eleusine coracana*
Kodomillet	Kodo, Varagu	*Paspalum scrobiculatum*

The common cereals and millets consumed in India are rice, wheat, maize, sorghum, finger millet (*Ragi*) (*Eleusine coracana*), pearl millet (*Pennisetum typhoides*), proso millet/common millet (*Panicum miliaceum*), foxtail millet/Italian millet (*Setaria italica*), barnyard millet (*Echinochloa frumentacea*), little millet (*Panicum miliare*) and kodo millet (*Paspalum scorbiculatum*). The grains are rich sources of starches or carbohydrates and form the main source of energy in Indian diet. These grains are also important sources of several other nutrients in Indian diets, such as proteins, calcium, iron and B-group vitamins. Cereals do not contain vitamin A and Vitamin C.

Cereals are the foods consumed in large quantity and at greater frequency by a vast majority of population in the world. They comprise the major segment of agricultural production of any country. In about 75 per cent of the countries of the world, cereals and millets form the staple food of diets.

In most developing countries, cereals and millets form the staple foods. As their cost of cultivation and production is low, cost : benefit ratio is high both in terms of yield and also nutrients. They can be stored easily and for long periods at a low cost, as their moisture level is low. They can be consumed in bulk; they provide blendness to the diet, hence can be incorporated in infant or invalid diets. Cereals and millets have a high satiety value, prolonged emptying time of the stomach and even if consumed in large quantities do not have deleterious effects on health provided they are not consumed at the cost of other foods. The starchy endosperm of cereals is the largest portion of the grain, and contributes to 76-90 per cent of the grain weight. About 10 per cent of weight is contributed by the pericarp, testa and aleurone put together and germ occupies 4-5 per cent of the structure of the grain. The detail chemical composition of millets is given in Tables 3.12,13 & 14.

Table 3.12: Composition of some oat, rye and barley products (g/100g)

Sl.No.	*Millet*	*Moisture*	*Protein*	*Fat*	*Carbohy-drates*	*Dietary Fibre*	*Non-starchy polysacch-arides*
1.	**Oats**						
	Oat meal	8.8	11.8	6.8	68.4	6.3	6.8
	Rolled	10.1	12.0	6.8	67.2	–	–
	Quick-cook	8.2	11.2	9.2	66.0	6.8	7.1
2.	**Rye**						
	Whole flour	15.0	8.2	2.0	66.0	–	11.7
	Bread	37.4	8.3	1.7	45.8	5.8	4.4
	Crisp bread	6.4	9.4	2.1	70.6	11.7	–
3.	**Barley**						
	Whole grains	11.7	10.6	2.1	64.0	–	14.8
	Pearled	10.6	7.9	1.7	83.6	5.9	–

Cereals and millets remain an important staple food in most countries and in many parts of rural Africa and Asia provide more than 70 per cent of the energy in the diet. As countries become more affluent the importance of cereals, millets and plant foods in general declines; notwithstanding, cereal foods provide approximately 30 per cent of the energy, 25 per cent of the protein and nearly 50 per cent of the available carbohydrate in the human diet. Even with the highest income groups, cereals and millets accounts for at least 50-60 per cent of the energy intake. Considering the amount of cereals and millets consumed it is estimated that fat present in cereals and millets in our diets can meet more than 50 per cent of our essential fatty acid requirement.

Table 3.13: Inorganic constituents in cereals/millets of different extraction (mg/100g)

Sl.No.	*Millet*	*Na*	*K*	*Ca*	*Mg*	*P*	*Fe*	*Zn*
1.	**Rye**	1	410	32	92	360	2.7	–
2.	**Oat**							
	Whole	28	450	94	138	385	6.2	2.0
	Rolled	33	370	55	110	380	4.1	–
3.	**Barley**							
	Whole	49	534	52	145	356	4.6	3.1
	Pearl	2	120	10	20	210	0.7	-
4.	**Sorghum**	11	277	30	148	305	0.7	3.0
5.	Pearl millet	11	454	36	149	379	11.0	2.5

Table 3.14: Amino acid composition of millets (mg amino acid/g Nitrogen)

Amino Acid	*Rye (hulled)*	*Oat (hulled)*	*Barley (hulled)*	*Sorghum*	*Pearl millet*
Isoleucine	220	240	220	238	269
Leucine	390	450	420	850	819
Lysine	210	230	160	125	106
Methionine	90	110	100	94	150
Cystine	120	170	140	69	113
Phenylalanine	280	310	320	306	350
Tyrosine	120	210	190	95	231
Threonine	210	210	210	194	194
Tryptophan	70	80	100	63	88
Valine	300	320	310	312	338
Arginine	290	390	300	163	206
Histidine	140	130	130	131	144
Alanine	270	280	260	593	706
Aspartic acid	450	480	350	394	400
Glumatic acid	1510	1310	1470	1356	1388
Glycine	270	290	240	194	143
Proline	590	320	680	494	431
Serine	240	290	250	269	431

Production

Small millets or minor millets are important feed crops for a large section of people in rural, tribal and hilly areas. Besides, the straw is often a precious fodder for cattle. Much of the production of these crops is by small landholders and farmers in the most marginal environments. Millets are crucial to the food security of many of the poor people in semi arid-tropics. Production of these grains has either remained stable over the past three decades or there is a tendency of slow decline. India is the largest producer of millet grains, producing about 33 to 37 per cent of the total about 28 million tones of the world produce. Small millets together occupy an area of around 4.5 million hectares, accounting for nearly 15-18 per cent of the area under coarse cereals. The annual production is 3.6 to 3.8 million tones accounting for nearly 18 per cent of the total production. On account of being grown in rain-fed marginal lands with low or no agricultural inputs, the yield of these millets is very low in comparison to other cereals and ranges between 0.5 to 1 tons per hectare.

Uses

All the coarse cereals besides being used as staple food by a larger section of people offer a good number of opportunities for industrial use as well. All these coarse cereals as well as millets can be utilized as human and animal feed also. Dr. Norman E. Borlogue stated that after the revolution in wheat and rice in the last two decades, next few decades will be known for maize. It is a truth that during the severe drought in Bihar during 1887, people harvested and ate green ears of maize-crop and survived under serious famine situation. Most coarse cereals can be puffed or popped. The pearling and polishing of millets can improve the appearance and reduce the fiber content. The flour of popped grain of millets provides a fine aroma to ready-to-eat products. The marketing of flour of coarse cereals is an important cottage industry. Refined flour is useful in the preparation of many snack foods. Especially, the extrusion cooking of millets appears to be highly promising in the preparation of value-added traditional and novel food products. In hilly areas, four small millets *viz.* finger millet, barnyard millet, proso millet and foxtail millet have special status and form staple food of the poor particularly during the period of food scarcity. The foods prepared from the small millets have special significance as they provide warmth to the body during winter and also have special medicinal value and taste. Since ancient times, many traditional foods and drinks have been made from the pros a millet and foxtail millet. Popped grains, especially of finger millet possess pleasant aroma and acceptable taste. Use of finger millet malt in low bulk, nutrient dense weaning foods has been well demonstrated and the technology is well adopted. Value-added products such as parboiled, flaked, popped, and expanded, extruded-deep fried products and noodles can be prepared from proso millet. The starch of the grain is suitable as sizing agent in the textile industry.

The de-husked grain of little millet is cooked like rice and eaten or boiled in milk and served as sweet dish. Sometimes the grain is parboiled like rice. Often roti and porridge are made and consumed. It is also made into flour and used for making puddings and cakes. In parts of Northern India, de-husked grains are cooked in plain water and used either alone or in combination with finger millet or rice for preparation of Haria (fermented drink) or the grain is fried and eaten as *sattu.* Finger-millet flour can be enriched with the addition of 5 to 10 per cent de-fatted soy-flour. Composite flour can also be prepared by mixing *ragi* flour with wheat flour/rice flour/sorghum flour. Special nutritional blends can also be made by mixing malted *ragi* flour with green gram flour to be used as weaning food.

Status of Millets

Millets are major subsistence food crops in many developing countries. These are more important than what statistics indicate because in the areas where they are grown for food, these are the only crops that people can grow as a source of energy and protein. Maize and wheat cannot replace millets as crops because millets can be grown even in extreme drought conditions of the semi arid tropics. Small millets are traditionally the indispensable component of the dry land farming system. Their time of maturation is comparatively earlier than major cereal crops. They are ideally suited to diverse agro-climatic conditions of soil, water, temperature and humidity. Their cultivation stretches from the sea-level to almost 8000 feet above sea level. Under favourable climatic conditions more than one crop of these millets can be taken in one year. The grains are cooked like rice while in some areas the grains are ground to flour and used in the form of *Chapatis.* These also provide substantial quantities of palatable fodder for the cattle. Nutritionally, the grains of small millets are rich in protein, minerals and vitamins and are comparable or even superior to major cereals. The grains of small millets, unlike other cereals, are practically devoid of stored pests besides retaining viability for long periods even under poor storage conditions. All these have conferred a special distinction for small millets and appropriately referred to as famine reserves.

Small millets being eco-friendly crops demand very little application of chemical pesticides. These are the crops of fragile and vulnerable agro-ecosystems and should be considered as preferred species for a sustainable green agriculture. Small millets can be a stabilizing force in building up super cereals in industrial and food products. These crops thus can make significant contribution in natural resource management and ultimately lead to a more holistic approach in the maintenance of agro-diversity.

Place in Human Dietary

In spite of many advantages, millet grains hold only a secondary place in actual dietary consumption patterns and trends. As a class, they are considered coarse grains compared to rice and wheat. They are used as articles of human food only in the situations where other food grains cannot be raised or purchased at economic prices and they have largely remained the food of the poorer and less privileged sections of the population. In developed countries, where the production is quite large, they are used mostly as animal feed and for industrial purposes. Hardly 5 per cent of millets accounts for food uses and only refined and processed products derived from these have become popular. In developing countries, for compelling reasons of food economy, millets are used either as sole sources of grains or for supplementing rice or wheat. Even in these countries, the people who are accustomed to rice and wheat do not relish the millets. Some major features that mitigate their use are indicated below:

1. The outermost layers of bran/glumes are coarse and fibrous and if used as whole meal, are unpalatable. Being tough in nature make these irritating to the tongue and hence are not readily accepted by the people accustomed to wheat and rice.
2. These cannot be easily cooked like soft grains such as rice, as the sub-aleurone layers are hard and quite resistant to hydration thus rendering soft cooking in water difficult.
3. Millets take very long time to cook.
4. Their grains do not have the gluten characteristics of wheat and hence cannot be put to typical baking uses; hence preparation of thin *roti/chapatti* becomes difficult.
5. Most of these also have very strong characteristic flavour.

Processing of coarse grains other than maize for improving consumer acceptability has not received wide attention because these have remained the food of the poorer people of the society in developing countries. Further, the cost involved in processing is high enough to make the processed product beyond the reach of people.

Importance and Nutritional Quality of Millets

Small millets though occupy a relatively lower position among food crops in Indian agriculture; these are quite important from the point of food security at regional and farm level. The millets further contribute to the widening of food basket which at present is narrowing because of excessive dependence on a fewer food crops. The researches at the Central Food Technological Research Institute (CFTRI), Mysore, and the National Institute of Nutrition (NIN), Hyderabad, have found that the millets are superior to other major food grains including rice and wheat in many ways. All millets are distinctly superior to fine cereals in mineral composition and can contribute to effective nutritional security and human well-being.

Like other cereal grains, millets and pseudo-cereals also contain large proportion of carbohydrates about (65 per cent, mainly starch) and thus provide bulk of energy in diets based on these. A high proportion of this carbohydrate is in the form of non-starchy poly-saccharides and dietary fiber, which helps in the prevention of constipation, lowering of blood cholesterol and slow release of glucose to the blood stream during digestion. The composition of millets is shown in Table 3.15.

By virtue of their composition, some of the millets are even better with respect to the average protein content compared to rice. The pearl millet is rich in fat while the protein content of sorghum and pearl millet is comparable with wheat. Protein in millets ranges from 6-12.5 per cent and fat content from 1.1 to 5 per cent. *Ragi* has the highest calcium content among all the cereal grains (300-350 mg/100g). It develops a high level of diastatic capacity upon germination. The capacity of elaboration of diastatic enzyme in *rigi* is next only to barley and similar to that of wheat. The essential amino-acid profile of the millets is also far better than that of wheat sorghum or maize (Table 3.16). Millets have a well balanced amino acid profile and are high especially in methionine, cystine and content.

Table 3.15: Comparative analysis of nutrient composition of cereals and millets

Food Grain	*Nutrient Composition (per 100 g)*							
	Protein g	*Carbohydrate g*	*Fat g*	*Crude Fiber g*	*Minerals g*	*Ca mg*	*P mg*	*Fe mg*
			Cereals					
Wheat	11.8	71.2	1.5	1.2	1.5	41	306	5.3
Rice	6.8	78.2	0.5	0.2	0.6	10	160	3.2
Barley	11.5	69.6	1.3	3.9	1.2	26	215	1.67
Maize	11.1	66.2	3.6	2.7	1.5	10	348	2.3
Sorghum	10.4	72.6	1.9	1.6	1.6	25	222	4.1
			Millets					
Pearl millet	11.6	67.5	5.0	1.2	2.3	42	296	74.9
Finger millet	7.3	72.0	1.3	3.6	2.7	344	283	3.9
Proso millet	12.5	70.4	1.1	2.2	1.9	14	206	10.0
Kodomillet	8.3	65.9	1.4	9.0	2.6	27	188	12.0
Foxtail millet	12.3	60.9	4.3	8.0	3.3	31	290	5.0
Barnyard millet	6.2	65.5	2.2	9.8	4.4	11	280	15.0
Little millet	7.7	67.0	4.7	7.6	1.5	17	220	6.0

Table 3.16: Essential amino acids in cereals and millets

Food Grains	*Value of Amino-acids (g/100g protein)*										
	Isoleucine	*Leucine*	*Lysine*	*Methionine*	*Cystine*	*Phenyl alanine*	*Tyrosine*	*Threonine*	*Tryptophan*	*Valine*	*Histidine*
					Cereals						
Wheat	3.3	6.7	2.8	1.5	2.2	4.5	3.0	2.8	1.5	4.4	2.3
Rice	3.8	8.2	3.8	2.3	1.4	5.2	3.9	4.1	1.4	5.5	2.4
Barley	3.5	9.8	2.6	1.6	1.6	5.1	3.6	3.5	1.4	5.8	2.1
Maize	3.7	12.5	2.7	1.9	1.6	4.9	3.8	3.6	0.7	4.9	2.7
Sorghum	3.9	13.3	2.0	1.4	1.4	4.9	2.7	3.1	1.1	5.0	2.1
					Millets						
Pearl	4.1	9.6	3.4	2.5	1.8	4.8	3.3	3.1	2.0	5.5	2.5
Finger	4.4	9.5	2.9	3.1	2.2	5.2	3.6	3.8	1.6	6.6	2.2
Kodo	3.0	6.7	3.0	1.5	2.6	6.0	3.5	3.2	0.8	3.8	1.5
Proso	8.1	12.2	3.0	2.6	1.0	4.9	4.0	3.0	0.8	6.5	1.9
Foxtail	7.6	16.7	2.2	2.8	1.6	6.7	2.2	2.7	1.0	6.9	2.1
Barnyard	8.8	16.6	2.9	1.9	2.8	2.2	2.4	2.2	1.0	6.4	1.9

These essential amino acids are of special benefit to those who depend on plant food for their protein nourishment. Millet grains are also rich in important vitamins *viz.* thiamine, riboflavin, folic acid or folate and niacin. It is quite interesting to note that lower incidences of cardiovascular diseases, duodenal ulcer and hyper glycaemia (diabetes) are reported among regular millet consumers. Millets are also rich in iron and phosphorus. Most of the millets are generally used for the preparation of the *roti* (Indian pan cake), textured products *(badi)*, sweet dishes *(sattu)* and porridge *(dalia)* (Table 3.17) contains the nutritional value of some of the common Indian dishes made from the small millets and various formulations developed for feeding programmes.

Table 3.17: Nutritional value of some Indian dishes made from small millets (per 100 g)

Food Grains	Nutritional Value, per 100 9										
	Energy, Kcal	Pro-tein, g	Fat, g g	Carbo-hyd-rates	Cal-cium g	Phos-phorus g	Iron, g	Vita min-A, IU	Thia-mine, mg	Nico-tinic Acid mg	Ribo-flavin mg
Ragibade	133	2.4	2.3	25.8	0.10	0.09	1.8	68	0.15	0.36	0.04
Roti of ragi	249	4.3	4.9	47	0.22	0.16	3.2	137	0.27	0.70	0.07
Roti of maize	221	6.8	3.9	39.7	0.007	0.20	1.3	43	0.20	0.8	0.05
Roti of jawar	168	5.0	0.87	35	0.01	0.13	3.0	60	0.16	0.87	0.05
Ragi puttu	289	3.3	5.1	57.7	0.14	0.14	5.5	193	0.14	0.4	0.04
Dalia of ragi	94	2.5	2.2	16	0.11	0.07	1.16	115	0.07	0.07	0.12

Traditional Processing Techniques

Traditional food processing methods play an important role in the utilization of locally available raw materials. It is unfortunate that not many attempts have been made to improve these existing traditional food-processing techniques or to improve the quality of indigenously processed foods.

Processing involves the partial separation and/or modification of the three major constituents of the cereal grain-the germ, the starch-containing endosperm and the protective pericarp. Various traditional methods of processing are still widely used particularly in those parts of the semi-arid tropics where small millets are grown primarily for human consumption. Most traditional processing techniques are laborious, monotonous and carried out by hand. These are almost entirely left for women to do. To some extent, the methods that are used have been developed to make traditional foods to suit local tastes and are appropriate for these purposes. Traditional techniques that are commonly used include decorticating (usually by pounding followed by winnowing or sometimes sifting), malting, fermentation, roasting, flaking and grinding. These methods are mostly labour intensive and give a poor-quality product. Small millets would probably be more widely used if processing techniques are improved and if sufficient good-quality flour is made available to meet the demand.

In general, industrial methods of processing small millets are not as well developed as the methods used for processing wheat and rice. The potential for industrial processing of small millets is good. Custom milling may have a significant impact in our country like Nigeria, where about 80 per cent of sorghum and millets are now custom milled into whole flour.

In India many different kinds of traditional foods are made from small millet grains and they form the staple diet for many rural and urban households. *Ragi* is eaten in the form of *ratio* many other traditional foods are made from popped *ragi* flour mixed with sugar/jaggery/ghee/milk/butter milk and salt. In several rural households a vast variety of traditional snacks are made from *ragi* and other small millets.

People in different countries prepare various products of finger millet in different ways. In India, the crop is usually ground into flour for cakes, puddings or porridge. A fermented drink or beer is made from the grain. The grain can also be malted. Flour from the malted grain makes a healthy food for babies and people who are sick or weak. In Africa, finger millet is used to make porridge, bread, malt and beer. It is also used for brewing. On adding a small amount of malted grain to a bowl of hot, starchy porridge, it turns into a watery liquid that can be fed to small or weak babies. If you eat a lot of starchy foods like wheat, rice, maize and potatoes *etc.* grain from finger millet helps your body to digest these foods. Some people use the straw from finger millet as food for working and milking animals. It is also used for thatching, for making walls for small granaries and for making dishes.

In times of distress or emergency when these coarse cereals are offered as part of the normal ration, people accustomed to rice or wheat finds it *very* difficult to accept and utilize them as food. It is therefore essential to give some type of simple and unsophisticated processing to these grains to *improve* their acceptability. It must be emphasized that the refinement introduced during their processing should be moderate and not *over-done* so as to affect their nutritive *value* adversely.

Crushing or grinding to produce coarse grits or fine flour can reduce the particle size of the endosperm fraction. Women in their normal work atmosphere do this unpleasant hard work. Traditional milling stones used to grind whole or decorticated grain to flour usually consist of a small stone, which is held in the hand and a larger flat stone which is placed on the ground. Grain, which should be fairly dry, is crushed and pulverized by the backward and forward *movement* of the hand-held stone on the lower stone. The work is *very* laborious, and one can hardly grind more than 2 kg of flour in an hour. In a traditional process used, decorticated grain is crushed to coarse flour either with a pestle and mortar or between stones. Grain is also ground to coarse or fine flour in mechanized disk mills now located in many villages.

Flat breads are made by baking batters made with flour and water on a hot pan or griddle. Almost any flour can be used. The batter can be based on sorghum, millet or any other cereal. These flat breads are known by many local names like *roti* and *chapatti*.

Various traditional local crops produced in the southern part of Rajasthan are Barnyard millet *(batti)*, finger millet *(mal)*, foxtail millet *(kangni)*, proso millet *(cheena)*, barnyard millet *tkuris, kodra, hamlai, bhadla*. The tribal people consume these millets in various forms like *roti, dalia/thuli, kadhi, rabri*, etc. The tribes of southem Rajasthan also consume these millets as a substitute of rice after de-branning by pounding and boiling in water. *Jhajharia, rab, ghat, kheench, maize curry, ghugari, baji, missi roii, dhokle* and *papdi* are the various forms maize and sorghum, which are prepared and consumed in Rajasthan. Sorghum is occasionally popped.

Small Millets as Animal-Poultry Feed

The small millet grain can become good substitute for other cereal grains in the preparation of concentrate mixture of live stock and poultry thereby reducing not only dependence on a few cereal crops but also the cost of feed. Hard pericarp and small seed size make them more resistant to insect damage and enhance their storage period. They contain about 7 to 12 per cent crude protein, 1 to 4 per cent fat, 70 to 85 per cent carbohydrate and 2 to 4 per cent minerals. Finger millet is low in fiber content (2 to 4 per cent) while other millets contain 7 to 10 per cent fiber. Pro so millet is rich in amino acid lysine while the finger millet possesses high level of sulphur amino acid and thus these two millets appear to be promising in meeting the limiting amino-acid requirement of high yielding dairy cattle and sheep reared for high quality wool.

The small millet grains except finger millet have an outer husk, bran and starchy endosperm whereas in finger millet, seed coat is tightly fused with soft endosperm. However, from the chemical composition they appear to be equivalent to other cereal grains, which are commonly used in the ration of poultry, pigs and ruminants. These crops can also be raised for fodder purpose and harvested at flowering stage. These yield good quality green fodder in a short duration of 40 to 80 days. The fodder is rich in crude protein (8 to 10 per cent), vitamins and minerals. These are highly palatable due to higher content of sugar and carbohydrates and form good material for making silage and hay. The straw/stovers of small millet form a source of dry roughage for ruminants. It contains 3 to 6 per cent crude protein and 30 to 40 per cent crude fiber. However, the digestibility of nutrients from these crop residues is generally low and is therefore, considered as poor quality roughage. There is a scope for selection of dual-purpose varieties which could give high grain yield as well as good quality straw. The threshed seeds of millets are coarsely grounded. The grounded millets, water and *Mahua's* flower are mixed in the ratio of 2.0:1.0:0.5 and given to the cows, buffaloes and oxen at the rate of 2 to 2.5 kg per day.

Storage of Millets

Storage of crops is an essential component of the whole production system. It facilitates to fulfill several farm objectives, *viz.* storing food for the future and avoiding food shortage, providing seed during the next growing season and allowing the farmer to sell the crop at a time when the price is good. The millets are grown primarily for own consumption by the farmers and for local trade. Therefore they are stored largely at domestic or household level.

Scientific attention to the storage of sorghum and millets has been considerably less than that for other cereals. The main reason is that sorghum and millets are regarded as minor grain crops despite their relative importance as staple food in many growing countries. The other notable reason is that farmers in the arid and semi-arid countries where millets are grown achieve quite impressive performance in grain storage by employing relatively simple traditional methods.

Most millet have excellent storage properties and can be kept for 4-5 years or even up to 30 years in simple storage facilities such as traditional granaries. This is because the seeds remain protected from the insect attack by the hard hull covering the endosperm and because grain is usually harvested and stored in dry weather conditions. Thus, despite large variations in production, year-by- year stocks are easily built up over the years.

Millet grains are known for good shelf-life. The grains dried to 10-12 per cent moisture can be stored for many years in the farmhouses. Proso millet *(cheena)* was found in a farm house of a tribe in southern part of Rajasthan stored for more than four decades without any adverse effect on grain quality and that too without any proper storage measures. The millet flour and their products also show good shelf-life. Besides India, in many countries of Eurasia including China and Japan, foxtail millet, proso millet and barnyard millet have been popular food grains for many centuries. Millets may be stored, after drying and threshing, as loose grains in bags or loose containers. They are commonly left on the field, prior to threshing, in stacks or piles of arrested plants. The detached heads may also be stored away from the field, in exposed stacks or in traditional storage containers. However, the essential pre-requisites for storage of millets are the same as those for other grains.

Storage life of millets is inversely related to temperature and relative humidity during storage. Quality can be maintained by reducing storage temperature and humidity or moisture content (or all the three factors). Mould growth and intrinsic deterioration of millet in storage is negligible when the grains are sufficiently dry. At rural and domestic levels, the types of storage structures used vary between regions. Bag storage is a common practice. Storing millets in jute bag facilitates aeration of the commodity but at the same time it favours cross infestation. Wooden crates dunnage is an ideal storage as it keeps the stacks 12 cm above the floor facilitating free circulation of air underneath. But these are seldom in use due to increase in cost and lack of knowledge. The traditional storage structures are largely unsuitable for fumigation. There are no chemicals generally accepted for the control of mould growth in stored coarse cereals. Small millets are usually stored in small quantities in traditional containers, often at the farms. Large quantities are seldom accumulated and bulk storage is rather uncommon.

Storage containers vary from small traditional on-farm or domestic containers to silos, which are sometimes found on large farms. Small granaries are made by weaving together various plant materials such as bamboo stalks, bark or even small branches and finally sealing in between gaps with mud or dung. These structures may be built directly on the ground or on the platforms so as to raise them from the ground. Occasionally sorghum and millets are stored on the ground, usually un-threshed. The ear-heads are heaped in a pile (either indoors or outdoors) and covered with straw. As the grain is needed, ear-heads are removed and threshed. More often, grain is stored in gunny sacks, which are stacked either on the floor or on raised wooden platforms. Underground pits, which may be located underneath the house or outside, are also used. The pit is lined with paddy straw or sorghum straw. When it is full, the grain is covered with straw and soil. For longer-term storage, the top or roof of the pit is plastered with mud. Storage jars, silos and bins are made from a number of different materials. On the small scale, grain is stored in clay (earthen) pots. Larger containers are made from wood, brick or stone or from bamboo made into a basket, which is then sealed with clay or dung. When these containers are kept indoors they are sometimes left uncovered, but when they are kept outdoors they are covered with either a lid or a thatched roof. If the grain is to be stored for a long time, the top of the bin is plastered over with mud or dung. Occasional exposure to sunshine is the most commonly used measure for preventing insect infestation.

Flour is usually produced as it is needed and is not often stored for long periods because it tends to turn rancid. This is particularly evident in case of pearl millet flour, which has a very high fat content. Sorghum and millets, particularly pearl millet, are therefore best stored as whole grain.

Scope of Processing

The first objective of processing is usually to remove some of the hull or bran-the fibrous outer layers of the grain. Pounding followed by winnowing or sieving usually does this. The grain may first be moistened with about 10 per cent water or may be soaked overnight. When hard grains are pounded, the endosperm remains relatively intact and can be separated from the heavy grits by winnowing. With soft grains, the endosperm breaks into small particles and winnowing and screening can separate the pericarp.

Finger millet is a staple crop in many communities. It is a source of good, healthy food during difficult times, such as drought or crop failure. It's relatively cheaper, easy to grow and can be stored well for years. There are many other good reasons to grow finger millet. A single seed produces a lot of grains, so the seed is not expensive. It doesn't need any extra fertilizer or water to grow well. Finger millet can be grown almost anywhere, including hilly areas and can be planted with other crops. It can be stored well for long periods of time and that's why finger millet is often called a famine crop. Finger millet stores well because it has small seeds. This means that the seeds dry quickly. Insects can't get inside the grains to damage these during storage. Finger millet generally does not have problem of pests. Even in a humid area, the millet seeds can be stored much longer than the seeds of cereals. It doesn't rot when it's stored during the wet months. Finger millet is also a nutritious food. It contains many essential nutrients that people need to stay healthy. One of the best things about finger millet is that various nutritious food products can be cooked quickly. The small millet seeds take less time to cook.

When suitably prepared grain is pounded, the bran fraction contains most of the pericarp, along with some germ and endosperm. This fraction is usually fed to domestic animals. The other fraction, containing most of the endosperm and much of the germ along with some pericarp, is retained for human consumption. Retaining the germ in the flour improves its nutritional quality, but at the same time increases the rate at which the flour becomes rancid. This is particularly important in case of pearl millet.

Dry, moistened or wet grain is normally pounded with a wooden pestle in a wooden or stone mortar. Moistening the grain by adding about 10 per cent water facilitates not only the removal of the fibrous bran, but also the separation of the germ and the endosperm, if desired. Although this practice produces slightly moist flour, many people temper the grain in this way before they pound it. Pounding moist or dry grain by hand is very laborious, time consuming and inefficient. A woman can decorticate hardly 1.5 kg of the grains per hour with pestle and mortar. Pounding gives a non-uniform product that has poor keeping quality.

The millet grains offer many opportunities for diversified utilization and for value-addition. With proper processing it is possible to make many different kinds of food products by adopting appropriate milling, popping and other technologies. Except finger millet, other millets resemble rice in grain morphology containing husk, bran and endosperm. Traditionally the husk and bran are separated by hand pounding. However, in recent years milling technology has been improved to enhance the grain quality and to save time, as well as energy. Millet mill is available today both for cottage level and large scale processing. Milled millet can be further processed towards various food uses such as flakes, quick cooking cereals, ready-to-eat snacks, supplementary foods, extrusion cooking, and malt based products, weaning foods and more importantly health foods.

Finger millet flour is easy to make since the endosperm and bran are pulverized freely and in such flour fiber content is normally higher. However, it is possible to reduce fibre content by adopting simple sieving methods. Malting of *ragi* for food uses has been in practice from the times immemorial. *Ragi* has superior malting properties compared to other cereal grains like rice, maize, jowar and *bajra*. *Ragi* contains high level of calcium and its protein is rich in methionine and sulphur containing amino-acids. While malting finger millet does not pose problems of mould growth, off odour etc., finger millet malt has acceptable taste, very good aroma and a longer shelf-life.

Extrusion is being used increasingly for the manufacture of snack foods. In extrusion processes, cereals are cooked at high temperature for a short time. Starch is gelatinized and protein is denatured which improves their digestibility. Anti-nutritional factors get inactivated. Micro-organisms are largely destroyed and the product's shelf-life is thereby extended. The products can easily be fortified with additives.

Alternative Uses of Sorghum and Millet

These grains can be used for traditional as well as novel foods. However, there is a need to look into the possibilities of alternative uses. Though sorghum and millets have good potential for industrial uses, they have to compete with wheat, rice and maize. Sorghum in particular could be in great demand in the future if the technologies for specific industrial uses are developed. Although pearl millet has some potential for industrial use, other millets have limited potential because of their small grain size and the associated difficulties of adopting a suitable dehulling technology. However, these can be used for animal and poultry feed. There is a need to compare their performance as feed with maize.

Sorghum and millets can be adopted for other food products by using appropriate processing methods. It may be possible to select grain types with improved milling quality that will make these crops competitive with other cereals in terms of utilization. Wheat milling technology with suitable modification can be effectively used for grinding sorghum and millets. Although bread can be produced from whole sorghum flour, the quality of the bread can be improved by using sorghum flour from which the bran fraction has been removed by passage through sieves. Sorghum malt can be used to make biscuits, weaning foods and beer. Addition of sorghum malt up to 40 per cent in biscuits causes reduction in stack height and increase in spread because of increased water absorption.

The use of sorghum in common foods such as *idli* (a steamed product), *dosa* (a leavened product) end *ponganum* (a shallow-fat- fried product) can be popularized for wider use in sorghum-growing areas. A few important sun-dried or extruded products from sorghum are *papad, badi* and *kurdigai.* These products usually have a shelf- life of over one year. These can be popularized through marketing channels similar to those used for rice products. To conclude, efforts should be made to market these foods commercially to popularize these more and more amongst a greater population.

4

Processing Technology and Value Addition in Pulses

Introduction

Grain legumes or pulses occupy an important place in the world food and nutrition economy. These are important constituents in the diet of a very large number of people, especially in the developing countries and are good sources of protein, which help to supplement cereal diets, improving their protein nutritive value. These also provide substantial quantities of minerals and vitamins in the diet. Although most legumes are consumed as dry grains, immature green pods or green seeds are also used as vegetables. The availability of grain legumes over the last a few years has dropped because their production has not been very profitable as compared with other crops. The consumer demand for legumes has, however, not fallen and prices have increased considerably. Improved conservation and processing to reduce postharvest losses and the manufacture of economically priced products based on grain legumes would help to increase the supplies. The development of this industry would provide additional rural employment, improve nutritional standards, offer a better price to the grower and ensure quality supplies at lower prices to the consumer.

In Asia and Africa, a substantial protein of the legumes is consumed after having been milled (affecting removal of the husk and splitting) or after some of processing. However, most of the commercial technologies available for this purpose are either obsolete or inadequate and result in heavy losses due to breakage and powdering of the grain. Some successful efforts have been made to develop improved technologies to reduce milling losses and improve product quality. Similarly, there is a need for development and utilization of improved technologies for the manufacture of products based on grain legumes. The reduction of pollution, recovery of cotyledon material from the waste and improved utilization of energy are some of important areas of concern.

Milling of Pulses

Generally pulses are consumed in India after dehulling and splitting and converting them into *"dal"*. Removal of the seed coat (husk) reduces roughage, improves storability and palatability for consumption in various forms. It also improves cooking quality and digestibility. Conversion of pulses into *"dal"* is the third largest food processing industry after rice and wheat milling industries. It is estimated that about 75 per cent of the pulses produced in the country are converted to *dal*. Milling of pulses has been practiced as a tiny and small-scale rural operation from the times immemorial but lately it has been converted into a fairly large commercial operation. One estimate indicates that about 30 per cent of the production of pulses is still carried out on a cottage scale level by the farmers in rural areas using traditional techniques.

There are several small-scale millers who process 0.1-0.5 tonne per day either by custom milling or for trade purposes as a home-scale operation. However, with the advent of organized milling systems, most of these processing systems are disappearing as the product is inferior in quality and yield. For this reason, the *dal* available in the market mostly comes from the large-scale commercial mills. There are about 7,000 *dal* mills working in various parts of the country processing different pulses throughout the year.

Traditional and Conventional Methods

Pulses go through several primary processes juice scouring hulling (dehusking), roasting, puffing, grinding, splitting, etc. before they can be used in different food preparations. Hulling, practiced widely in Asia and Africa either on a home-scale or as a cottage industry, produces refined cotyledons with good appearance, texture, and cooking qualities. Dehusked grains are easily digested and efficiently utilized by the body. As the husk tightly envelops the endosperm, usually through a thin layer of gums and mucilage, the primary step in hulling involves laborious procedures.

The oldest and most common home-scale technique for hulling grain legumes is to pound them in a mortar with a pestle, either after spreading the grains in the sun for a few hours or after mixing them with a little water. The husk is then winnowed off to get the clean cotyledons. Methods followed in the home or village industry or in commercial mills are usually similar in principle, but different in the use of technique for better yield, operational efficiency, and large-scale application. Home-scale hulling consists generally of two steps: (*i*) loosening of the husk by wet or dry methods and (*ii*) removal of the husk and cleaning. In South Asia, the first step is achieved by sun-drying the raw mature grains as such, or after they have been treated with oil and/or water. In some areas, grain is steeped in water for two to eight hours prior to sun drying. Grain varieties whose husks are tightly attached to the cotyledons are soaked and then treated with red-earth paste before being sun-dried. The steeping technique to loosen the husk is also practiced in several Southeast Asian and African countries.

Pounding the grain in a mortar with a pestle or grinding in a hand-operated wooden/stone sheller accomplishes dry-method husking. Winnowing thereafter separates the husk. This is a common practice where dry cotyledons or grain-legume flours are used in food preparations. When a batter or dough is to be prepared, the soaked grain is either rubbed by hand to remove the husk, after which it is separated by flotation or wet-ground in a stone grinder. In several Southeast Asian countries, this technique is used for extracting starch from mung beans, the husk being removed by straining dilute slurry through coarse muslin or other cloth. Home-scale methods are generally laborious and not always hygienic. In village industries, the techniques employed for loosening the husk are:

1. Prolonged sun-drying until the husk is loosened
2. Application of small quantities of oil, followed by several hours or even days of sun-drying and tempering
3. Soaking in water for several hours, followed by coating with red-earth slurry and sun-drying
4. Soaking in water for several hours to loosen the husk before manufacture of food products and
5. A combination of these techniques.

There are no standard procedures developed for any specific variety of pulse, which really calls for characterization of pulse varieties in terms of physical and thermal properties and their processability-a fact which is often lost sight of by the breeders.

Removal of the loosened husks from the grain in the dry-milling technique is commonly done in small machines. Hand or power-operated under-runner disc-shellers or grinders with emery or stone contact surfaces are used. A plate mill with a blunt contact surface is sometimes used both to husk and split the soaked and dried grains. After aspirating or winnowing off the husk, sieving separates the split cotyledons. Remaining unsplit whole grains are similarly processed until almost all the grains are husked. In certain parts of India, oil-treated and sun-dried grains are husked in an Engelberg-type rice-huller after being mixed with 2-3 per cent stone powder. Sound kernels are removed by sieving, while the husk, powder, and small brokers remain in the stone powder.

Hulling methods are not one-step operations. About 50 per cent removal is achieved in the first operation. After separation of the husked split-cotyledons *(dal)*, the process is repeated several times until almost all the grains are converted into *dal*. In the process, excessive breakage with powdering of grains occurs because of repeated splitting and husking operations. Complete removal of the husk from the grain is not always achieved in cottage-scale processing, particularly with some varieties and hence unpolished (split and unhusked) *dal* is sometimes put on the market. The cooking quality of *dal* prepared by wet methods is usually poor. This is especially true of the pigeon pea for which cooking time increases with the duration of soaking. However, such *dal* has an attractive appearance and more desirable processing. Hulling of legumes on a commercial scale is generally based on dry-processing techniques. Many of the operations, particularly husking and splitting, are mechanized. The drying is done in large yards and is completely dependent on sunshine. Legumes such as pigeon pea, black gram, and mung bean, which are more difficult to husk, require more of the oil or water treatments followed by prolonged sun-drying (pre-milling treatments), while grains such as chickpea, lentil, pea etc., with more easily removable husks, require short periods of sun-drying and fewer oil or water treatments. Sometimes these grains are given an initial "pitting" in the roller mill to crack the husk and improve the absorption of oil or water. In the case of black gram, the coating of wax and dust is removed by initial scouring in a roller mill, facilitating absorption of water or oil. Husking and splitting are done either in a single operation or, more advantageously, as independent operations. Moisture addition adversely affects husking, but it helps to split the grain. Addition of water prior to husking helps to induce simultaneous splitting, but this often leaves patches of husk that have to be removed by scouring in polishing machines. As the husk forms 10-16 per cent of the grain, a maximum theoretical yield of 84-90 per cent of kernels should be possible during hulling. In practice, the yields vary from 68 to 76 per cent as a result of breakage, powdering, and other milling losses. In wet methods, water-soluble nutrients are also lost in the soak-water.

Improved Technologies for Hulling of Grain Legumes

The improved method developed at the Central Food Technological Research Institute, Mysore for hulling grain legumes commonly consumed as *dal* or split-grain legumes in India, Pakistan, Bangladesh, Nepal and some other countries involves a conditioning technique of moisture adjustment of the grain to a critical level in order to loosen the husk. In this process, the grain is exposed to heated air, at a specific temperature appropriate for each variety, for a predetermined time and equilibrated to the critical moisture level with gradual aeration in tempering bins. The husk is removed in an improved abrasion-type hulling machine that accomplishes almost complete removal in a single pass with the least possible scouring or breakage of the endosperm. The whole husked grain can be split in a splitting machine after suitable conditioning for which the technology and equipment have already been developed. The method is independent of the influences of climate and variations among the varieties. This improved technology has been shown to increase the yield by 5-10 per cent. The time taken for processing is also less and the cost of operation is lower. Small commercial models of complete automatic milling plants of 0.5-2 tonne per hour capacity have been designed and put into operation. The process, originally developed for pigeon pea has been adapted with suitable modifications for processing other legumes such as chick pea, mung bean, black gram, lentil, pea, soybean and some other beans.

In the Prairie Regional Laboratory, Saskatoon, Canada, a hill-threshing unit consisting of carborandum discs has been successfully used for dehulling cowpea. Initial findings indicate that the unit is capable of mechanically hulling brown Nigerian cowpea that can subsequently be converted to flour. The husk is removed by the abrasive action of rotating discs; the amount removed being dependent upon the through-put and retention time. An acceptable product would necessitate removal of about 27 per cent as polish from the cowpea, whose husk content is 3-5 per cent (22-25 per cent loss of kernel). This unit is being put into rural-scale operation for processing cowpea in some African countries.

The hulling characteristics of grain legumes are influenced by the variety, season, when harvested, and agro-climatic situation. Larger of bold-grain varieties are easier to hull, give a higher yield and are preferred by millers, while the smaller varieties require repeated and severe pre-hulling treatments and complex procedures. Freshly harvested grain and winter crops are more difficult to process, possibly because of their higher moisture content. Such grain is either stored for some time to reduce moisture or treated with limewater or a solution of sodium carbonate to loosen the husk. Whole legumes or husked splits are either ground dry into flour or ground wet into a batter for a number of sweet and savoury preparations, either alone or in combination with cereals and millets. The eating quality of many of these products, particularly the texture, depends on the composition of the flour, degree of fineness of grinding relative proportion of particles of different mesh grades and cooking conditions. Chickpea, pea, black gram and cowpea are the common grain legumes ground wet or dry.

Different Pre-milling Treatments

Wet Method

This method of dehulling generally involves water soaking and sun-drying and is common in many parts of India. The wet-method has the advantage of facilitating good dehusking and splitting of the cotyledons, giving less breakage though it may adversely affect the cooking quality. The method is also labour intensive and is completely dependent upon climatic conditions for drying the soaked seeds. The entire process usually takes about five to six days and only limited quantities can be processed at any given time. *Dal* produced by wet-method tastes better but takes longer time to cook. The soaking of pigeon pea in water for the time ranging from 2 to 14 hours is a common practice in Maharashtra, Uttar Pradesh and Madhya Pradesh. Soaking for longer periods is preferred in certain region where pulses are processed in summer season. Soaking is generally practiced at village level. In large scale operation spraying the seeds with water increases the seed moisture level. Comparatively smaller dehusking units practice alternate wetting and drying known as conditioning. Dehulling can be rendered easier by prolonged soaking in water for 12h or more but *dal* remains uncooked and tough even on prolonged boiling.

Dry Method

The cleaned and graded grains are passed through an emery-coated roller for initial pitting or scratching of the husk. Pitted grains are thoroughly mixed with about 1 per cent oil and spread in a thick layer for sun drying in the drying yards for 2-5 days. Grains are heaped during the night to preserve the heat. About 2-3 per cent of water is added to the grains before milling and the grains are subsequently passed through the roller for dehusking by abrasion. In this process, about 40-50 per cent of the grains are de-husked and major portion of these gets split simultaneously. The loosening process may be slow but the husk can be totally loosened if the treatment is extended to several days. This method is said to produce *dal* that cooks faster than the *dal* produced by the wet-method. The major disadvantage of this method is high dehulling losses due to breakage and powdering.

Sodium Bicarbonate Treatment

Krishnamurthy *et al.* (1972) tried the use of sodium bicarbonate, sodium carbonate, sodium hydroxide, acetic acid and ammonia as a replacement for vegetable oil in the traditional process and reported a considerable improvement in *dal* yield when sodium bicarbonate was used. Reddy (1981) used sodium bicarbonate (5 per cent solution) and reported a *dal* yield of 75 per cent. Saxena *et al. (1981)* treated pigeon pea grain with aqueous solution of calcium hydroxide, sodium carbonate and sodium chloride of different normalities. Sodium bicarbonate solution was reported to be the most effective, resulting in *dal* yield of 78 per cent in addition to loosening of husk. It also reduced the cooking time of the resulting *dal.*

Srivastava *et al.* (1988) showed that the pigeon-pea grain soaked in 6 per cent sodium bicarbonate gave improved dehusking efficiency they however observed an increasing loss of crude protein due to leaching as the concentration of sodium bicarbonate solution was increased. They reported the total of crude protein in the range of 6-8 per cent at 6 per cent $NaHCO_3$ concentration. Mayande (1987) reported a maximum hulling efficiency of 90 per cent for grains treated with 10 per cent $NaHCO_3$ solution at grain to solution ratio of 20:1 for 5 hours. Broken percentage was the highest in control and was comparatively less in grain to solution ratio of 20:1. The powder was the minimum in control and did not show any specific trend due to the effect of time or ratio.

Enzyme Treatment

Verma *et al.* (1993) employed edible enzyme as a husk-loosening agent. He reported a maximum hulling efficiency of 88.93 per cent at an enzyme concentration of 0.08g protein per 260g pigeon-pea grain. Grains were treated with the enzyme and allowed to incubate. During this period of incubation, enzymatic hydrolysis took place which brought about the biodegradation of complex molecules of the grain. The complex gums were degraded which resulted in easy dehusking. It established that a lesser force was required to bring about the dehusking of enzyme treated grain. The action of the enzyme also disturbed the microstructure of the grain affecting its strength. He further reported an increase in the protein digestibility and 37.03 per cent reduction in cooking time. Further, this *dal* was reported to cause less flatulence due to fermentation which broke down the poly-saccharides responsible for causing flatulence, in many people.

Advance *Dal* Milling Technologies

Processing

A large number of varieties of pulses are grown, processed and consumed in India such as pigeon pea, mung-bean, urd-bean chickpea, lentil, pea, etc. The requirements of processing differ not only as per the variety but also often differ for the same variety of pulse grown and harvested in the areas having different soil, climatic conditions. Almost all the pulses have to undergo the processing sequences as shown in the flow chart (Figure 4.1).

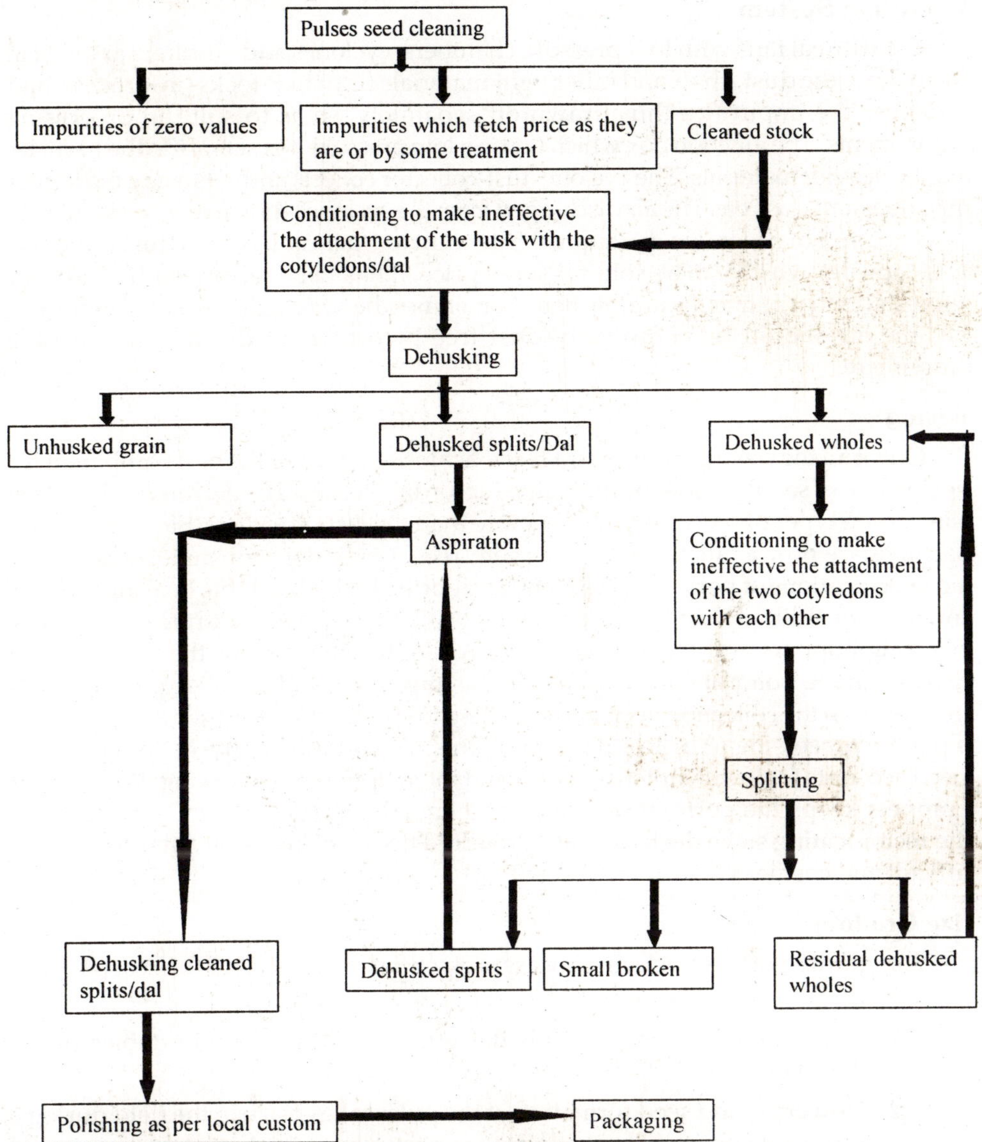

Figure 4.1: Sequences in processing of pulses (*Source*: Matanhelia, 1980)

Cleaning and Size-grading

Cleaning

Elimination of all impurities, foreign matters, and a damaged and/or unwanted grain is the first step in the processing of pulses.

Aspiration System

Centrifugal fans with low pressure chambers, cyclones and air filters are well in use to draw the dust, chaff and other light materials from the stocks- in-process, and to deliver the impurities thus drawn off separately, so as to suit the processing requirements. The junction box which is also a low pressure box sets aside the particles and the heavier materials. The cyclone dust-collector collects and sets aside the lighter impurities and dust, etc. The air discharged from the cyclone is however not completely dust- free and may contain dust upto 0.5 per cent to 1 per cent. If such dust is the fine pulse powder which can be sold for some price, it has to be recovered from the air before its discharge into the atmosphere. Sometimes the state laws make it compulsory that the air should be completely dust-free before being discharged into the atmosphere.

Sieving

Out of all the sieving machines, a reel machine seems to be the best equipment for the purpose of setting aside the impurities according to size. The *dal* mills in India are using crude types of reel machines for more than the last fifty years. Since improved machines are now available, they are preferred. A cylinder with suitable numbers and specifications of perforated sets and supported by two suitable bearings rotates slowly inside the body of the machine having suitable number of partition chambers, collecting hoppers and inspection doors. While the stocks enter the cylinder and travel from the point of entry to the exit, the impurities and the size graded stocks drop down in the collection hoppers according to size from where they are moved off as per the processing or disposal requirements. The rotation of the cylinder at slow speed accounts for low maintenance cost and interruption-free working. Provision of scrappers keeps the perforations clean and ensures high efficiency continuously. The reciprocating sieve decks are less preferred because the repair/maintenance is difficult and costlier.

Size Grading

Size grading of the stocks in process is essential on account of the reasons such as:

1. A size graded stock offers better efficiency at the gravity tables or the destoners.
2. Flowering and seed formation of the pulse takes place in the field during a prolonged period. The seeds grown in the later periods are less mature and differ in characteristics requiring different adjustments at the conditioning stage, and
3. Product of the graded raw material has better appearance and fetches better price.

Destoners

The stocks-in-process carry with it impurities such as stones, hardened mud-particles, metal and glass pieces. These impurities when equal in size to the stocks to be processed do not separate by sieving. However since these are of higher specific ıvity, a destoner can conveniently separate them. Wet destoners and dry destoners ꞉ available and the latter is preferable. Pulses being hygroscopic in nature attract of moisture, even to unwanted extent, when the stocks are passed through water a wet destoner and the miller does not have a control on it. When a dry destoner is ed, water only as per the conditioning requirements can be added in a paddle ewworm. Gravity tables can also be used with advantage but they are more costly d intricate.

eumatic Separators

Even after cleaning aspiration, screening and destoning, the stocks-in-process ıy contain some damaged, weevilled, under-developed or shriveled grains which lighter in weight (specific gravity) than the sound grains. They can conveniently separated by pneumatic separators.

ckles Cylinders or Carter Discs

These are used to separate grains which do not differ in size or gravity but differ shape. Presence of such unwanted grains is particularly significant in the case of til. Their separation, when done by sieving, removes grains along with which can erwise be converted into dehusked wholes or dehusked splits to be produced as the marketing requirement. These impurities can conveniently be separated by kles cylinder machines. The unwanted grains are generally round in shape while lentils are flat. Spiral towers can also be used for separation.

gnetic Separators

Permanent magnets of good standard quality should be used to separate the n particles from the stocks-in-process at the points of their entry into the pitting chines, de-huskers or the shellers otherwise these will badly damage the emery ls or the emery disc and may even cause fire.

nditioning

Conditioning loosens the binding of the husk with the cotyledons and the binding he two cotyledons with each other. Conditioning requirements differ not only m pulse to pulse but also for the same variety grown in different areas depending on the type of soil, climate and harvesting season. Three alternative systems of ditioning, dehusking and splitting are being suggested (Figures 4.2A, 2B and 2C). de-husking and splitting of the heated stocks take place simultaneously in a ller (fitted with emery dis of the improved design and coated with emery or borandurn of the required granulites), when the material reaches the point of tact at the circumference of the discs. The stocks after being heated in the sheller or screening whereas the de-husked splits are at the heated temperature and have e cooled before packing.

Type-A conditioning system is recommended for pigeon pea, mung-bean and l-bean and alike varieties of pulses which are most difficult to de-husk and split. se require more intensified conditioning effect to loosen the bond of the husk to cotyledons and the bond between the two cotyledons. Heating and tempering are vided twice to get the desired effects. Type-B conditioning system which provides

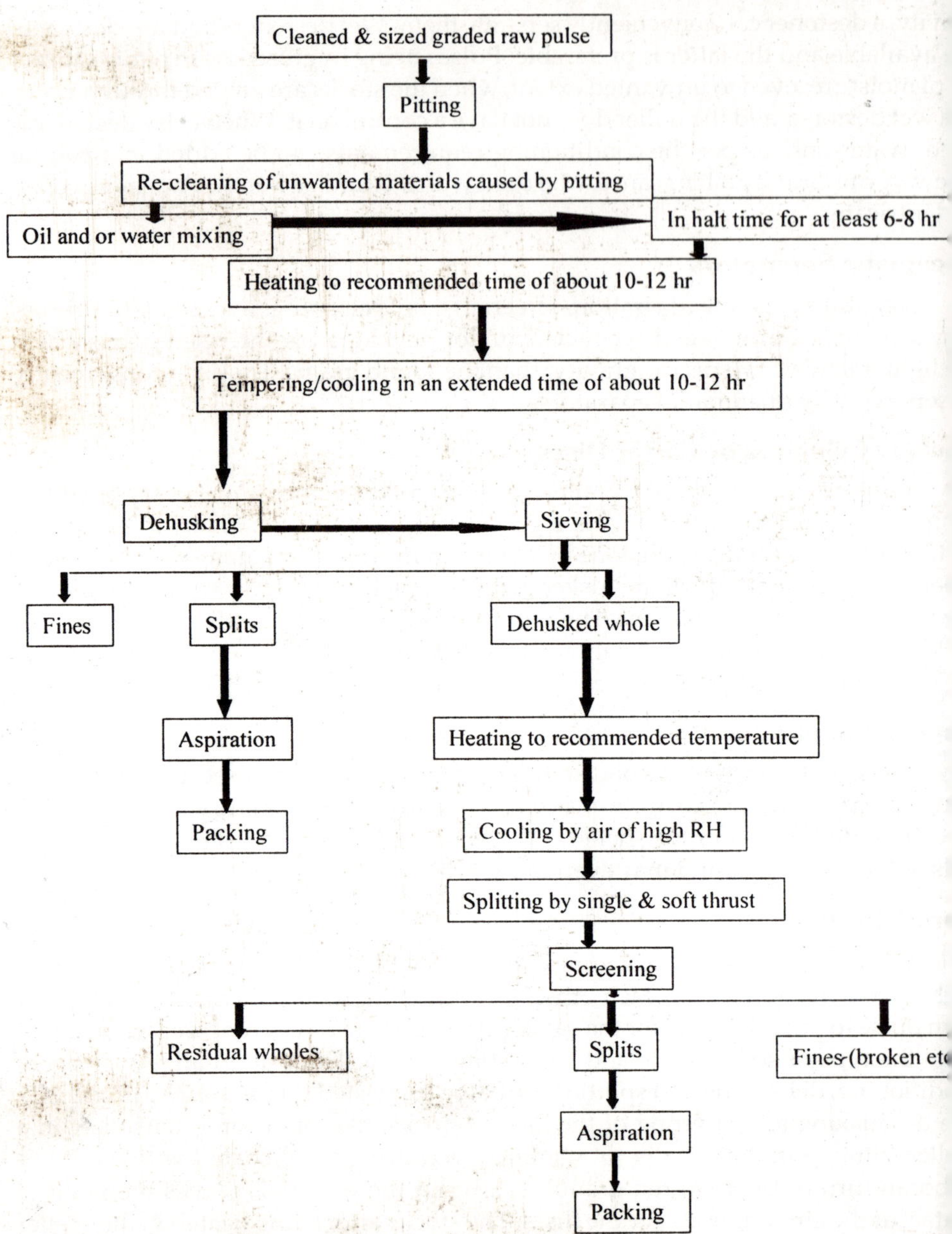

Figure 4.2A: Conditioning system type-A (*Source*: Matanhelia, 1980)

a single heating and single tempering system is suitable for de-husking and splitti of dun pea, faba beans, broad bean, chickpea, soybean and alike varieties of puls Conditioning system of type-C is especially suitable for de-husking and splitting chickpea. An over- treatment to some of the grains can damage the quality, wh under- treatment may result in husk- patches on dehusked splits.

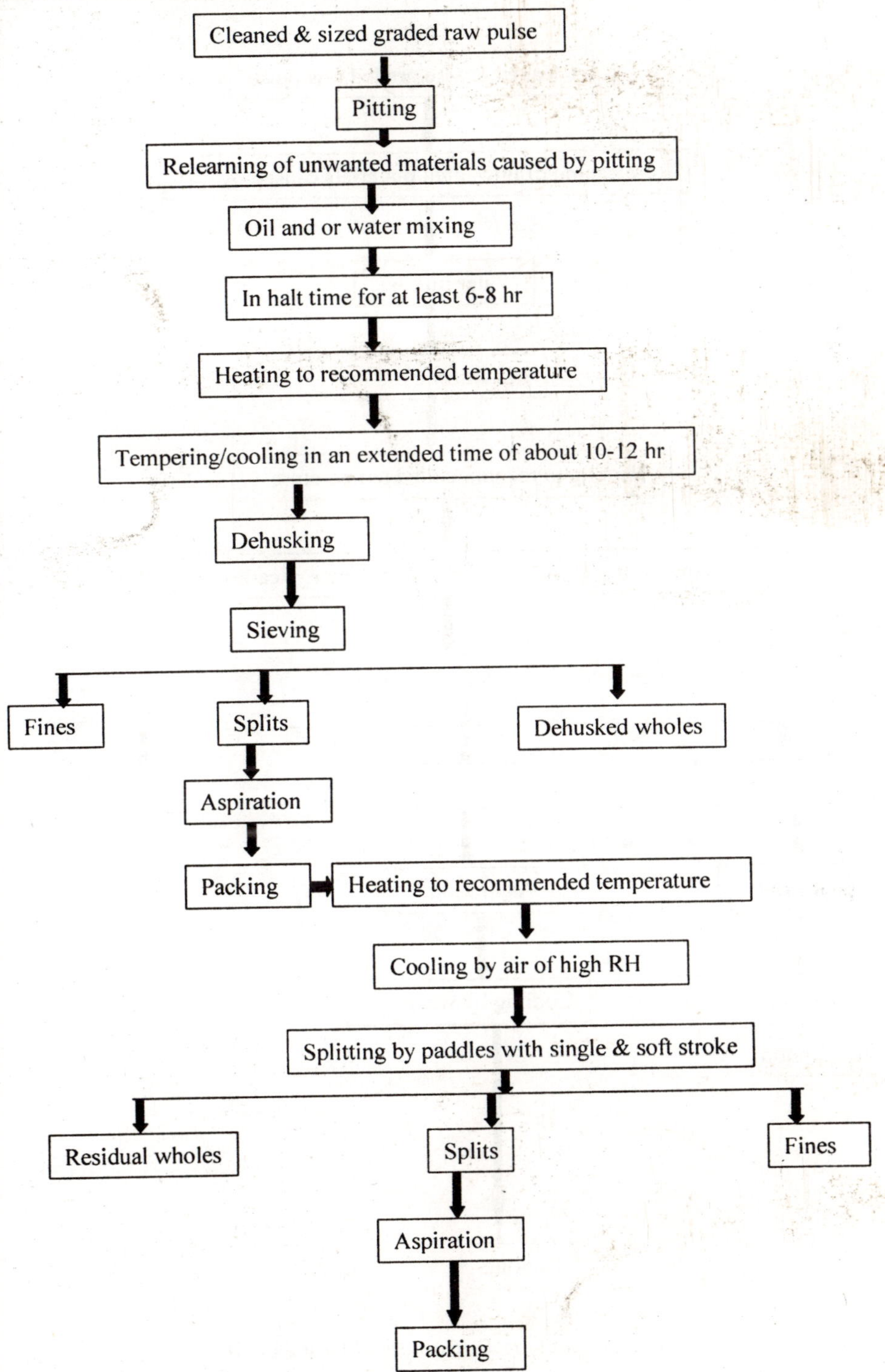

Figure 4.2B: Conditioning system type-**B** (*Source*: Matanhelia, 1980)

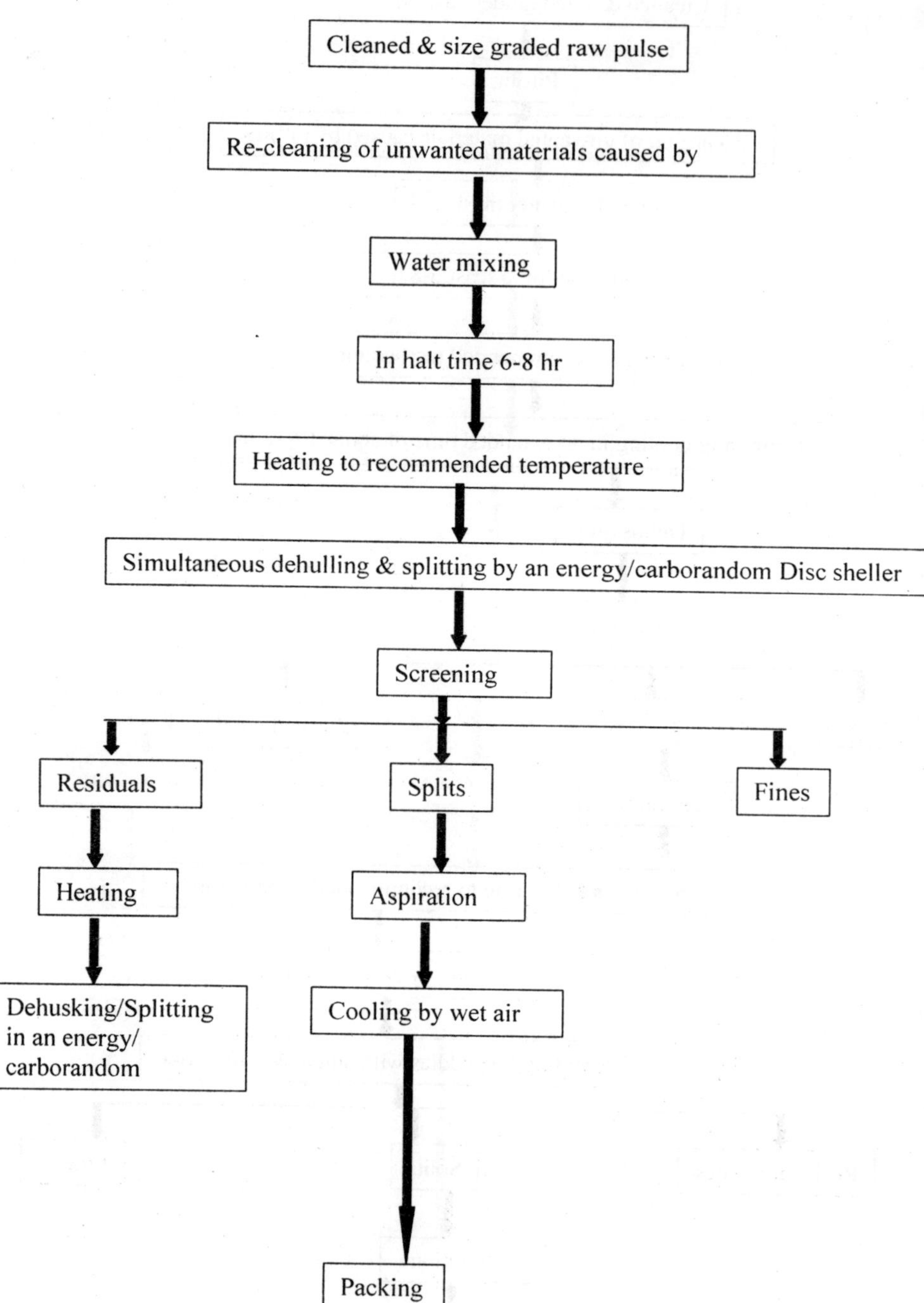

Figure 4.2C: Conditioning system type-**C** (*Source*: Matanhelia, 1980)

Pitting

Pitting is done to cause some cracks and pores in the seed coat to enable better penetration of the various conditioning effects such as those given during oil and/or water mixing, heating and tempering, etc. The ideal pitting would be such that the cracks or the pores may not be visible by naked eye. De-husking or splitting is to be avoided as far as possible at this stage although some amount of dehusking and/or splitting is quite unavoidable on account of the condition effects.

Oil/Water Mixing

Mixing of oil and/or water followed by an in-halt time of 6 to 8 hours sometimes produces very good effect for the better conditioning of the stocks-in- process and this can be well adopted to supplement the effect of conditioning at subsequent stages whenever necessary. While the mixing of water helps de-husking, it also leads to excessive splitting and is sometimes avoided for specific product requirement.

Heating and Tempering

It has very well been established that the heating of almost all the pulses to the recommended temperature followed by tempering *i.e.*, cooling in extended time, has a very positive effect in losing the binding of the husk to the cotyledons. This process of heating is done once or twice depending on the intensity of conditioning required. The temperature of the air used for heating as well as the temperature, to which the stock is heated, will vary from pulse to pulse. Similarly, it has been observed that allowing more time for cooling of the heated stock during tempering has a better effect but the equipment-cost as well as the cost of the material which will remain in process, puts a limit for the time which can be allowed at this stage. The equipment is accordingly designed. Air is preferred as the media for carrying the heat from the source of heating to the grains for heating and also from the grains to the atmosphere for tempering *i.e.*, cooling. It also helps in bringing down the moisture to the required level. For heating of the air, electrical air- heaters or fuel-oil burners using furnace oil, H.S.D. or LSD oils or the LPG have all been very successfully tried. In the case of heating of the air by fuel-oil burners, the heating system should be indirect so that the gases do not injuriously affect the material.

De-husking

Emery/carborandum coated rolls or cones placed and operated in vertical and horizontal position have been tried with varying results. The rolls and cones placed vertically have been observed to be totally unsuitable. The material from the feed point drops instantaneously to the lower part where it forms practically a jam. While there is too much overloading and power consumption, not all the grains get the proper contact with the abrasive surface of the emery roll. It also causes too much powdering and the dehusking is poor. On the other hand, emery/carborandurn coated rolls of conical shape and horizontally placed give the best results. Almost all the grains get dehusked. There is no rubbing or powdering or quibbling. The quality of the product is naturally better and the yield higher. Emery roll-coated with the most appropriate grains of emery and/or carborandum and accurately balanced rotates inside a shell of which the clearance to the roll can be adjusted while the machine is in operation to suit the specific requirements. The roll is fixed to a shaft provided with ball bearing of the best quality suitable to withstand the working loads. The shell is fitted with covers of mild steel and perforated sheets of sufficient thickness. The unit is placed in a body fabricated from steel and provided with covers and inspection doors. A powerful magnet is provided at the inlet to eliminate the entry of iron pieces in the machine along with the stocks to be treated so as to avoid the possibility of damage to the roller or shell cover. The inlet has also arrangement to regulate the feed of the stocks to the machine as well as for instantaneous shut-off. The delivery is provided with mechanism to regulate the pressure inside the shell. All these arrangements enable best dehusking results.

For splitting after making the binding of the two cotyledons ineffective by the conditioning system, the grains just need a single thrust and not harsh multiple thrusts to split. There is no quibbling and this again increases the yield.

Husk Aspiration System

The de-husked splits received by screening after dehusking as well as after splitting contain husk of pretty big size and are therefore to be cleaned by an efficient system of aspiration. There being a little difference between the weight of the husk and the splits, the system requires provisions for every minute adjustment so that no splits are carried away by the air with husk and at the same time no husk should remain with the splits. The new type of aspiration equipment are well designed to adjust the air to the required amount and pressure and when the same is passed at cross currents with the flow of the stocks to be treated, it efficiently draws off the husk and lets the cleaned dehusked splits to the packing. The equipment comprises of (*i*) cascade where the strokes meet at cross current air at required pressure and speed; (*ii*) junction box, which is also a low pressure box to recover the small quibbling which are carried away by the air; (*iii*) cyclone dust collector; (*iv*) air locks and (*v*) air filters, etc. all suitably arranged. For still finer separations, vibro-aspirators are recommended instead of the cascades.

Prospects

Modernization of *Dal* Milling Industry

Traditional technologies for making *dal* are laborious, time-consuming and completely dependent on climate and low yielding. As such there is a need to modernize the processing steps and machinery involved for better and quick recovery to reduce the cost of processing and make the process independent of climatic conditions. In modernizing the *dal-milling* industry, the CFTRI, Mysore, has employed the "conditioning technique" to loosen the husk without resorting to sun-drying and oil and water application. This step has been mechanized with the introduction of conditioning units where the grains are incipiently roasted by a counter-current through-flow technique and tempered in specially designed tempering bins where moisture is reduced to the desired levels. The husk thus loosened is then removed in an abrasion-type dehusking machine wherein almost complete removal of the husk is achieved in a single operation. The scouring and breakage losses are minimized. As the milling characteristics and physical properties of pulses vary, appropriate processing conditions and steps have been developed with the "conditioning echnique" which is almost 5-10 units more than that obtained by using the traditional methods. The resulting products are also better in cooking quality.

Milling of Pulses

Introduction

Pulses are rich in proteins and are mainly consumed in the form of dehusked split pulses. Pulses are the main source of protein in vegetarian diet. There are about 4000 pulses mills *(Dhal mills)* in India. The average processing capacities of pulses mills in India vary from 10 to 20 tons/day. Milling of pulses means removal of the outer husk and splitting the grain into two equal halves. Generally, the husk is much more tightly held by the kernel of some pulses than most cereals Therefore, dehusking of some pulses poses a problem. The method of alternate wetting and drying is used to facilitate dehusking and splitting of pulses. In India the dehusked split pulses are produced by traditional methods of milling. In traditional pulses milling methods, the loosening of husk by conditioning is insufficient. Therefore, a large amount of abrasive force is applied for the complete dehusking of the grains which results in high losses in the form of broken and powder. Consequently, the yield of split pulses in traditional mills is only 65 to 70 per cent in comparison to 82 to 85 per cent potential yield. It is, therefore, necessary to improve the traditional methods of pulses milling to increase the total yield of dehusked and split pulses and reduce the losses.

Milling of Pulses

In India there are two conventional pulses milling methods: wet milling method and dry milling method. The latter is more popular and used in commercial mills.

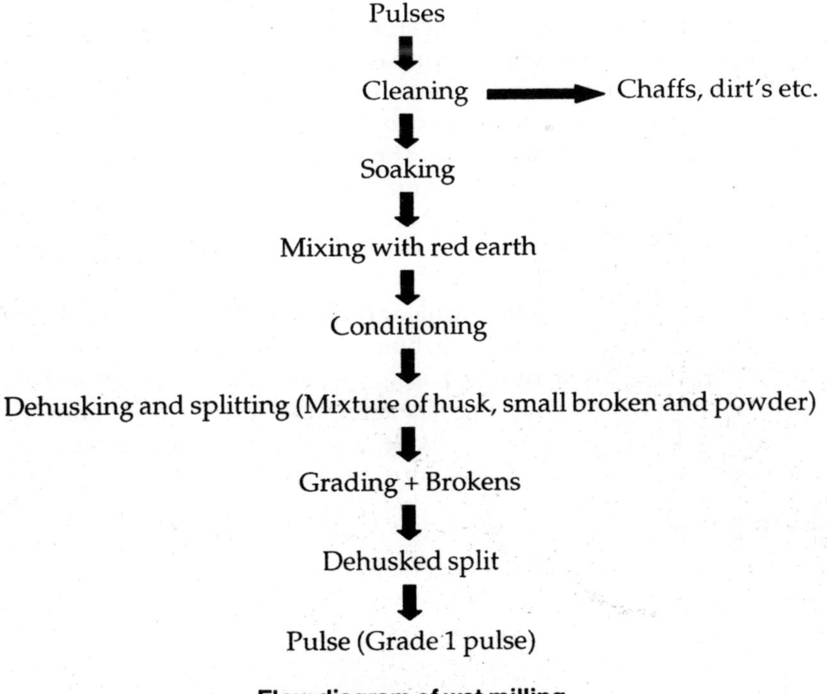

Flow diagram of wet milling

Traditional Dry Milling Method (Dhal Milling)

There is no common processing method for all types of pulses. However, som general operations of dry milling method such as cleaning and grading, rolling c pitting, oiling, moistening, drying and milling have been described in subsequer paragraphs.

Cleaning and Grading

Pulses are cleaned from dust, chaff, grits, etc., and graded according to size by reel type or rotating sieve type cleaner.

Pitting

The clean pulses are passed through an emery roller machine. In this unit, hus is cracked and scratched. This is to facilitate the subsequent oil penetration proces for the loosening of husk. The clearance between the emery roller and cage (housing gradually narrows from inlet to outlet. As the material is passed through the narrowin clearance, mainly cracking and scratching of husk takes place by friction betweer pulses and emery. Some of the pulses are dehusked and split during this operatior which are then separated by sieving.

Pretreatment with Oil

The scratched or pitted pulses are passed through a screw conveyor and mixec with some edible oil like linseed oil (1.5 to 2.5 kg/tonne of pulses). Then they are kep on the floor for about 12 hours for diffusion of the oil.

Conditioning

Conditioning of pulses is done by alternate wetting and drying. After sun drying for a certain period, 3- 5 per cent moisture is added to the pulses and tempered fo about eight hours and again dried in the sun. Addition of moisture to the pulses car be accomplished by allowing water to drop from an overhead tank on the pulses being passed through a screw conveyor. The whole process of alternate wetting and drying is continued for two to four days until all pulses are sufficiently conditioned. Pulses are finally dried to about 10 to 12 per cent moisture content.

Dehusking and Splitting

Emery rollers, known as Gota· machine are used for the dehusking of conditioned pulses. About 50 per cent pulses are dehusked in a single operation (in one pass). Dehusked pulses are split into two parts also. The husk is aspirated off and dehusked, split pulses are separated by sieving. The tail pulses and unsplit dehusked pulses are again conditioned and milled as above. The whole process is repeated two to three times until the remaining pulses are dehusked and split.

Polishing

Polish is given to the dehusked and split pulses by treating them with a small quantity of oil and/or water.

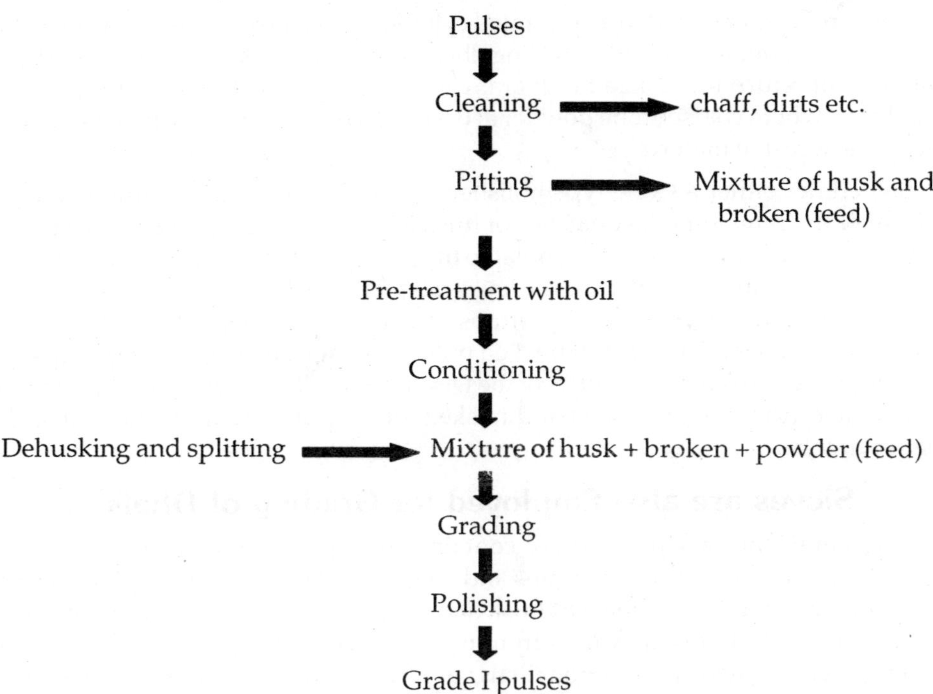

Flow diagram of dry milling of pulses

Commercial Milling of Pulses by Traditional Methods

It is discussed earlier that the traditional milling of pulses are-divided into two heads, namely, dries milling and wet milling. But both the processes involved two basic steps: (*i*) Preconditioning of pulses by alternate wetting and sundrying for loosening husk and (*ii*) subsequent milling by dehusking and splitting of the grains into two cotyledons followed by aspiration and size separation using suitable machines. 100 per cent-dehusking and splitting of pulses are seldom achieved particularly in cases of certain pulses like tur, black gram and green gram. Of them tur is the most difficult pulses to dehusk and split. Only about 40 to 50 per cent tur grains are dehusked and split in the first pass of preconditioning and milling. As sundrying is practiced the traditional method is not only weather dependent but also it requires a large drying yard to match with the milling capacity. As a result it takes 3 to 7 days for complete processing of a batch of 20 to 30 tons of pulses into dhals. Moreover milling losses are also quite high in the traditional method of milling of pulses.

In general, simple reciprocating or rotary sieve cleaners are used for cleaning while bucket elevators are used for elevating pulses. Pitting or scratching of pulses is done in a roller machine. A worm mixer is used for oiling as well as watering of the pitted pulses.

The machines used for dehusking are either power driven disc type Sheller 'chakki' or emery-coated roller machine, which is commonly known as 'gota' machine.

The emery roller is engaged in a perforated cylinder. The whole assembly is normally fixed at a horizontal position. The Engelberg type rice hullers are also used for dehusking of return unhusked black gram and green gram pulses in some parts of south India, where coarse stone powder at 0.5 to 0.75 per cent level is mixed with the grains as a abrasive material.

Sometimes either a Cone type polisher or a buffing machine is employed for removal of the remaining last patches of husk and for giving a fine polisher to the finished dhal. The cone polisher is similar to the polishing machine used for polishing of rice (*i.e.* for removal of bran from brown rice). The buffing machine is equipped with a rotating paddle having leather straps which can remove the last patch of husk and can give a fine polish to the dehusked pulses. Blowers are used for aspiration of husk and powder from the products of the Disc sheller or Roller machine. Split dhals are separated from the unhusked and husked whole pulses with the help of sieve type separators.

Sieves are also Employed for Grading of Dhals

In general, the raw pulses may contain 2 to 5 per cent impurities (foreign materials), some insect infested grains and some extra moisture. Though the clean pulses contain about 10-15 per cent husk and 2-5 per cent germs, the yield of dhals from commercial dhal mills varies from 68-75 per cent. It may be noted that the average potential yields of common dhals vary from 85 to 89 per cent. These milling losses in the commercial pulses mills can be attributed to small broken and fine powders formed during scouring and simultaneous dehusking and splitting operations. Some of the commercial milling methods commonly followed for different pulses is briefly described in the subsequent paragraphs.

Dry Milling of Tur

The detail flow sheet of dry milling of tur is given below:

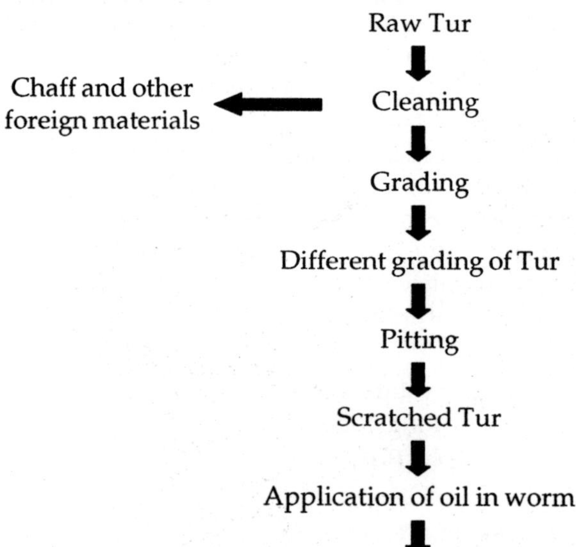

Scratched and oil coated Tur

↓

Sundrying and overnight tempering for 2 to 4 days

↓

Addition of about 5 per cent water by spraying and overnight moisture equilibration

↓

Dehusking and Splitting

↓

Aspiration + (Husk+ Powder)

↓

Sieving

↓

Dhal

Dry Milling of Tur

The dry milling of tur is generally practiced in M.P. and Uttar Pradesh as it can give higher turnover in terms of capacity of a mill having the same area of drying facilities. In this method the pulses are subjected to pitting in a roller and then they were subjected to oil treatment by applying 0.5 to 2 per cent linseed or any other edible oil in worm mixers. The pulses are then spread in the drying yard for sundrying for 2 to 4 days. The pulses are tempered by heaping and covering during the nights in between these days. After thorough sundrying the pulses are once again moistened uniformly with about 5 per cent water and kept as such on heaps overnight for uniform moisture equilibration. Then these grains are passed through the rollers for dehusking and splitting. About 50 per cent grains are dehusked and split in first operation. After removal of the husk by aspiration the split dhal is separated from a mixture of husked and unhusked whole pulses. The mixture is once again moistened by spraying water and dried in the sun and then dehusked and split as before either in roller or in an under runner disc sheller where around 30 per cent of the grains are dehusked and split. The above process of alternate wetting and drying is repeated until almost all the remaining pulses are converted into split dhal. The average yield of dhal ranges from 68 to 75 per cent.

Wet Milling of Tur

The detail flow diagram of the wet milling of Tur is given below.

Raw Tur

Soaking in water for 3-12 hrs

Mixing of soaked pulses with wet earth (5 per cent)

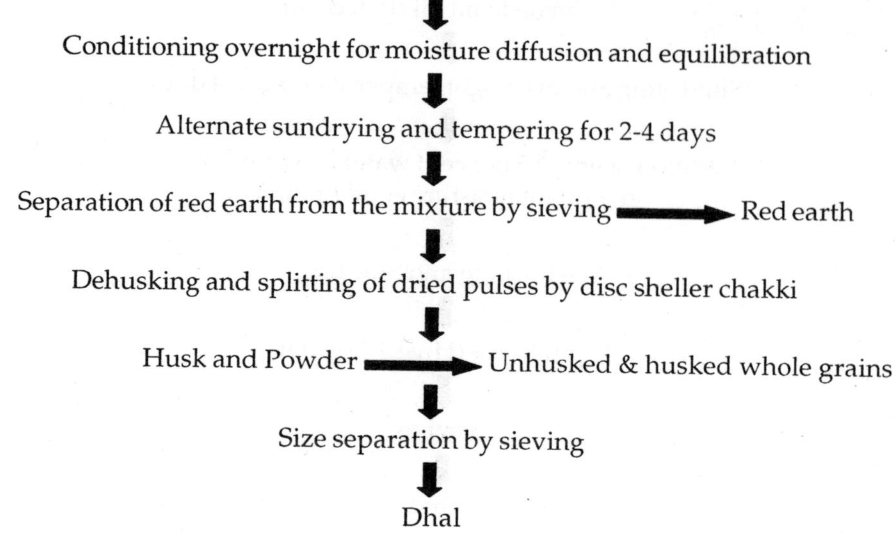

Wet milling of Tur

In wet milling of tur the grains are soaked in water for a period of 3 to 12 hours. The soaked pulses are thoroughly mixed with wet red earth at about 5 per cent level. The mixture is kept in heaps overnight. The whole mixture is then dried in the sun for 2 to 4 days until the husk of all grains are loosened. The pulses are tempered overnight in between these days. The red earth is then separated from the pulses by sieving. The sundried grains are dehusked and split in a disc sheller (chakki). Dhal and other fractions are separated as usual. It is claimed that about 95 per cent of the grains can be dehusked and split in' a single milling operation. The split dhal is separated from the mixture usually. The rest of the unhusked and husked whole grains are preconditioned and milled as above for conversion of these grains into dhal. Though the above wet milling of tur is popular in South India, the purpose of using earth is not well understood. However, it is believed that the red earth facilitates in increasing the rate of drying and in consequence in loosening the husk. So also the earth may act as a milling aid on account of its abrasive nature. It is also considered that dhals produced by following wet method are attractive in colour and good taste. The wet method requires 5 to 7 days for complete processing of a batch of grains.

Dry Milling of Black Gram

After cleaning the black grams are subjected to pricking in a rough roller mill for some scratching as well as partial removal of the waxy coating on the black grams. The scratched grains are then coated with 1 to 2 per cent oil in a worm mixer and then heaped over night for diffusion of the oil in the grains. The scratched and oil coated pulses are sprayed in drying yards for sundrying for 4 to 6 hours. The partially dried grains are moistened with a spray of 4 to 5 per cent water and kept overnight for moisture equilibration. The wetted pulses are then dried for 3 to 4 days in the sun and tempered over nights in between these drying periods. The thoroughly dried pulses e dehusked in a roller. About 40 to 50 per cent pulses are dehusked and split in first illing operation. The husk and powder are then aspirated off. Then the split 'dhal' separated from the dehusked whole dhal and unhusked pulses by sieving. Both ısked and unhusked whole grains are again dried in the sun and milled as above ıd the same process is repeated until the desired milling of pulses is "achieved. The erage yield of dhal is 70-71 per cent. Sometimes the last part of the unsplit grains ıd partially husked grains are allowed to pass through sheller and polisher achines for splitting and removal of the husk, which result in a large amount of sses due to formation of powder and brokens. In some cases polishing is done in a ıffing machine. In order to give a white finish and to protect from insect attack a ating of soapstone powder is generally given to these 'dhals'

Dry Milling of Bengal Gram, Lentil and Peas

It is comparatively easy to dehusk and split Bengal gram, Lentil and Peas as eir husks are loosely attached to the cotyledons. It requires shorter period of econditioning prior to milling these pulses. After cleaning, the pulses are pitted in roller machine. The pitted grains are then wetted with water (5 to 10 per cent) in a orm mixer and then these are kept in heaps for a few hours for diffusion of water to the grains. These grains are dried in the sun for a day or two, with overnight mpering in between these days. About 60 to 70 per cent dried pulses are then husked and split in the first pass of a roller machine. The husk and powder are pirated off. The split pulses are separated from the unhusked and husked whole ains by sieving. The alternate wetting with 5 per cent water and sundrying and ıbsequent milling operations are repeated till the most of the pulses are converted to ıal. The preconditioning and milling of Lentils and Peas are comparable with engal gram. The same initial pitting, wetting, conditioning, sundrying and ıbsequent milling by dehusking and splitting in a roller and aspiration of husk ith a blower and separation of split dhal from the mixture of unhusked and husked hole grains with a sieve, are being followed. The whole process of preconditioning ıd milling are repeated till most of the pulses are converted into dhal. However, the nversion of these pulses into dhals is easy compared to tur a takes about 3 to 5 days r complete processing of a batch of pulses.

Dry Milling of Green Gram

In dry milling of green gram, both oil and water treatments are given to the ains. The wetted grains are dried in the sun. Then the dried pulses are multaneously dehusked and split using a dehusking machine. After removal of ısk split dhal is separated from the mixture as usual. The yield of dhal is poor hich varies from 62 to 65 per cent only.

5

Processing Technology and Value Addition in Oilseeds

Introduction

The importance of oils and fats in human nutrition is well recognized. Thes form a vital component of many cell constituents, are an important source of energ and act as a carrier of fat-soluble vitamins. Besides, they contribute significantly a functional ingredient in improving the sensory characteristics of several processe food products. Oils and fats are mainly derived from plant and animal source About 71 per cent of edible oils/fats are derived from plant sources. Oilseeds a those crops in which energy is stored mainly in the form of oil. Some oil crops such a groundnut (peanut) can be used directly as a food, but others are exclusively processe to obtain fat or oil, and cake or meal. The production of oilseed crops has expande rapidly in response to the growing world population and the rising living standard In addition, technological advances have lead to higher production levels an improvements in product quality and versatility. This has also paved a way for th development of technologies for the processing of nonfood products using oilseec as the raw material. The cultivation of oil seeds is documented since ancient time Soybean has been an important food in China for thousands of years. The names c many oilseeds such as sesame and rapeseed appear in the Indian Sanskrit literatur The oil crops are now grown all over the world. There are three major groups of o crops:

1. Those are annual or biennial such as soybean, sunflower, groundnut an rapeseed.
2. The Perennial crops such as coconut, babassu nuts and oil palms.
3. Crops such as cotton and corn germ, where the embryo, a rich source of oi is a byproduct.

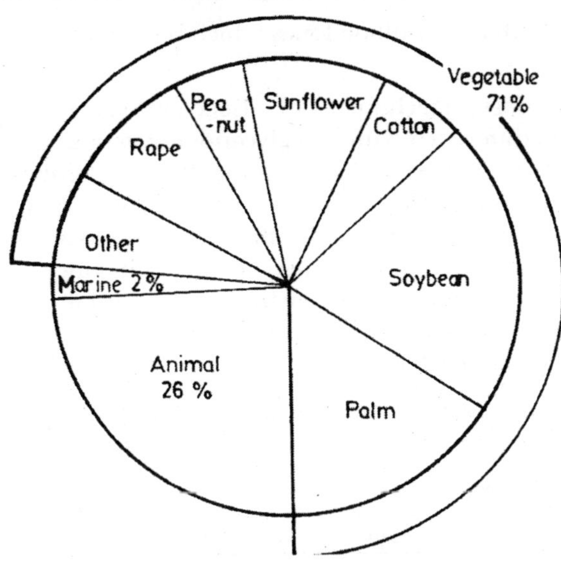

Depending upon the use of oil, oilseeds can be classified into seeds which contain edible vegetable oil such as soybean, peanut, cottonseed, rapeseed, sunflower, safflower, sesame and those which contain nonedible oils such as castor beans. Nonedible oil called tall oil is a byproduct of pine tree pulping in kraft paper mills. Although not a vegetable oil in the same sense as those from seeds, tall oil is nevertheless a dominant source of industrial oil. There are forty different oilseeds whose oil can be used for human consumption. However, only a small number of such crops are significant in the world's edible/nonedible oil supply. Among the oilseeds, soybean accounts for more than 30 per cent of the world's oilseed output. The major oilseed producing areas are in temperate regions of the world.

Based on the figures of 17 major oils, the annual world production increased by 52 per cent between 1975 and 2010. The sources of edible oils shifted in the last decade toward an increased dependency on plant products. Rapeseed and palm oil have undergone significant increases in production between 1975 and 1985, but not all seeds are processed to obtain oil. A part of the produce is used as seed, fed unprocessed to animals, or used directly for human consumption. There are losses also during storage and handling. In general, 15 to 20 per cent of the seed produced remains unprocessed.

For the last two decades, India is meeting the shortage of oils and fats by importing these commodities at huge costs in foreign exchange. Based on an income elasticity of 1.32 and a national income growth rate of 3.5 per cent, the demand for vegetable oils in 2000 A.D. will be 8.75 million tones as against an expected production of 5.15 million tones. In order to augment the oils and fat resources in India the following steps should be taken: To make up the deficit of oil, rice bran oil, corn oil and other non-conventional oils are to be made available to the maximum extent. High Yielding varieties of plam, sunflower, com, cotton etc. are to be grown widely. Collection of forest oilseeds such as 'sal', 'mowrah' etc. needs to be intensified. The potential of these sources is as high as 15 million tones which is equivalent to 2 million tones of oils and fats.

Improved post harvest technology of oilseeds is to be introduced. The oils and fats are composed of different mixtures of glycerides of various fatty acids. The waxes are mixtures of higher polyhydric alcohols (other than glycerol) with fatty acids. The fatty acid compositions and their contents of different oils are given in Table 5.1. It may be noted that the main constituents of the vegetable oils are the 16 and 18 carbon acids. Esters containing more unsaturated acids and having lower melting points are oils. The more saturated Esters constitute fats. All fats and oils are broadly classified into edible and non-edible. Groundnut cottonseed, linseed, soybean are some of the sources of the edible oil.

Table 5.1: Per cent composition of fatty acid of different vegetable oils

No. of C Atoms	*Vegetable Oils*					
	Fatty Acid	*Chemical Formula*	*Linseed Oil*	*Cotton Seed Oil*	*Soybean Oil*	*Coconut Oil*
8	Caprylic	$C_7H_{15}COOH$	—	—	—	87.8
10	Capric	$C_9H_{19}COOH$	—	—	—	7.3
12	Lauric	$C_{11}H_{23}COOH$	—	—	—	48.0
14	Myristic	$C_{13}H_{27}COOH$	—	1.6	0.5	17.5
14	Unsaturated	$C_{13}H_{25}COOH$	—	—	—	—
16	Palmitic	$C_{15}H_{31}COOH$	6.0	23.0	10.2	9.0
16	Unsaturated	—	—	2.0	1.2	—
18	Stearic	$C_{17}H_{35}COOH$	2.9	1.3	2.5	2.2
18	Oleic	$C_{17}H_{33}COOH$	19.0	22.8	23.5	5.7
18	Linoleic	$C_{17}H_{31}COOH$	241.4	47.9	51.0	2.5
18	Linolenic	$C_{17}H_{29}COOH$	47.0		8.5	—
20	Arachidic	$C_{19}H_{39}COOH$	0.5	1.41	2.6	—
24	Lignoceric	$C_{23}H_{47}COOH$	0.2	—	—	—

The edible oils are mainly used for cooking purposes, table uses and salad dressings. Oils are also used for inedible purposes. These include soap industry, drying oil industries including paints and varnishes, plasticizers. The waxes are used as components in the manufacture of floor and shoe polishes, carbon paper, candles etc.

The mechanical extraction as well as solvent extraction methods is employed for the manufacture of oil from the oilseeds. Almost all oils contain some free fatty acids, colouring and odouring matters and in some cases gums and waxes (particularly in rice bran oil). The crude vegetable oils are, therefore, to be refined for removal of these undesirable materials prior to their uses for edible purposes. Though steps followed in extraction and refining of oil vary to some extent from one oilseed to other. The specific methods of production of crude and refined oil from different oilseeds are described below.

Soybean Processing

Cooking

It is known that soybeans do not support the growth of rats unless they are cooked in a steam bath. The degree of improvement in nutritive value depends upon the temperature, duration of heating and moisture. Maximum nutritive value of soybean proteins is achieved by treatment with live steam for about 30 min or by autoclaving at 15 lb pressure for 15-20 min. The improvement in the nutritive value of soybean proteins appears to be related to the destruction of trypsin inhibitors and possibly other biologically active components. The excessive amount of heat may adversely affect the nutritive value of proteins and the damage thus inflicted can usually be overcome by supplementation with lysine and sulfur containing amino acids. Excessive heating results in destruction or inactivation of cystine and lysine. Cystine is particularly sensitive to heat. Lysine not only undergoes destruction when soybean protein is overheated but is also rendered unavailable. This is due to the fact that amino groups of lysine interact with sugars; this is referred to as browning or Maillard reaction. Lysine so modified is no longer physiologically available since the peptide bond containing the modified lysine is not susceptible to tryptic cleavage. Thus, the digestibility of soybean protein by proteolytic enzyme is considerably reduced if the protein has been subjected to excessive heat treatment. A direct consequence of impaired digestion is retardation in the rate at which all amino acids are released from protein during digestion, since methionine is the limiting amino acid of soybean proteins, a delay in digestion leading to excretion of methionine would only serve to accentuate a deficiency of this amino acid. The destruction of cystine further intensifies the deficiency of sulfur-containing amino acids in excessively heated soybeans. In addition to cystine and lysine, a number of other amino acids including arginine, tryptophan, histidine and serine are either partially destroyed or inactivated by excessive heating of soybean meal.

Processing for Oil

Soybean oil is an important product of soybean-processing industries. Almost all of the soybeans harvested are processed to oil and meal products through the solvent extraction process in the developed countries. There are essentially three steps involved in this process: (1) bean preparation, (2) oil extraction, and (3) solvent stripping and reclamation. The processing of crude soybean oil into a variety of products is presented in Figure 5.1. About 30 per cent of soybean oil is produced in the U.S. is degummed. The purpose of degumming is the removal of phospholipids and other non triglyceride material. A valuable product of degumming is lecithin, which is used in many products as an emulsifying agent. Crude or degummed soybean oil is treated with caustic soda to neutralize free fatty acids, hydrolyze phosphatides and remove some colored pigments and unsaponifiable matters. Bleaching is an important technique for absorbing pigments using activated earth. Deodorization is the last major processing step in the refining of soybean oil. This operation is designed

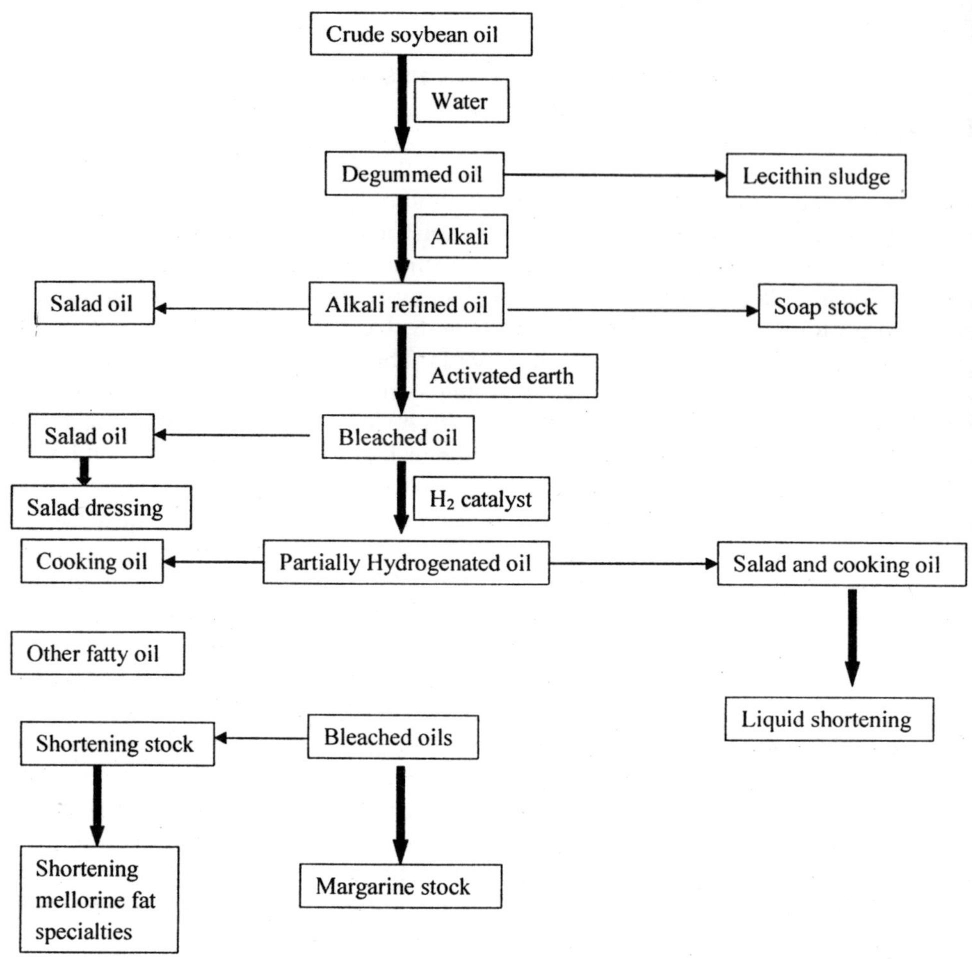

Figure 5.1: Flow diagram showing the processing of soybean oil

to remove all oxidative cleavage products that impart an odor or flavor to the oil. It may also remove tocopherols, sterols and other useful minor constituents of free fatty acids and undesirable foreign materials. The average composition of crude and refined soybean oil is presented in Table 5.2. Soybean oil is used as a cooking medium or in the formulation of margarines and shortenings.

Soybean Processing into Protein Products

Flours

Soy flours are screened and graded products obtained after expelling or extracting most of the oil from dehulled soybeans, except the full-fat soy flours, which contain all of the oil originally present in soybeans. Flours are ground fine enough to pass through 100 mesh or smaller screen. Full-fat soy flours are essentially dehulled ground soybeans, thus containing 18 to 20 per cent fat. Defatted soy flours are produced by nearly complete removal of oil from soybeans by the use of hexane. They usually contain about 1 per cent fat. Low-fat soy flours are produced either by partial removal of the oil from soybeans or by adding back soybean oil and/or lecithin to soy flours to a specified level such as 5 to 6 per cent. High-fat soy flours are produced by adding back soybean oil and/or lecithin to defatted soy flours usually in the range of 15 per cent.

Table 5.2: Average compositions for crude and refined soybean oil

Composition	*Crude oil*	*Refined oil*
Triglycerides (per cent)	95-97	> 99
Phosphatides (per cent)	1.5-2.5	0.003-0.015
Unsaponifiable matters (per cent)	0.6	0.3
Plant sterols (per cent)	0.33	0.13
Tocopherols (per cent)	0.15-0.21	0.11-0.18
Hydrocarbons (squalene) (per cent)	0.014	0.01
Free fatty acids (per cent)	0.3-0.7	0.05
Trace metals		
Iron (ppm)	1-3	0.1-0.3
Copper (ppm)	0.03-0.05	0.02-0.06

Protein Concentrates

Protein concentrates are prepared from high-quality clean dehulled soybeans by removing most of the oil and water-soluble nonprotein constituents (mainly sucrose, raffinose and stachyose) and contain not less than 70 per cent protein on a moisture-free basis. Defatted flakes or flours are starting materials used for the production of soybean protein concentrates. The basic process involves extraction of defatted meal under conditions where bulks of the proteins are insoluble but sugars are dissolved and removed. Three solvents are employed for this purpose. Aqueous ethanol is the most commonly used solvent. The proteins and polysaccharides are insoluble in this solvent whereas sugars, ash and minor constituents are soluble. After solvent removal, a concentrate is obtained. In a recent modification, defatted flakes are extracted with hexane-ethanol to remove residual lipids prior to alcohol treatment. Water is another solvent used for preparation of protein concentrates several years ago. In this, moist-heat treatment is applied to the flakes to denature and insolubilize the proteins. Subsequently washing with water removes mainly sugars and a concentrate is obtained by drying the washed product. This yields full-fat soybean concentrate. Yields of the protein concentrate by three solvents range from 60 to 70 per cent based on the weight of the starting defatted flakes or flours.

Protein Isolates

The protein isolates contain more than 90 per cent proteins (N x 6.25) on a moisture-free basis. These are extracted from undenatured, defatted flakes by extraction with dilute alkali. The slurry is centrifuged to remove the spent flakes, the resulting extract is adjusted to pH 4.5, where most of the proteins coagulate and precipitate as curd. After washing, the curd may be spray-dried directly to yield an isoelectric protein or more commonly, the curd is redispersed in water adjusted to pH 6.8 to 7.2 and then spray-dried. The resulting soy proteinate is more water-soluble than isoelectric form and, hence, more readily incorporated into wet food systems. Yields of isolates are about 30 per cent of the weight of the starting defatted flakes (Figure 5.2).

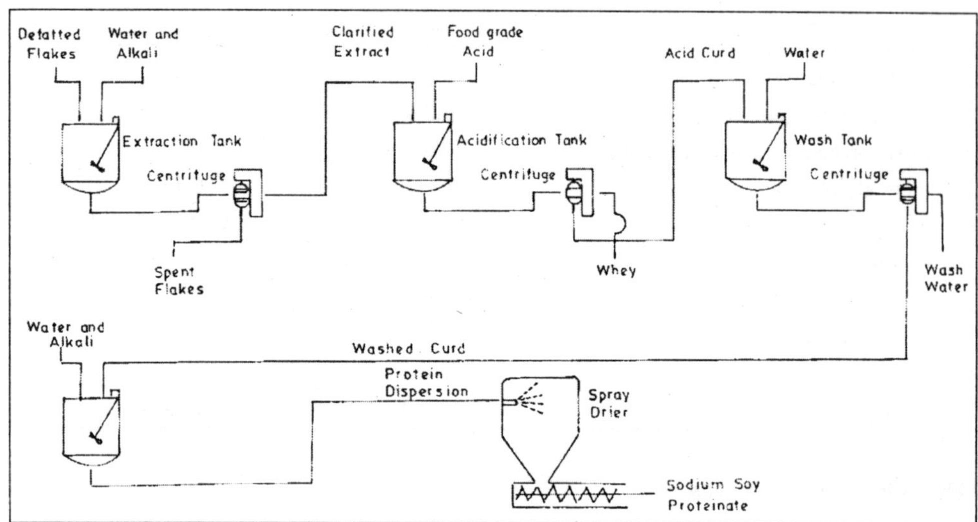

Figure 5.2: Processing of soybean for preparation of protein isolates

Textured Protein Products

The textured products can be prepared from flours, concentrates and isolates. They are available in textured forms with fibrous and chewy properties of muscle proteins of meat. The majority of products are made by thermoplastic extrusion. In this process, defatted flour is mixed with water plus additives to form dough, which is then fed to an extruder where the material is subjected to high temperature, pressure and mechanical stress. On emerging through the die, the dough puffs and assumes a fibrous texture simulating certain meat products. The size and shape of the textured dough can be controlled at the extruder die. Protein concentrates can be textured by analogous processes and are available commercially. Protein isolates are also available in two textured forms. In the first type, an isolate is prepared, mixed with alkali to form a spinning dope and then pumped through a spinnerette into a coagulating bath containing acid and salt. Under these conditions, the alkali is neutralized and the protein is coagulated to produce tiny protein filaments that are gathered into bundles or tows. The tows are washed, stretched, blended with fats, flavors, colors and other additives and, finally, fabricated into a variety of sizes and shapes of meat-like products. Textured soy protein isolates are also produced by the "dry spinning" process, in which isolated slurries are pumped through a heat exchanger under high pressure at temperatures in the range of 116 to 157°C. Fibers are obtained by expelling the heated slurries through a small circular or slot-type nozzle. The expelled fibrous isolate is cooled by dropping it through ambient air into a collecting vessel. Excess water is removed by centrifugation.

Germination

Soybean sprouts have been used as food in the Orient since ancient times. The commercial equipment for growing bean sprouts has been described by Chen and Chen (1956). The beans are washed well, placed in ajar covered to screen from light and sprinkled with water three times a day. The water should contain a small amount of calcium hypochlorite to discourage the growth of micro-organisms. The seeds are allowed to germinate at 30°C. After the beans are harvested, sprouts may be considered as a fresh vegetable and can be used throughout the year. The sprouts may be used uncooked in a salad, boiled in water with suitable seasoning or may be fried in fat and used alone or mixed with other vegetables. Germination results in changes in the nutrient contents of the seeds. A significant increase in ascorbic acid was noted during germination of soybean seeds. The nutritive value of soybeans for rats increases upon germination. Several scientists concluded that nutritional value of soybean sprouts compares very favorably with other soybean products which are used as food in the Orient.

Rapeseed Processing

Drying and Handling

Rapeseed is normally harvested when the seed contains less than 10.5 per cent moisture. If the moisture content is higher, the produce is artificially dried to less than 10.5 per cent moisture immediately after harvest. Rapeseed with a moisture content of 9 to 10 per cent can be stored safely for periods of up to 3 months under most ambient conditions. For longer storage, the seeds should contain less than 9.5 per cent moisture. Rapeseeds for processing can be dried at temperatures as high as 85°C without loss of oil quality if the seed is processed soon after drying. The quality of oil from rapeseed which has been dried at temperatures high enough to destroy the seed viability deteriorates during several months' storage of the seed. The relationship between seed moisture content and seed fragility would suggest that moisture contents of 6 to 8 per cent allow safest handling of rapeseed. Cracked or broken seeds contain high amounts of free fatty acids which decrease the quality of the oil. Accordingly, it is essential that the seeds be handled carefully and mechanical parts that come in contact with the seed should be free from sharp edges. In order to produce oil and meal of high quality, the seed entering a crushing plant must be well-matured and sound. Crushing of a high-quality seed allows oil extraction and refining process to proceed smoothly and continuously. A poor quality seed can cause difficulties in almost every step in the processing and significant losses of oil.

Seed Cleaning and Preparation

The modern rapeseed processing involves several steps in the preparation of rapeseed for oil removal, *i.e.* cleaning, flaking, and cooking. After cleaning, the moisture content is adjusted to about 8.5 per cent. In winter, the seed may be heat-conditioned. The seed entering the flaking rolls may be tested for color, free fatty acids, and moisture. Flaking rolls may be smooth or corrugated and may operate at 0 to 5 per cent speed differential. The flaked seeds (0.20 to 0.22 mm for direct solvent, 0.22 to 0.25 mm for prepress solvent) pass to the cooker where, in four to five steps, the seed temperature is raised from 30°C to 80-100°C in 30 min. The initial seed moisture of 8 to 10 per cent is decreased to 4 to 5 per cent. Cooking of rapeseed is important in order to:

1. Dry seed to a suitable moisture content
2. Complete the breakdown of oil cells
3. Coagulate the proteins to facilitate oil separation
4. Reduce the affinity of the oil for solid surfaces
5. Insolubilize phosphatides
6. Increase oil fluidity
7. Destroy moulds and bacteria
8. Inactivate enzymes

The inactivation of enzymes is particularly important in the case of rapeseed. The thioglucosidase enzyme, which mediates the hydrolysis of glucosinolates, is particularly active between 40 and 70°C. For optimum inactivation, the temperature of the flaked seed must be increased rapidly to a temperature of 80 to 100°C. The hydrolysis products, which appear if the enzyme is not inactivated, remain in the oil and cause problems in hydrogenation. The cooking conditions used depend on the subsequent extraction procedure. In the top kettle, the moisture content of flaked seed should be between 6 and 10 per cent and the temperature between 85 and 90°C. The temperature rises in the lower kettles as the seed becomes drier. For direct solvent extraction, the flakes should reach a final temperature of 105°C and a moisture content of 4 to 6 per cent. Somewhat higher temperatures are required for flakes which are to be prepressed. Seeds processed by the expeller procedure (with no solvent extraction) are cooked to 105 to 110°C with 5 min at 130 to 135°C. In all cases, the final moisture content of the flakes is about 4 to 6 per cent. For pressing, the flakes go directly from the cooker to the press. For direct solvent extraction, the cooked flakes are crisped in an open screw conveyer to leave porous granules, rerolled on a smooth roller mill to a uniform flake thickness.

Extraction and Processing of Oil

Extraction

Extraction of oil from flaked rapeseed can proceed by one of the three processes: by direct screw pressing, by direct solvent extraction, or by prepress solvent extraction. Prepress solvent extraction utilizes a combination of two processes and is the most common and probably the most economical process. Several factors influence the efficiency of the extraction process, and quality and stability of the extracted oil. Flakes (at 105 to 110°C and 4 to 6 per cent moisture) are generally conveyed directly from the cooker to the press. Prepressing leaves a residual cake with 12 to 20 per cent oil (compared with 4 to 12 per cent from direct pressing). Pressures of 1000 to 1400 kg/cm^2 force the oil from the flakes through the 0.25 mm bar spacings. After pressing, the press cake is passed through a flaking mill to give flakes of 0.3 to 0.4 mm thickness. Although this thickness does not give the most efficient extraction, the small size and fragility of rapeseed makes this a necessary compromise.

Solvent extraction may take place in either percolation-type extractors, where the solvent is allowed to percolate through the seed bed, or filtration- type extractors, where the solvent and seed mass are slurried and filtered. Solvent extraction is most advantageous for low-oil-content seeds (20 per cent oil) but can also be used for direct extraction of rapeseed in high-capacity plants. In order to remove the maximum amount of oil with a minimum of solvent, extraction usually takes place in a continuous countercurrent process. Flakes containing most of the oil are extracted with the miscella (solvent plus oil) from about the halfway point in the process while the final stage involves extraction with pure solvent. In modern extractions, the ratio of solvent to meal is 1.1 to 1.3, and the concentration of oil in the final miscella is 10 to 50 per cent. The meal leaving filtration-type extractors has 25 to 30 per cent solvent.

Solvent is removed from the miscella by distillation procedures involving heating with vent gases, direct steam-heating, and the use of a "stripping column". The oil usually contains about 0.1 per cent residual solvent. Solvent is removed from the meal in a desolventizer toaster in which the meal is subjected to a temperature gradient from 80°C to 100-110°C with the addition of moisture. The heat and moisture in the later stages denature the proteins of the meal. The resulting meal is dark in color, has 10 to 12 per cent moisture, 1 to 2 per cent oil, and low protein solubility. Solvent is recovered from vent gases by condensation and scrubbing. Solvent loss from a modern plant is less than 0.15 per cent.

Degumming

Solvent-extracted rapeseed oil usually contains gums amounting to about 2 per cent of the oil. The major constituents of the gum fraction are the phosphatides. This phosphatide material can cause difficulties by settling out in storage tanks, leading to large refining losses. The oil is often degummed by treatment with hot water (85-95°C) or steam or water plus phosphoric acid, citric acid, or other acidic materials. The degumming treatment precipitates the gums; the precipitate is removed by continuous centrifugation. The degummed solvent-extracted oil (now free of solvent) is then mixed with the press oil and the gums usually are added to the meal as it enters the desolventizer toaster. Some crushing plants mix the two oil streams prior to degumming. There is little market for rapeseed lecithin (main component of the gum fraction) as soybean lecithin dominates the market.

Refining

Refining, the term used to describe the purification of the crude oil, is ambiguous. North American "refining" refers to caustic refining only, while European "refining" refers to all processes from degumming through deodorization. Caustic (alkali) refining appears to be the only type of refining applied to rapeseed oil. Alkali-refining can be carried out as a batch or more commonly as a continuous process. Generally, refining losses are 3 to 4 per cent for continuous refining and 4 to 6 per cent for batch refining.

Bleaching

Bleaching is the process in which 0.25 to 2 per cent of finely divided clay (neutral or acid-activated) is added to the oil to improve the color of oil. Bleaching also may have an effect on oxidation of the oil. Rapeseed oil contains significant amounts of chlorophyll and is usually more difficult to bleach than other oils. Oil from damaged or immature seed may be extremely difficult or impossible to bleach. Acid clays are especially useful as bleaching agents since they are particularly effective in removing chlorophyll. After bleaching, the oil is filtered, usually in pressure leaf filters or plate and frame filters (in older plants).

Hydrogenation

This process modifies the physical properties of the fat or oil as well as rendering it more resistant to oxidative and thermal damage. Some improvement in their color is sometimes obtained by destruction of carotenoid pigments. Hydrogenation results in the addition of hydrogen to the double bonds of the unsaturated fatty acids and through the conversion of some of the *cis* double bonds to the higher melting *trans* forms. Hydrogenation may be more or less selective depending upon the conditions (high temperature, high catalyst, low pressure is selective conditions). Although for rapeseed hydrogenation usually is carried out under selective conditions. A problem sometimes encountered in the hydrogenation of rapeseed oil is the presence of sulfur compounds, which poison the catalyst even when present at very low levels. Several scientists found three to five ppm sulfur in commercially refined and bleached canola oil was reduced to one ppm by hydrogenation. The decrease was accounted for by a corresponding increase in the sulfur content of the nickel catalyst.

Deodorization

This process removes odoriferous and flavor compounds. Both bleached and hydrogenated oils require further processing to remove flavor-active compounds. Deodorization is achieved by low- pressure steam distillation. A number of efficient semi- continuous deodorizers are available for commercial use. Often 0.005 to 0.01 per cent citric acid is added to the oil as an iron scavenger and the deodorized oil may be stored and packaged under nitrogen to further preserve its natural flavor. Rapeseed oil which has been slightly hydrogenated to decrease the linolenic acid content and improve flavor stability may also be "winterized" (stabilized by removing some of the high melting triglycerides).

Protein Concentrates

Rapeseed meal obtained after oil extraction can be used in the preparation of protein concentrates and textured protein products. The available methods of protein concentrate and isolate production may be classified into two categories, namely (1) physical separation of protein-rich fraction, and (2) solubilization of proteins using suitable solvent system followed by precipitation and drying. Sometimes a combination of the two methods is employed. When solvents such as an alkali are used for solubilization, several changes in proteins may result such as destruction of certain amino acids, racemization of amino acids, and formation of new compounds and aggregation of proteins which may reduce protein solubility. However, preparation of protein concentrate results in reduction or elimination of antinutritional factors.

Protein Isolates

Protein isolates have a higher production cost but their properties can be tailored to specific needs. Protein isolates are appropriately used in beverages, toned and artificial milk and ice-cream. Blaicher prepared defatted rapeseed meal by countercurrent extractions at pH 9.5 and pH 12 in successive stages. The resulting extract, upon consecutive precipitation at pH 6 and 3.6, provided two protein isolates containing about 60 per cent and 12 per cent of meal protein, respectively. The color of the protein isolate was improved by using sodium bisulfite in the solvent for protein extraction. Rapeseed protein isolates containing about 10 per cent phytic acid exhibited a PER value of 2.2 compared to a value of 2.5 for casein. Paulson and Tung prepared protein isolates after succinylation of canola protein and studied their functional properties. Both emulsification activity and stability increased by succinylation but extensive succinylation was not required to significantly improve these properties. Tzeng prepared protein isolates using aqueous extraction with sodium hexametaphosphate, ultrafiltration, diafiltration and purification by ion exchange. About 60 per cent of the nitrogen in meal was recovered as an isolate free of glucosinolates, low in phytates, and fibers. It was light in color and bland in taste. All these results suggest that protein concentrates and isolates low in antinutritional factors with improved nutritional quality can be prepared from rapeseed meal.

Sunflower Processing

Drying

Drying of seeds is the first step in sunflower seed processing. The moisture content of freshly harvested sunflower seeds may be as high as 20 per cent. To ensure safe storage, the seeds must be dried to less than 10 per cent moisture. Most drying methods use forced air through a mass of seeds. The rate of drying can be enhanced by increasing the air flow, increasing the air temperature or by reducing the relative humidity of the air. The drying systems commonly followed for sunflower seeds are: (1) in storage-bin drying, (2) batch drying, (3) batch-in-bin drying, and (4) continuous air flow are drying. The method and effectiveness of drying directly influence the oil quality. Incomplete drying tends to promote microbial activity and can lead to spoilage. The microbial activity increases the amount of soluble pigments and free fatty acids in the extracted oil.

Dehulling

Method

For dehulling, the seeds are passed through air separation chamber to remove foreign matter. The seeds are dehulled in a drum-shaped chamber which contains a revolving plate that impels seeds against the smooth inner wall of the drum by centrifugal force. The hulls and kernels are separated by air gravity. Kernels are then passed over to mechanical color sorters and then go over a vibrating table for visual inspection and removal of unacceptable kernels obtained almost complete hull-free kernels from high oil sunflower seeds by means of a continuous dehulling separator using air jet impact huller. The seed coat of non oilseed-type of sunflowers is loosely bound to the kernels and can be easily removed. The recovery of kernels may be 50-55 per cent. Kernel fragmentation occurs during dehulling process, particularly if the moisture content of seeds is below 10 per cent. The addition of dehulling step in oil extraction may increase the energy requirement by 5-10 per cent. However, there is a reduction in the unnecessary movement of mass through the system, in the wear of expeller and in the wax content of oil and fiber content of meal.

Effects on Chemical Composition

The removal of hulls during dehulling has a significant influence on the chemical composition of sunflower seeds. Complete dehulling of sunflower seeds before oil extraction reduces the transfer of pigments from hulls to flour and reduces the fiber content in the finished product. Meals from size-separated seeds contain more protein and less cellulose than those from without size separation.

Byproducts

Hulls are the byproduct of dehulling process. The hull content among the genotypes varies from 10 to 60 per cent. Sunflower seeds of oil-type varieties contain less hulls (20-25 per cent) than the confectionery types (40-50 per cent). The hull contains heavily lignified cells. Oil-type seed hulls are thinner and more difficult to remove. Most commercial de hulling processes are 90 per cent effective at the best. The hulls have low bulk density (0.24-0.42 g/ml) and, therefore, present storage and disposal problems. Sunflower hulls contain about 4 per cent crude protein, 0.5 to 2 per cent fat, 50 per cent crude fiber and 2.5 per cent ash. The hull flour from oil-type sunflower contains less crude fiber and more protein, oil and ash than the hull-flour from confectionery-types. Cellulose and lignin are the principal constituents of the hull and account for almost 50 per cent of the hull weight. Reducing sugars (26 per cent) are the second major constituents of the hull and include 52 per cent xylose, 27 per cent arabinose and 9 per cent galactose. Lipids represented 5.2 per cent of the total hull weight, including 2.96 per cent wax, in sunflower. The wax is composed of long chain (C_{14}-C_{28} mainly C_{20}) fatty acids and fatty alcohols (C_{12}-C_{30}, mainly C_{22}, C_{24} and C_{28}). The wax of sunflower hull was reported to contain high levels of arachidic acid (46.3 per cent), followed by behenic (16.3 per cent), oleic (4.7 per cent), and lignoceric (4.5 per cent) acids. The fatty acid composition of sunflower hull lipids is similar to that of kernel oil.

Oil Extraction and Purification

Method of Extraction

Extraction of oil from sunflower seeds or kernels can be done using general equipments and operating conditions used for soybean or other oilseeds. The extraction of oil from sunflower is done by mechanical extraction, prepress solvent extraction and direct solvent extraction methods. Cleaning, drying and dehulling prior to extraction are the common steps in all the methods.

Mechanical Extraction

In mechanical screw press or expeller method, the kernels or seeds are crushed or rolled and cooked. Cooking facilitates the disruption of oil-bearing tissues and also inactivates phospholipases and lipases, thereby reducing the concentration of non-hydratable phosphatides and minimizing the acid value. The crushing, screw pressing and filtering operations reduce the oil content to 15-18 per cent.

Prepress Solvent Extraction

This is the most common method for sunflower oil extraction. In this method, the kernels are screw-pressed to obtain cake, with about 16 per cent oil content, as in the mechanical extraction method. The cake obtained is granulated or flaked and prepressed in a screw press or an expeller in single or two stages. In the two-stage press, the first stage has a shallow pitch. The second stage has a sharper pitch and operates at higher temperatures and pressures than the first stage. The pressed cake is granulated prior to solvent extraction. The solvent extractor may be of horizontal or vertical type. The oil cake may be moving or remain stationary. The solvent and the flaked oil cake move in counter flow. Hexane is widely used because of its low cost and low toxicity than the other solvents. The solvent is recovered from the miscella by evaporation using excess heat from the meal desolventizer-toaster process. The vaporized solvent is condensed and then recycled. The oil goes through the stripping column, where steam removes the last traces of the solvent. The toasting operation may be omitted for sunflower meal as it lacks toxic factors and its color does not improve upon heating. Prolonged heating of meal reduces the solubility of protein and lowers the concentration of some essential amino acids. The prepress solvent extraction reduces the oil content in the sunflower meal to about 1-2.5 per cent. The screw press operation consumes much of the electrical energy (about 40 per cent) during oil extraction. Some alternative methods have been suggested to reduce the power requirements for screw press operation and reduce the cost of production. These are: (1) immersion in solvent, (2) comminuting the seed, heating in the presence of water, drying and extracting in normal manner, and (3) separating oil, protein and fiber using a combination of solvent and water, followed by centrifugation.

Direct Solvent Extraction

In this method, kernels are conditioned, flaked and oil is extracted directly instead of expelling or screw pressing and then flaking prior to oil extraction. The resultant meal contains about 2.5 to 3.5 per cent oil.

Refining

Sunflower oil usually does not require extensive refining as it contains relatively low levels of free fatty acids, phospholipids, tocopherols, pigments and sterols. The oil is refined by degumming, neutralization, bleaching, deodorization and winterization.

Degumming

The oil is degummed to remove hydrophilic phosphatides and to reduce subsequent refining losses. The oil is degummed by treating it with phosphoric acid (or phosphate salt) and 1-2 per cent water to break the phosphatidic linkages. The hydrated phosphatides are removed by continuous centrifugation. Some scientists used citric acid, phosphoric acid, oxalic acid and acetic anhydride for their effectiveness in degumming crude sunflower oil. Of these, oxalic acid was found to be more effective than other reagents.

Neutralization

After degumming, the oil is treated with dilute sodium hydroxide to neutralize the free fatty acids and to react with the residual phosphatides, pigments and other minor constituents. The mixture is heated to break the emulsion and separated by continuous centrifugation. The oil is then washed with small quantities of water to remove traces of sodium hydroxide and soap.

Bleaching

Sunflower oil normally does not contain chlorophyll but contains small quantities of carotenoids and xanthophylls. Bleaching is, therefore, unnecessary for sunflower oil. However, oil from poor-quality seeds, or the seeds stored for a long time require bleaching. Many crushers routinely bleach the sunflower oil to ensure a constant color in the finished oil. Bleaching of sunflower oil is done by batch or continuous operation using 1 per cent activated clay or bleaching earth.

Deodorization

Deodorization of sunflower oil is accomplished by injecting steam to remove the volatile products such as free fatty acids, sterols and some unsaponifiables to improve the flavor and color of the oil. It also decomposes peroxides which may have formed during refining. Deodorization is usually done at temperatures of up to 270°C under reduced pressure.

Winterization

The winterization treatment is given to sunflower oil to improve its clarity so that it remains clear without clouding or precipitating during storage at low temperatures. The winterized oil remains clear for at least 5 h when held at 0°C. The wax content of the oil may be as high as 1 per cent of the total lipids, depending upon the efficiency of the dehulling and refining processes. During winterization, the oil is slowly chilled to promote the formation of crystals, held at a low temperature to allow crystals to grow and then separated by filtration. The slow process of filtering viscous oil can be avoided by winterizing the oil as a miscella containing up to 6 per cent hexane. In addition, refined oils can be cooled to 4°C, treated with dilute alkali, held and then centrifuged to produce clear oil. Morrison and Thomas were able to reduce the phosphatides tenfold, waxes from 0.021 per cent to 0.009 per cent and to produce finished sunflower oil that remained clear for more than 7 days.

Hydrogenation

Hydrogenation of fats involves the addition of hydrogen to double bonds in the chains of fatty acids in triacylglycerols. The process is of major importance in the fats and oils industry since it accomplishes two main objectives. First, it allows the conversion of liquid oils into semisolid or elastic fats more suitable for specific applications, such as in shortenings and margarine and second, it improves the oxidative stability of the oil. In practice, the oil is first mixed with a suitable catalyst (usually nickel), heated to the desired temperature (140-225°C) then exposed to hydrogen at pressure up to 60 psig and agitated. Agitation is necessary to aid in dissolving the hydrogen, to achieve uniform mixing of the catalyst with oil, and to help dissipate the heat of the reaction. The starting oil must be refined, bleached, and low in soap and dry, the hydrogen gas must be dry and free of sulphur, CO_2 or ammonia, and the catalyst must exhibit long-term activity, function in the desired manner with respect to selectivity of hydrogen and isomer formation, and be easily removable by filtration. The course of the hydrogen reaction is usually monitored by determining the change in refractive index, which is related to the degree of saturation of the oil. When the desired end point is reached, the hydrogenated oil is cooled and the catalyst removed by filtration.

Other Methods of Refining

The oil can also be refined in the presence of solvents. Miscella-refining eliminates wash water, vacuum drying and reduces refining losses. Physical refining, used for oils with high contents of free fatty acids, is also suitable for sunflower oil. This process eliminates the alkali neutralization step and combines degumming and bleaching. The oil is purified further in a deodorizer column which removes free fatty acids, destroys pigments and deodorizes the oil. The overall process reduces refining losses and waste byproducts and is more economical to operate.

Sunflower Meal

Sunflower meal is obtained as a byproduct of oil extraction process. Four types of meal may be obtained, *viz.* (1) mechanically extracted whole seed meal, (2) solvent-extracted whole seed meal, (3) dehulled mechanically extracted meal, and (4) dehulled solvent extracted meal. The chemical composition of the dehulled meal is significantly different from that of the whole seeds. The meal contains more protein and nitrogen-free extract and less of ash and fiber than the whole seeds. The oil content of solvent extracted meal is almost negligible. The chemical composition of sunflower meal compares favorably with most oilseed meals. The main exceptions are the higher fiber and ash contents, which tend to reduce the metabolizable energy. The composition of meal varies with the efficiency of the dehulling and extraction procedures. When the processing operations are harsh or when excess heat is used to desolventize the meal, destruction of lysine, arginine and tryptophan occurs. Cegla and Bell scientists isolated 8.3 per cent sugars from dehulled and defatted sunflower meal. These sugars included 0.6 per cent glucose, 2.3 per cent sucrose, 3.2 per cent raffinose and 0.8 per cent trehalose. Sunflower meal compares favorably with most oilseed meals as a source of calcium and phosphorus. Other minerals are neither deficient nor present in toxic proportions. Sunflower meal is a good source of B-complex vitamins: thiamin, nicotinic acid, pantothenic acid, riboflavin and biotin.

Sunflower Butter

Sunflower butter is made similar to peanut butter and has similar consistency. It is rather darker in color than peanut butter. Nutritionally, it is rich in iron and vitamin E. Dreher scientist studied the nutritional, sensory and physical properties of sunflower butter. They observed that the taste score of sunflower butter was less than that of peanut butter. The protein quality, calorie content and phytic acid levels in sunflower and peanut butters were approximately the same. Sunflower butter can be used like peanut butter. However, the chlorogenic acid present in sunflower products may cause a greenish grey color and a distinct bitter flavor.

Heat Processing

Roasting

The large-sized non-oilseed-type sunflower seeds (> 8 mm dia) are usually roasted and sold as in-shell sunflowers for cracking. The medium-sized seeds are dehulled, roasted and sold as peanut substitutes and confection or snack item. In-shell roasted sunflower seeds are considered to be of high quality if they are clean, large and of uniform size. For roasting of non-oilseed-type kernels, the moisture is reduced to 5-6 per cent. Roasting is done at 149-191°C for 3 to 6 min/batch, an antioxidant (BHA, BHT or TBHQ) is usually added to the oil used for roasting. Roasted kernels usually contain about 0.5 per cent moisture.

Dry Heating and Autoclaving

Heat treatment of sunflower meal affects its amino acid composition and nutritional quality. Bandemer and Evans scientists reported that heating sunflower seed at 121°C for 45 min caused a large decrease in basic amino acids. Some researcher observed a destruction of lysine in sunflower meal after dry heating at 121°C. Sunflower meal heated at 110°C for various periods showed a decrease in the available lysine content from 3.4 per cent to 0.9 per cent. Autoclaving of sunflower meal at 151b pressure significantly decreased the performance of chicks and rats. Heating of sunflower seeds before oil extraction improved the nutritional value of the defatted meal.

Salting and Frying

Salted sunflower kernels contain about 1.25 per cent salt. The salt used often contains antioxidants like BHA and BHT. The roasted/salted kernels are consumed as snack. Sunflower oil is a good frying medium. However, prolonged heating of sunflower oil results in darker color and oxidative changes. Sebedlo isolated the cyclic fatty acid monomers (CF AM) formed during heat treatment of sunflower oil when it was heated at 275°C for 12 h under nitrogen and at 200°C for 48 h in a commercial fryer using a 2 h daily cycle.

Germination

Canella and Bernardi (1983) studied changes in phenolics and oligosaccharides during sunflower seed germination from the view point of increasing the use of germinated seeds. There was a rapid fall in chlorogenic acid and a slight rise in caffeic and quinic acids during germination. Sucrose and raffinose concentrations decreased with concomitant rise in glucose and fructose contents.

Protein Products

Commercial sunflower meal contains about 40 per cent protein. This meal is not suitable for human consumption because of high fiber content and the presence of polyphenols. Efforts have, therefore, been made to obtain protein products with high protein contents and a better nutritional quality. Various protein products from sunflower may be broadly classified into three categories:

1. Dehulled, defatted high-protein meals containing 50 per cent or more protein.
2. Protein concentrates containing 70 per cent or more protein.
3. Protein isolates containing about 90 per cent or more protein.

High Protein Meals

Bau *et al.* (1983) prepared colorless protein products from sunflower seed by various technological treatments like dehulling, defatting, soaking or boiling in citric acid or in sodium bisulfite solution. The processed products containing more than 60 per cent protein were white in color, low in dietary fiber and phenolic contents. The diets containing whole seed meal presented a low PER value in rats, which was slightly improved upon dehulling. Lysine supplementation of protein products significantly improved their PER values. Citric acid treated products (CADHP and CAADP) exhibited higher in vitro digestibility than the untreated meal.

Protein Concentrates

Sosulski *et al.* (1973) produced a stable white protein concentrate by hot aqueous diffusion of cracked sunflower kernels. A creamy or light grayish colored protein, readily soluble in water and with an improved amino acid pattern, by a single step and successive extraction with aqueous calcium hydroxide and water. The protein had an isoelectric pH of 5.0. Creamy white protein concentrate, low in chlorogenic acid, could be obtained by a coholic extraction.

Protein Isolates

A number of workers have prepared protein isolates from sunflower, mostly using alkaline extraction. A flow chart for the preparation of protein isolates is shown in Figure 5.3. The above methods caused denaturation of proteins and an incomplete removal of color-forming phenols. Saeed and Cheryan (1988) have prepared low-polyphenol concentrate and low-polyphenol reduced-phytate isolate from sunflower seeds using acidic butanol, alkali and acid extractions.

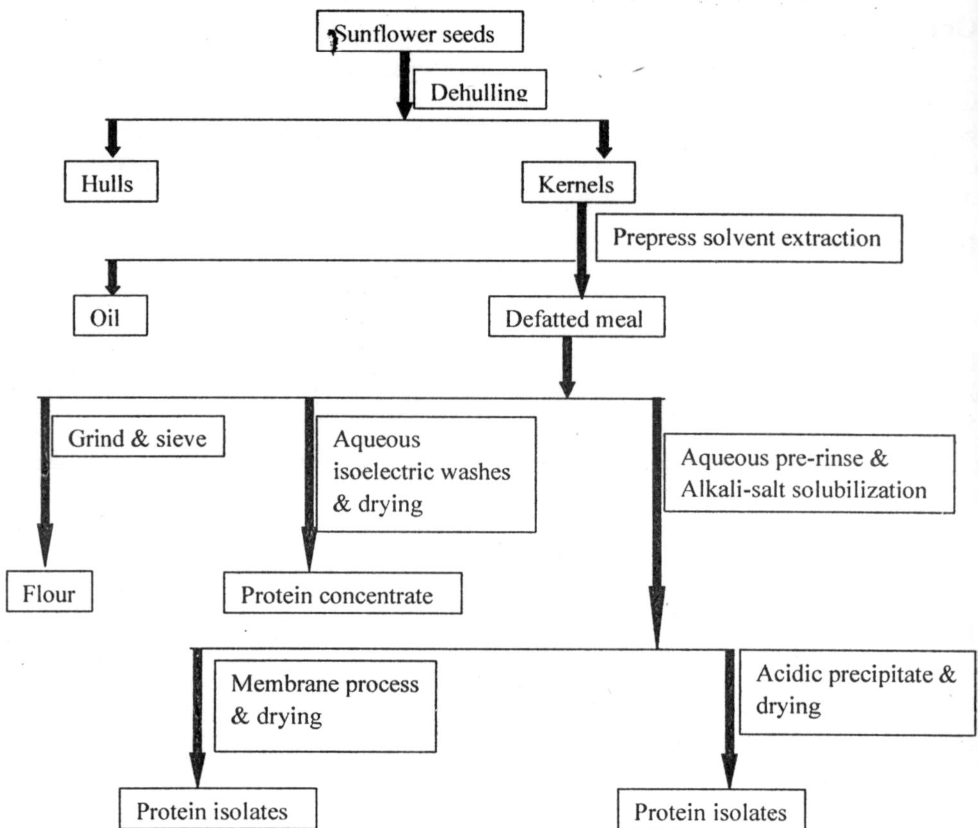

Figure 5.3: Flow diagram showing the preparation of sunflower protein concentrate and isolates

Peanut/Groundnut Processing

Peanuts are processed into a wide variety of edible products. Edible oil, peanut butter, salted peanuts and various confections are the major products. Peanuts are also processed to obtain peanut flour, peanut protein isolates and concentrates, peanut milk, curd, cheese, peanut paste for seasoning vegetables, boiled unshelled fresh peanuts, defatted peanuts, peanut butter ice-cream, peanut butter milkshakes, peanut butter powder, peanut butter sand witches, and peanut bread. The nonfood items include soaps, cosmetics, medicines, shaving creams, lubricants, and synthetic fibers. The quality of the processed products depends on the initial quality of the peanuts. The criteria of quality may vary for farmers, merchants, millers and processors. For farmers, peanuts that are resistant to insect pests and giving higher yields are important, while for merchants and decorticators quality relates to shelling out-turn and natural test weight of pods and kernels. To the crusher, the oil and the free fatty acid contents are most important. For manufacturers of peanut butter, salted nuts and confections, quality is related to desirable sensory properties such as color, appearance, aroma, taste and shelf life of the products. For any kind of processing, peanuts need to be free from foreign material, off-odors, broken and mold infestation. In order to achieve the best quality of the processed product, the in-shell peanuts need to be properly graded on the basis of the percentages of foreign material and loose shelled kernels, contamination of *A. flavus* mold, shelling percentage, moisture contents in-the-shell kernels, and various types of damages such as brokens, internal or concealed damage, rancidity, decay, freeze damage, discoloration etc. For shelled peanuts, the grading may involve sound split and broken kernels, split and broken kernels with damage, split and broken kernels with minor defects, sound whole kernels, whole kernel with minor defects and unshelled kernels.

Shelling of pods should follow the careful grading of farmers' stocks. The shelling involves receiving, precleaning, shelling, screening, separating, sorting, sizing and packaging operations. The shelled peanuts are stored in cold storage at a temperature of 2-5 °C and RH of 65 per cent until used for product manufacture. In most developing countries, both grading and shelling operations are paid least attention. The operations are of traditional type and inefficient; this often results in the inferior nutritional and sensory characteristics, and short life of the processed products.

Extraction, and Refining of Oil

Edible oil is the major product produced from peanuts in the world. About two-third of the world's peanut production is crushed to obtain edible oil. According to USDA estimates for 1987-88, peanut (5 per cent) is ranked seventh after soybean (29.1 per cent), palm (16 per cent), rapeseed (13.2 per cent), sunflower (13.1 per cent), cotton seed (6.2 per cent) and coconut (5.1 per cent) in the world's production of major vegetable oils. On the basis of its cooking and flavor qualities it is among the three premium oils (peanut, safflower and rapeseed) consumed in India. Oils from soya, palm, sunflower, cotton seed and coconut need refining before they are used for edible purposes. However, the crude oil from peanut, safflower and rapeseed produced by expeller or hydraulic press is often directly used for cooking purposes in the developing countries. These oils are, however, refined before consumption in the developed world. In India, the major peanut oil is produced by expeller press and consumed in the form of crude oil. In most of the developing countries, a major oil seed produce is crushed to extract edible oil. In India, over 80 per cent of the peanuts are subjected to crushing through traditional screw pressing to extract edible oil, while in the western countries including the U.S. quality peanuts are processed for other food uses and only the farmers' stock peanuts that do not meet the requirements of food uses after preliminary cleaning are shelled and crushed for oil.

Usually, three methods are employed to recover the oil: expeller, hydraulic press and solvent extractor, either separately or in combination. The residual oil in the cake is about 7, 5 and 1 per cent, respectively. Most peanut mills are equipped with shelling and crushing facilities. After removal of foreign materials and shelling through shaker and destoner, the loose shelled kernels are removed in suction and metallic contaminants by electromagnets.

Expeller Pressing

This method is most widely used to recover the oil, particularly in the developing world. The equipments employed range from the traditional bullock drawn *ghani* to the highly mechanized electrically operated screw press. Peanuts containing about 6 per cent moisture are crushed and fed into the large conical end of the expeller chamber. Pressure is exerted as the screw turns, forcing the material towards small end. The oil is expelled through screen due to friction heat and pressure. The oil is collected beneath the expeller chamber, while the residue is extruded from the small end. More than one expeller units are often arranged side by side on commercial oil mills, where the oil is collected through a common channel at one place, filtered and packed in tin cans or barrels of suitable size as crude peanut oil. The residual oil in the cake ranges from 7 to 15 per cent depending upon the quality of peanuts and the efficiency of the expeller unit. Over 15 per cent residual oil remains in the cake with bullock-drawn traditional wooden expellers. The cake can be powdered to edible meal if the peanuts are of edible grade and the operations are conducted under hygienic conditions. However, this is seldom possible. Peanuts crushed for oil extraction are often not graded. On the contrary, some proportions of unshelled peanuts or peanut hulls are intentionally mixed before crushing to facilitate friction for maximum oil recovery. The red skins are intact before crushing. The hulls and skins contribute excessive fiber and phenolic constituents in the cake. The bitter compounds present in the germ also come in the cake. Thus, the cake obtained in expeller pressing is full of fiber, bitter saponins and phenolic compounds. The crude oil obtained in expeller units is filtered to remove suspended material, allowed to stand for settling and stored in tin cans. It is used as cooking oil directly or subjected to further refining. The press cake can be further subjected to solvent extraction to recover residual oil.

Hydraulic Pressing

The shelled peanuts with about 6 per cent moisture are crushed, heated with live steam and spread on press cloth with edges folded to prevent running. These are formed by a modified rack-and-cloth operation. The racks are placed in tiers and about 14,000 psi pressure is applied. The crude oil is collected beneath the tiers. Similar to expeller pressing, a small amount of unshelled peanuts or peanut meal is mixed with the ground peanuts as a press aid. The efficiency of oil extraction is lower than expeller pressing and the press cake is often not suitable for human consumption. The cake is usually subjected to solvent extraction to remove the residual oil.

Solvent Extraction

The residual oil in the cake produced in hydraulic or expeller press is usually extracted by employing a suitable solvent. The system is rarely used for full fat peanut grits or meal. It is a closed system in which the partially defatted cake grits or meal is washed with hexane to remove oil. Other solvents include absolute ethanol and 95 per cent ethanol. However, n-hexane is most commonly used. The solvent is removed by evaporation and recycled while the oil is further refined for edible purposes. Almost completely defatted cake meal obtained is used as cattle feed and manure. The meal can, however, be used for human consumption if both expeller pressing and subsequent solvent extraction operations are performed under hygienic conditions and if good quality peanuts are used.

There are several types of extractors that may be used for solvent extraction of oil. The systems are devised to extract maximum oil with minimum losses in both meal and solvent, and to obtain high-quality protein for industrial uses. An aqueous system for simultaneous recovery of peanut oil and food-grade protein concentrates and isolates directly from raw peanuts has been described. Under optimized conditions, about 96 per cent of the oil and 94 per cent and 92 per cent of the proteins were recovered in protein concentrates and isolates, respectively. A method for direct solvent extraction of oil from peanuts to produce a low-fat proteinaceous material has been patented by Woodroof (1983). The steps involved include wet-heat conditioning of the kernels at 92-205°C for 10-20 min (6-12 per cent moisture), flaking of the conditioned seeds, rapid dry-heat conditioning of flakes with forced heating system (1.9 to 6 per cent moisture), and solvent extraction of oil from such flakes by stationary bed, counter flow or cross-flow extraction, either under vacuum or by using a hot solvent (24-60°C). Solvents may include hexane, acetone, ethyl alcohol, isopropyl alcohol, tetrachloroethylene, chlorinated hydrocarbons or their mixtures.

Refining of Oil

The fresh crude oil obtained in hydraulic or expeller units is always subjected to refining before consumption. The steps involved in oil refining are mechanical removal of impurities such as meal and moisture by sedimentation, filtration or centrifugation of phosphatides by degumming, free fatty acids by alkali treatment, colors (bleaching) by treatment with activated charcoal and deodorization by steam-heating under vacuum. The process also removes natural flavor, antioxidants and fat-soluble vitamins. Hence, antioxidants and certain fat-soluble vitamins are often added to refined oil to compensate these losses. The refined oil has a pale yellow color, a bland taste and good storage stability. Freshly refined peanut oil has a smoke point of about 227°C, melting point of 0.56-2.22°C, refractive index of about 1.4692; iodine number of 90-94, free fatty acids 0.0137-0.0422 per cent, peroxide value of 3.5-8.0 meq/kg and saponification number of 188-191.

Palm Oil Processing

Oil palm fruit is processed to obtain two types of oil. Palm oil is obtained from the mesocarp of the fruit while palm kernel oil is extracted from the kernels. Palm fruits must be processed locally after the harvest for extraction of palm oil as palm fruits cannot be transported due to economic and quality reasons. Palm kernels are, however, mostly exported. Palm oil is the major economic product from oil palm fruits, representing about 20 per cent of the harvested palm fruit bunch while palm kernel oil represents only 4 per cent.

Palm Oil

Palm oil is extracted by traditional methods in Africa, although improved methods of oil extraction are available. In the traditional methods, the oil may be extracted by a "soft oil" or "hard oil" process. Mechanical methods include the hand-operated presses, power-operated centrifuges and presses, and solvent extraction (Raymond

1961; Hartley 1967). Solvent extraction is, however, not popular for the extraction of palm oil due to some technical and economic considerations.

Traditional Methods

In the "soft oil" method, the fruits are boiled for about 4 h and left for 3 days. The pulp is then disintegrated by pounding the fruits in a pestle and mortar. The disintegrated pulp is immersed in water and stirred. The oil rising at the surface is skimmed off. The water in the oil is expelled by heating the crude oil in a shallow vessel. In the "hard oil" process the pulp is softened by fermentation in wooden troughs. Fermentation takes place in successive stages alternating with moistening and pounding for several days. The oil is allowed to drain and mixed with water. The oil rising at the surface is skimmed off and boiled to expel water. The extraction rate by the traditional methods is very low, *i.e.* 30-50 per cent in the soft oil process and 20-30 per cent in the hard oil process. The free fatty acid content may be 7-12 per cent and as high as 30-50 per cent in soft and hard oil processes, respectively.

Mechanical Methods

The efficiency of extraction by mechanical methods is significantly higher (80-85 per cent) than that by the traditional methods. In the hand-operated process palm fruit bunches are sterilized to destroy lipolytic enzymes. The fruits are separated and digested at a high temperature. The kernels are removed and the oil is extracted from the pulp by a screw or hydraulic press. The screw press is similar to that used for the extraction of fruit juices. In the hydraulic hand press, a ram moves into a perforated press cage when the hydraulic fluid pressure is increased by hand operation of the piston pump. The power-operated palm oil mills were started toward the end of the nineteenth century. They consist of several sections and operations including sterilization, stripping, digestion, extraction, clarification, nut separation, nut drying, nut grading and cracking, kernel separation and kernel drying. The object of sterilization is to loosen the fruits on the bunch, which otherwise would require about a week at room temperature. During this period, free fatty acids would increase significantly. Loosening of the fruits is achieved by sterilization of bunches for 60-75 min. Vertical type sterilizers are suitable for small mills while horizontal-type sterilizers are used in large mills. The sterilized fruits are separated from the bunch by beater arm type strippers or threshers in small mills and by rotary drum type strippers in large mills. The digesters break up the pulp physically and release the oil from the cells. The oil is extracted using a screw press. The extracted oil contains large amounts of water and some impurities of non oil solids such as fiber pieces, small fruits, and sand or soil particles. Larger impurities are removed by passing the crude oil through vibrating screens of stretched gauze. The nuts may be separated from the fibers by hydraulic, mechanical or pneumatic separators. Pneumatic separators are most commonly used. The clean nuts are dried and graded according to their size. Nutcrackers produce a mixture of kernels, shells and uncracked nuts. The kernels and shells are separated using a hydrocyclone machine. The separated kernels are dried and bagged.

Palm Kernel Oil

Palm kernel oil is not usually extracted in the producing countries, except for only small quantities extracted by primitive means. The bulk of palm kernels are exported and extracted in the importing countries. Palm kernels are subjected to industrial processes for extraction of oil. The kernels, being solid and hard, are crushed to meal before oil extraction. Roller mills are usually employed to rupture the oil-bearing cells. Cooking under pressure helps to release the oil still further. The oil is extracted by hydraulic press, screw press or by solvent extraction.

Refining of Palm Oil

With increasing competition from other edible oils, processing of crude palm oil has become necessary to yield more saleable products. The developments in the technology of refining, bleaching, deodorization, fractionation, hydrogenation and interesterification have helped to expand the use of palm oil for edible purposes. The crude palm oil contains several nonglyceride impurities and has a dark yellow or red color. The color of crude oil is primarily due to the presence of carotenoid pigments. For edible purposes, the palm oil is refined to remove free fatty acids, phospholipids, pigments and volatile components by neutralization, bleaching and deodorization treatments.

Neutralization

Klein and Crauer (1974) have described continuous caustic refining of palm oil containing upto 8 to 10 per cent free fatty acids. The following conditions were found to produce good refined oil yields and quality: preheat crude palm oil to 54-71°C, neutralize with just a sufficient quantity of caustic soda and mix to complete the chemical reaction, heat to 82-88°C to flock the soap phase and immediately centrifuge.

Bleaching

The neutralized palm oil contains some color and flavor compounds. It is, therefore, bleached to remove its natural reddish brown color to produce a colorless stable product for use in margarines, shortenings and bakery products. Jasperson and Pritchard (1965) have given typical user specifications of the color (Lovibond Red units, 5.25 cell) of palm oil for use in various products: margarine 4.0 to 5.0; first-grade frying fat 2.0 to 2.5; blended cooking fats 2.0 maximum and white fat 1.5 maximum. Bleaching of palm oil is achieved by using bleaching earths or by destruction of pigments by heat. Bleaching earths possess a large surface area and have specific affinity for pigment-type molecules. Activated carbon has also a large adsorbent area but is more expensive. Because of the high intensity of color in palm oil, it requires more bleaching earth and a longer period than for bleaching of other vegetable oils. Arumughan *et al.* (1985) reported that the addition of 3 per cent bleaching earth containing active carbon in the ratio of 9: 1 and bleaching at 150 °C for 1 h in a vacuum of 700mm Hg represents the optimum conditions for the bleaching of indigenous crude palm oil on a pilot-plant scale.

Deodorization

The refined and bleached palm oil contains some aldehydes, ketones, alcohols, low molecular weight fatty acids, hydrocarbons and other compounds derived from decomposition of peroxides and pigments. Although these compounds are present in very low concentration, they can be detected by their flavor or odor. For the use of oil for edible purposes, these compounds need to be removed. These compounds are more volatile at reduced pressures and elevated temperatures. Deodorization is essentially steam distillation under vacuum. Refining causes significant changes in the physicochemical properties of palm oil. Refined palm oil contains less free fatty acids, iron, copper, carotene, moisture and dirt and has better oxidative stability than the crude oil. Siew and Mohammad (1989) studied the effects of bleaching and deodorization on the oxidative properties and possible isomerization and interesterification of fatty acids in palm oil products. It was observed that the conjugated dienes and trienes formed in refined palm oil products were minimum. Very little isomerization occurred in refined products as indicated by the levels of *trans* fatty acids. During refining, a part of the lipoproteins is removed together with free fatty acids and a part is hydrolyzed, releasing cholesterol. As a result, there is an increase in the cholesterol level during alkaline and physical refining of palm oil.

Fractionation

Palm oil is semisolid at ambient temperatures. It contains triglycerides of varying melting points and solubilities. On standing long, palm oil separates into a deep red, liquid upper layer and a light yellow, viscous lower portion. The separation of the two layers can be clearly seen if the oil is cooled slowly. The separation of liquid and solid triglycerides of vegetable oils by controlled slow cooling is called winterization. The liquid and semisolid fractions are called "olein" and "stearin", respectively. Commercially, three systems are available for fractionation of palm oil. In the dry system, palm oil is cooled carefully and filtered to separate the fractions. In the wet system, the crystals of the stearin fraction are preferentially wetted by using surfactants or aqueous detergent solutions. In solvent systems, palm oil is diluted by solvents like hexane, acetone, isopropanol or n-nitropropane. It is cooled and then filtered. The yields of the olein and stearin fractions after solvent fractionation of palm oil were 80.5 per cent and 19.5 per cent, respectively. The palm olein and stearin fractions differ in their physicochemical properties, fatty acid composition and have different end uses. The carotene content of the liquid fraction was considerably higher (940-1184 ppm) than the semisolid fraction (566-768 ppm). The residual color after bleaching was lighter in the stearin than in the olein fraction.

Interesterification

Fats and oils are a mixture of triacylglycerols in which the acyl groups are distributed in a nonrandom manner. Under the influence of a basic catalyst, *e.g.* NaOH, the acyl groups are redistributed both intermolecularly and intramolecularly. This provides a method of transferring saturated fatty acids to predominantly unsaturated glycerides and vice versa. Interesterification leads to the production of a triglyceride mixture having a composition completely different from that of the original oil. Industrially, the interesterification procedures are used to improve the physical properties of lard and to produce cocoa-butter substitutes from cheaper oils. Palm oil of more desirable melting point (39°C) can be prepared by interesterification.

Hydrogenation

Hydrogenation of vegetable oils is carried out to increase their melting point (or convert liquid oil to solid fat at ambient temperatures) and to retard the oxidation and flavor deterioration. Hydrogenation makes the oil less absorbable, resulting in a less greasy fried food. Palm oil does not require artificial hardening or hydrogenation and can be utilized in its natural form. Further, during hydrogenation some of the fatty acids are rearranged from their natural "*cis*" form to an unnatural "trans" form. Unlike the normal "*cis*" polyunsaturates, the "*trans*" polyunsaturates fail to act as essential fatty acids. They are metabolized like saturated fatty acids. When palm oil is hydrogenated, the amount of *trans* isomers of fatty acids formed is relatively much less than other selectively hydrogenated vegetable oils.

Cotton Processing

Ginning

The seed cotton collected from field or storage contains both lint and seed. In addition, it may also contain extraneous impurities such as dried leaf fragments. The lint is separated in a ginning operation using either a *charkha* gin, roller gin or saw gin. Cotton lint is packed or compress- baled and seeds are separately collected. Waston and Helmer (1964) found that approximately 1 per cent seed damage was attributable to the cleaning, drying and conveying operations while an additional 5 per cent damage was associated directly with ginning process. Cottonseeds obtained after ginning are stored or used for the extraction of oil.

Production and Refining of Cottonseed Oil

The processes and operations for the production of crude oil are outlined in the following sequences:

Mechanical Expression of Cottonseed Oil

The cottonseeds are thoroughly cleaned, delinted and dehulled. These materials are allowed to pass through a roller mill to convert them into thin flakes. The flakes are then cooked to precipitate the phosphatides, to detoxify the gossypol, to coagulate the protein and to bring down the moisture content of the flakes from about 12 to 5 per cent. Hydraulic press is mostly used for expelling cottonseed oil. However, expeller or screw press is also employed in smaller proportion for the same operation. The cooked flakes are placed in the hydraulic press and gradually pressure is applied. The pressure is raised to a maximum of 113 to 141 kg/cm^2 till the oil begins to flow. Then the pressure is drained for 30 to 45 min. The cakes are discharged. The capacity of a press may be as high as 7 tones of cooked flakes per day. A continuous motor driven screw press consisting of a tapered screw operating inside a perforated barrel can also be employed. It can apply a pressure of IS SO to 1860 kg/cm^2 and can discharge oil through the barrel spacing and deoiled cake through the orifice of the tapered end of the barrel. The residual oil in the deoiled cake is 4 to S per cent only, but the power consumption is very high. The yields of the oil, deoiled cakes, hulls, and linters from 1 tons of raw cottonseed are 136 kg, 408 kg, 227 kg and 91 kg respectively. Now-a-days solvent extraction is also practiced for the extraction of oil from either full fatted cottonseed flakes or mechanically deoiled cakes. The cotton seed oil can be used for the production of refined cooking oil or salad oil, margarine and shortenings. The deoiled cake is used for cattle feed. The linters can be utilized as a source of highly pure cellulose. The hulls can also be used for live stock feeding.

Extraction of Oil

Godin and Spensley (1971) have described the pretreatment and extraction processes for the commercial production of cottonseed oil. The high-moisture cottonseeds are dried, cleaned, and delinted. The hulling or decortication is carried out using either a bar or disk-type huller. The separation of split hulls and meals is accomplished in an assembly consisting of shakers and air separators. The separated kernels are rolled into thin flakes between heavy, smooth iron rolls for oil recovery by mechanical expression or solvent extraction. Cooking is necessary before mechanical expression and sometimes before solvent extraction. The flaked meals are subjected to a heat treatment. For hydraulic pressing, the cooking time ranges from about 90 to 120 min, temperatures varying from 79 to 91°C in the top pan and to about 107 to 112°C in the bottom pan. The moisture content is reduced from 11-12 per cent to about 5-5.5 per cent. For direct solvent extraction, flakes are first cooked and cooled to surface-dry the particles and put them in a more granular state. When cooking is omitted before extraction, the meal must be heat-processed to destroy free gossypol. The oil is extracted from the cottonseed by the following methods: hydraulic press, screw expeller press, solvent extraction prepress, solvent extraction, and filtration extraction. In hydraulic pressing, the meals are discharged from the cooker into a mechanical cake former and the slabs are formed and enclosed in a press cloth. The cake thus produced contains 5.6 to 6.5 per cent oil. Continuous pressing may be employed where cottonseed is crushed in a screw press. The oil is settled to remove the bulk of the fine, then passed through a plate filter press, and pumped to storage tanks. The press cake is ground into meal and then ground cottonseed hulls are added to adjust the protein content given by the required standards. The meal contains about 2.5 to 4 per cent oil.

More recently, the solvent extraction method has largely been employed. The oil-solvent solution is filtered and clarified, unless a percolation-type extractor is used. The miscella is passed through a tube extractor which recovers 90 per cent of the solvent. The oil concentrate is then passed through a stripping column, where sparging steam is applied to remove the remaining solvent. The meal thus extracted has only about 1 per cent residual oil. The crude cottonseed oil is refined to free most of its nonglyceride constituents by treatment with alkali, followed by bleaching and deodorization.

Refining of Crude Cottonseed Oil

The crude oil is refined as follows: Free fatty acids are neutralized with either caustic soda (NaOH) or soda ash with the formation of soaps commonly known as alkali foots. The soaps are removed from the oil by centrifuges and filters. The neutral oil is washed twice or thrice with water for washing out the remaining soap in the oil. The waste water is separated by filters and centrifuges. The oils are then decolourised with the help of the adsorbent like activated carbon and or activated clay at a high temperature by a batch process or by a continuous process. The bleached oil is then deodorized by, heating the oil with super heated steam under high vacuum. For the manufacture of salad oil bleached oil is subjected to winterization which consists of cooling the oil to a low temperature for a long time and filtering the solid materials from the oil.

Flour and Meal

Cottonseed is increasing in significance in relation to the total monetary value of the cotton crop. Edible-grade glandless cottonseed and the liquid cyclone processed flour from glandless cottonseed are expected to play an increasing role in coping with the world protein shortage. Glanded cottonseed can be processed into edible flour or concentrate by the liquid cyclone process. The liquid cyclone process uses prime-quality whole and cracked kernels that are essentially hull-free. The kernels are first dried with air at 82°C to a moisture content of 1.5 to 2.0 per cent and then ground with a pin mill. The milled meals go into a fluidizer where they are slurried with hexane. The discharged slurry with a solids level of 22 per cent is sent to an agitated liquid cyclone feed surge tank. The liquid cyclone capitalizes upon the rates at which the different components of cottonseed settle out or classify in hexane by using the action of centrifugal force, which accentuates classification. The liquid cyclone classifies feed slurry into gland-free overflow slurry containing 13 to 15 per cent high-protein solids, and a gland-rich coarse meal underflow slurry containing 43 to 45 per cent solids. High-protein solvent-damp cottonseed cake is recovered from the over-flow slurry on a rotary vacuum drum filter. During the filtration step, the lipid level in the solvent-free cake is reduced to approximately 0.6 per cent. The filter cake is desolventized in a stainless steel rotary twin shell "V" type blender equipped for vacuum and solvent recovery operations. The cake is heated to 82°C, whereupon nitrogen gas is injected into the blender to strip solvent from the flour to a level below 50 ppm.

High-pressure screw processing and prepressing followed by solvent-extraction methods used in oilseed processing plants can produce edible-grade cottonseed meal. During these processes, gossypol is made physiologically inactive by heat. Cottonseed meal is employed as a source of protein in the manufacture of adhesives and fibers. Plywood glue is prepared by combining it with casein, soybean meal, and synthetic resin which is water resistant and nonabrasive. Plastics containing equal parts of phenolic resin, cottonseed hulls, and cottonseed meal have excellent flow properties, a short curing cycle, water resistance, and strength. Cottonseed meal is the cheapest source of raffinose (2 to 4 per cent yields).

Protein Concentrates

Beginning with cottonseed flour (liquid cyclone processed glanded or glandless cottonseed), protein concentrates (containing 70 per cent or higher protein, dry-weight basis) can be prepared either by air classification or by one of the several wet-extraction procedures. The composition and yield of cottonseed protein concentrates obtained by various methods are given in. The first two methods are wet operations. Ninety per cent ethanol gives an optimum extraction of residual lipid and sugars with minimum removal of nitrogen. Ethanol extraction increased the protein content from 66 to 72 per cent, decreased the lipid content from 1.2 to 0.1 per cent and decreased the sugar content from 5.1 to 2.1 per cent. Little shift in the amino acid composition of the product from that of the original flour is to be expected because of the minimal loss in nitrogen.

Aqueous extraction at essentially neutral pH is another wet procedure for the preparation of cottonseed protein concentrates. Very dilute divalent cationic salt solutions such as 0.008 M $CaCl_2$ can be used to extract the defatted flour. The increase in protein content depends on both the solubility characteristics of the constituents of the flour and the particular extraction system used. Although this procedure results in an increase in the fiber content, it also provides maximum removal of the total sugar. Selective removal of water-soluble, low molecular weight proteins and sugars results in a significant reduction in the total weight and the total nitrogen content and, consequently, a marked change in the average molecular weight and amino acid composition of the final product. In the water extraction process, the proteins are retained in their original structures and concentrated at the expense of the low molecular weight cytoplasmic proteins of the cell.

Wet operations suffer from two major difficulties: the processing of the extract and the byproduct in an economically feasible manner, and the drying of the major product. The first difficulty involves low-yield product recovery, solvent recovery, and pollution problems. The second difficulty involves denaturation of the major product during drying. This difficulty is usually more severe when alcohol is a part of the extracting solvent system.

The dry procedure of air classification has neither of these difficulties. A cottonseed meal with high nitrogen solubility can be ground in a pin mill and air-classified to give two products, one for food and the other for feed, without further processing. In this procedure of separation of particles, clusters of unruptured cells, cell wall fragments with adhering residual cytoplasm, and residual spherosome membranes are separated from free, intact protein bodies. This mechanical separation results in an increase in the nitrogen content at the expense of the lipid, sugar and crude fiber contents of the resulting concentrate. A parallel increase in the ash content with nitrogen content, evident in most of these concentrates, is consistent with the particulate nature of the intracellular constituents.

The nonaqueous, liquid classification procedure called the liquid cyclone procedure was developed primarily for the removal of pigment glands. Since this procedure removes hull particles and cell wall fragments as well as pigment glands, the final product is very similar in composition to the glandless air-classified concentrate. Approximately defatted flakes are ground in hexane to remove the cellular tissue from the pigment gland. The pigment glands are then separated from the flour particles in liquid centrifuge called a liquid cyclone. The final product, a light yellow bland concentrate with high nitrogen solubility, contains < 400 ppm free gossypol and < 3000 ppm total gossypol.

Protein Isolates

Cottonseed protein isolates (90 per cent protein or higher) are conventionally prepared by extracting protein from cottonseed flour. In the classical procedure, this is accomplished by a single extraction at alkaline pH. Other isolation procedures separate cottonseed proteins into non-storage protein (NSP) and storage protein (SP) isolates (Figure 5.4). NSP and SP possess differing functional and nutritional properties. Protein isolation by each of the conventional isoelectric precipitation methods generates a byproduct, whey-like stream. Processes have been developed in which cottonseed wheys are processed by ultrafiltration (UF) and reverse osmosis

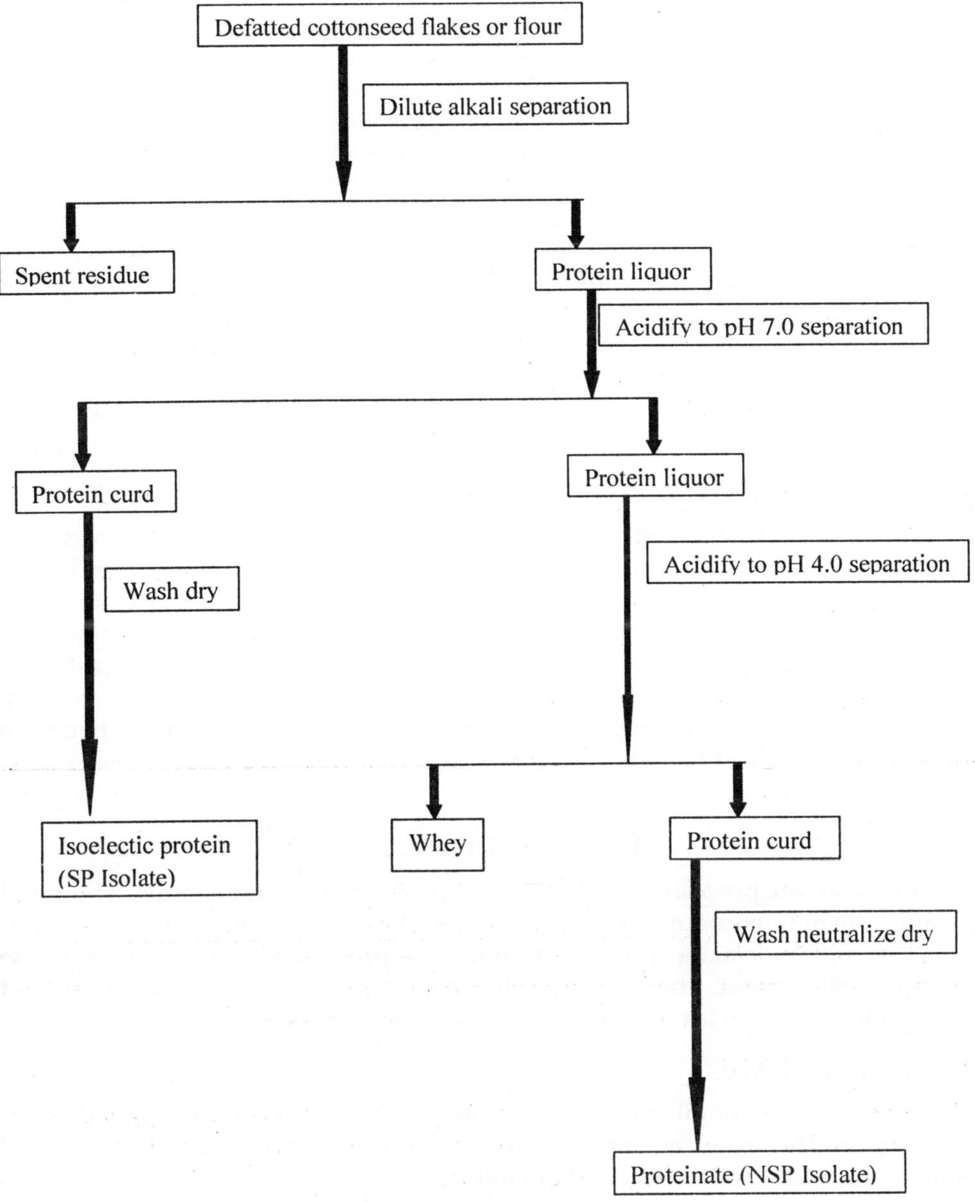

Figure 5.4: Processing of cottonseeds for protein isolates

(RO) to recover valuable constituents and prevent environmental pollution from whey disposal. Another approach to protein isolation employs industrial membrane systems. To avoid the generation of whey-like byproduct streams, liquid extracts from defatted oilseed flours are ultrafiltered instead of being acid-precipitated.

Cottonseed flour has been used in preparing co-isolates from flour blends made with any combination of vegetable proteins. In general, the method involves extraction of proteins from a flour blend with dilute aqueous sodium hydroxide, acidification of the extract to pH 2.5, and adjustment of the resulting mixture to pH 5.0 to precipitate protein curds. The protein curds are removed, resuspended in water, washed free of contaminants, neutralized, and lyophilized or spray-dried as co-isolate. Polyacrylamide gel electrophoresis studies suggest that the proteins in the extract dissociate into subunits at pH 2.5, and then reassociate into their original and new protein forms as the pH is adjusted to 7.0. The presence of some new protein forms, as well as original proteins in the coisolate, suggests that new derivatives are made that differ in compositional, solubility, functional, and nutritional properties from those of the isolate. The protein isolate had a low free gossypol content of 660 ppm with a PER of 2.1 and an available lysine content of 3.3-3.5 g/16 g N. Washing of flour prior to extraction improved the color but reduced the extractability of protein to 52 per cent. The color was further improved by hydrogen peroxide treatment. It was essential to supplement lysine and methionine to improve the nutritive value of the peroxidetreated protein.

Textured Protein Products

Cottonseed proteins can be texturized to prepare various products such as meat analogs. Various methods are employed to texturized cottonseed proteins. Patties prepared with 15 per cent flour were not as flavorful as 100 per cent beef patties but were similar in texture to them. A low-cost texturization process could promote the use of cottonseed products as food ingredients. The nonextrusion texturization of cottonseed flour with Korean hard press has been reported. The physical and chemical characteristics of the textured cotton flours made from glanded cottonseed flours are presented in. The results indicate that extrusion reduces the free gossypol content during texturization.

Coconut Processing

Coconuts are processed into coconut oil, desiccated coconuts, coconut milk and cream, skim milk, coconut protein, copra press cake and meal, and utilized in variety of bakery and confectionery products. The husk is processed into coir fiber while the shell into shell flour, charcoal, utensils and art products. The processing and utilization of coconut leaves, palm hearts, and stem has been described.

Extraction of Oil

The extraction of oil from copra is one of the oldest seed-crushing industries in the world. The processes employed to obtain oil from copra include the traditional milling, improved wet milling and dry milling.

Traditional Milling

It is essentially a wet-milling process. However, the recovery and quality of oil is low, and the byproduct is not useful for any food application. It is practiced only at the domestic level. Fresh coconuts are grated and mixed with hot water; the mixture is hand-kneaded into a white emulsion. The emulsion is poured into clay or wooden containers with a plugged opening at the bottom. The grated coconut is repeatedly pressed and the emulsion is collected. The combined emulsion in the container is allowed to settle for a few hours for separation of a creamy layer at the top and a watery layer at the bottom. The watery layer is removed through the bottom hole and the creamy layer is boiled to separate the oil from water, which is cooled and filtered through cloth. The residue on cloth contains about 16 per cent oil, 50 per cent protein and 12 per cent carbohydrates. The oil recovery in the traditional extraction process is only about 60 per cent. The residue of this process is either fed to animals or wasted. Thus, a significant portion of the raw material is wasted in this method. Since the domestic oil extraction practice is going to be continued in the developing countries, simple low-cost devices need to be designed to improve oil recovery. A hand-operated hydraulic press developed at the Royal Tropical Institute in the Netherlands is reported to extract 85 per cent of the oil in two pressings. Such a device can be used in villages for domestic oil extraction.

Wet Milling

The wet milling of coconut endosperm has been described by several investigators. The advantage in this process is the utilization of byproducts for food applications, besides the extraction of oil. Hagenmaier (1979) reported a new concept for industrial processing of fresh coconut with 95 per cent recovery of oil containing very low content of free fatty acids. The production of high-quality natural oil, coconut skim milk powder, dried coconut milk protein, pressed meal and coconut flour through wet milling has been described.

The steps employed by Hagenmaier *et al.* (1975) in wet milling of coconut are outlined in Figure 5.5. The technology has not been adopted for large-scale processing, mainly due to technical difficulties in obtaining a high oil percentage and the high cost involved in the installation of the plant. The problem exists in extraction of the emulsion from the fresh endosperm and breaking the emulsion to separate oil. The percentage of cell rupture influences the recovery of oil and proteins. The particle size of 40.1 mesh obtained by grinding the meal in a colloidal mill, slurring the mixture for 15 min at a water to meal ratio of 0.21 : 1 coupled with two pressings in a Hander expeller extracted about 55 per cent oil and 78 per cent proteins. However, the conditions to extract over 90 per cent oil and proteins need to be standardized.

The oil-in-water emulsion of coconut milk is opaque in color and is like dairy milk. However, the oil globules of coconut milk are surrounded or enclosed in a membrane made up of phospholipids cephalin as emulsifier, while salts and proteins act as stabilizers. Attempts have been made to standardize a process to break this emulsion and obtain higher oil recovery. These include heating and centrifugation, enzyme treatment of cream followed by freezing and thawing, freezing and thawing treatment of coconut milk, inversing the emulsion by shear, chilling the whole mass

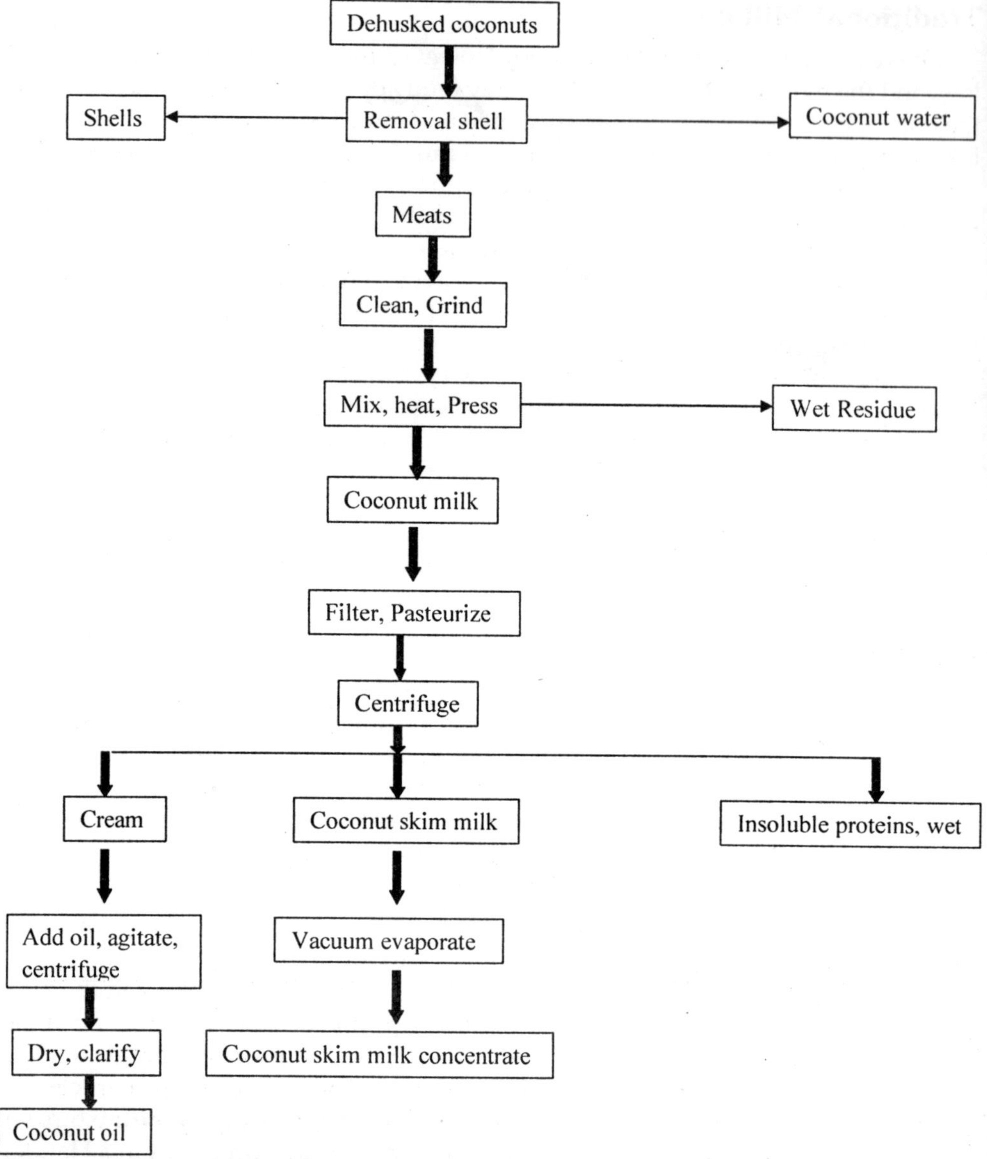

Figure 5.5: Flow diagram for wet-milling of coconut

of cream obtained by centrifugation to 17°C or below with subsequent thawing, agitation of cream and drying of fresh coconut meal, followed by grinding and pressing. Although the wet-milling process is advantageous due to a complete utilization of the endosperm and elimination of the drying costs, development and standardization of economical and commercially viable conditions to obtain maximum extraction of oil and proteins and of techniques to break the oil-water emulsion remains to be done.

Dry Milling

This involves the extraction of oil from copra. It is a conventional method of industrial coconut oil extraction on a commercial scale. The processing system is almost similar to that for other oilseeds. Copra is dried, if necessary, cleaned of stones, fibers and other foreign materials and crushed to a coarse and, subsequently, into a fine powder. This facilitates' the extraction of oil and improves its efficiency. The fine powder is heated to about 160°C using a traditional process, while only up to 120°C using modern processes. The optimum moisture content of copra to be treated in expellers is about 2-3 per cent. This moisture level is regulated during cooking, either by continued drying of copra or by the addition of water. Hydraulic presses require more moisture than expellers. The efficiency of oil extraction is generally low and depends on the nature of the expeller. The most perfect expellers extract 94 per cent, the less perfect ones extract 90 per cent, while the rural mills extract even less oil from copra. The cake is generally fed to the animals. If available in larger quantities, it can be subjected to a solvent extraction process.

The quality of oil and cake meal depends on the initial quality of the copra used for extraction. The unrefined oil is colorless to brownish yellow, depending upon the quality of the raw materials and the processing methods. The flavor of oil may vary from pleasant to rancid. After refining, it is odorless, tasteless and clear white. An intense roasted flavor is preferred in certain countries. Such oil has better storage stability than ordinary coconut oil. Heat treatment of disintegrated copra at 160°C for 15 min gives intensive roasted aroma and a better keeping quality to coconut oil.

Safflower Processing

Extraction of Oil

Almost the entire produce of safflower seeds is processed to obtain edible oil in the developing world. In the developed nations including the United States, the extracted oil is mainly used to manufacture alkyd resins and various types of drying oils. In India, the oil is extracted mostly by traditional methods, while more efficient expeller-pressing coupled with solvent extraction are employed in the United States. The seeds are often partially dehulled before oil extraction; however, crushing whole seeds without dehulling is also in practice.

Traditional Methods

The most popular method of extraction in India is by *ghani,* which is basically a large pestle and mortar. The mortar is an inverted cone, the pestle a heavy baulk of timber seated at the bottom of the cone and inclined at an angle, so that meal is crushed against the mortar during rotation. The timber baulk is pulled against the side of the mortar by weight, and is rotated by a pole inserted through it at right angles. Power is applied by one or two bullocks yoked to the pole who walk blindfolded in a circle round the mortar. There are different types of *ghani* used in India, which include improved bullock-driven *ghani,* overhead-drive *ghani* and portable power-operated *ghani.*

The steps employed for oil extraction by *ghani* generally include cleaning and grading, decortication, sieving and winnowing, watering and crushing, and oil extraction. The seed is sieved and made free of dust, stones and foreign matter. The decortication is carried out in a power-operated stone decorticator which can decorticate about 200 kg of seeds per hour. The decorticated material is sieved with the help of an electric screen set comprised of four types of sieves having 5, 6, 8 and 12 holes per 2.5cm^2, respectively. Some hull still remains in the broken and unbroken kernels which are separated by a winnower to free the kernels from hulls. About 48 to 50 kg kernels are obtained from 100 kg clean whole seeds. The clean kernels are mixed with sufficient water to make fairly hard balls and kept for 10 to 12h for softening. The wetted kernels are pulverized and powdered to some extent in the mortar. Some amount of water is sprinkled to act as a cementing material and to provide a grip to the pestle. Under normal conditions, pulverization requires one-third of the total time needed for complete oil extraction. High moisture content of kernels, overwatering, insufficient pressure on the pestle and irregular feeding of the meal may delay the pulverization. Under normal conditions, the quantity of water required for the entire process is about 6 per cent of the weight of the seeds taken for pressing. The pulverized meal is subsequently subjected to pressing, where the meal is cooked by the heat of friction and water. Cooking coagulates the protein and frees the oil for efficient pressing. Additional water is added during pressing to facilitate proper cooking. Heat is generated in the *ghani* by friction. To supplement this, some artisans heat the meal with a burning torch. Some others remove oil in the middle of the process, heat it and then pour it in the *ghani.* The oil is separated from the meal by pressure. After the stage of cooking is over and the cake begins to form, the oil droplets are expelled due to displacement by water. At this stage, pressure brings about cohesion among the various drops of oil. These drops come together during pressing; form themselves into bigger ones and flow out of the cake. A thick cake retains more oil, takes a longer time to lose moisture and becomes rancid soon. Hence, a thickness of about 0.5 inch at the top and about an inch at the bottom is recommended. The extracted oil is allowed to stand for about 24 h to settle the sediments, filtered to remove any remaining insoluble particles and packed in tin containers for sale as edible-grade crude safflower oil.

Improved Methods

The improved methods used for the extraction of safflower oil include a mini expeller (40 to 80 kg seeds/h), mechanical pressing (expelling), and a combination of continuous mechanical pressing followed by continuous solvent extraction. The steps involved in these extraction methods, namely hulling, cooking and drying, extraction, and cleaning and conditioning are essentially similar to the traditional *ghani* and operate on the same principles. To provide high-quality edible oil, dehulling is recommended. It also has the advantage of reducing the fiber content of the meal. However, dehulling increases the cost of production. When the meal is intended only for manure purposes, dehulling is not necessary. Partial dehulling is carried out if the meal is to be used for cattle or poultry feed. The oil recovery is often low with whole seeds unless additional pressure is employed. The thin-hulled seeds may avoid the additional costs in terms of dehulling or increased mechanical pressure.

The traditional thick-hull seeds are subjected to partial dehulling. It consists of cracking of the seeds and passing them through an air blast adjusted to remove only lighter hull particles. Disc or roller hullers are used for this purpose. A disc huller with 60 cm discs running at 1200 rpm can hull up to 135 kg of safflower seed per hour, and may also be set to crack the kernels. Safflower seed is difficult to decorticate owing to the hull strength and kernel softness. A Ripple Flow Mill developed by CSIR in Australia is suitable for large or small amounts of seeds.

Cooking of the pulverized meal or kernels is carried out at a moisture content of 10 to 12 per cent and 88 to 93°C to coagulate the proteins and free the oil for efficient pressing, followed by drying at 110 to 115°C to approximately 2 to 3 per cent moisture before entering the screw press. The drying temperature is important as safflower oil is heat-sensitive to burning. A screw press operating in a cage-like perforated cylinder through which the oil diffuses is commonly employed for pressing the cooked and dried meal. An extremely high pressure is developed by the restricted discharge in screw press, the oil being expelled through narrow slots around the press barrel.

For direct solvent extraction, safflower seeds are dehulled, rough-milled and then extracted in a batch or continuous process. This process is based on a counter-current flow of the solvent and an oil-bearing material in the extraction vessel, solids running in one direction while being washed by solvent passing in the other. The extraction rates using this process are usually higher than those from mechanical methods. The common solvent used is hexane of 65°C boiling points. In a combined process, following cleaning and conditioning, seeds are prepressed in expellers, the resulting cake conveyed directly to extractors without flaking or granulation. The miscella produced directly from the extractor is clean and brilliant. It is distilled without further filtration, and after conventional refining the oil is nearly equivalent to high-grade nonbreak processed oil.

After extraction, the oil is subjected to solvent recovery and then it is cleaned, bleached and filtered. Oil from a press usually contains suspended meal called "foots", which is removed by settling, screening and filtering. The filtered crude oil contains impurities, phosphatides, gums, resins, free fatty acids, and coloring matter, which need to be removed. Alkali-refining removes the gums, free fatty acids and some of the coloring matter. Thus, the edible oil is produced from crude filtered oil by refining, bleaching, and deodorization in a modern process.

Sesame Processing

Dehulling

Sesame seeds are mostly processed or used without removing the cuticle or the seed coat. The presence of cuticle contributes to the color, bitterness and high fiber and oxalate contents of the seeds. The bitterness may be due to the removal of calcium by oxalic acid. Dehulling of the sesame seed is, therefore, essential to improve its quality and utilization as a source of human food. Dehulling is an integral part of the modern oil extraction plants. It is also recommended to produce a high-quality oil and meal. Several wet processing methods and mechanical treatments have been tried for dehulling. The most commonly used method of decuticulizing of seeds is to soak the seeds and remove the cuticle manually by light pounding or by rubbing on a stone or wooden block. In this process, seeds are cleaned and given a hot lye (0.6 per cent) treatment for 1 min. The seeds are washed with excess cold water. The ruptured seed coats are separated by scrubbing in suitable equipment. The dehulled seeds (kernels) are then dried. The chemical compositions of the different parts of sesame seed are not identical. Therefore, removal of hulls results in a significant change in the chemical composition of the seeds. The dehulled seeds contain significantly more fat and less crude fiber, calcium, iron, thiamin, and riboflavin and slightly less phosphorus than the whole seeds. Oxalic acid, being present mostly in the seed coat, is significantly decreased after dehulling treatment observed that the digestibility of proteins was improved as a result of dehulling of sesame seeds. Heat treatment during dehulling as well as subsequent processing of the flour did not lower the available lysine.

Oil Extraction and Purification

Extraction

The most popular method of oil extraction in India is by *ghani*. *Ghani* is basically a large pestle and mortar. In earlier days, the *ghani* was made from wood and driven by bullock. Subsequently, power-driven steel *ghani* came into existence. The extraction of oil by *ghani* is not complete and the yield of oils about 40-45 per cent (Weiss 1983). The Burmese *hsi-zon* is similar to Indian *ghani* but it is now replaced by power-driven mills. The oil yield from sesame seed by *hsi-zon* is about 33 per cent. In Central Africa, sesame seeds are boiled to make them soft, and then squeezed in a sausage made from tree fibers to extract the oil. In the developing countries, some of these traditional methods are still being followed.

Commercially, sesame seeds are extracted using a continuous screw press, hydraulic press, prepress solvent extraction or direct solvent extraction methods as in the case of other oilseeds. The sesame seeds produced by the farmers are not of uniform size, color, or maturity, being an admixture of varieties. They are also contaminated with soil particles. Because of the small size of the seeds, it becomes difficult to clean the seeds. The oil quality is affected if the seeds are not properly cleaned. Similarly, prolonged storage under unsuitable conditions results in a loss of oil quality.

In Europe and Asia, the oil is usually extracted in three stages. The first pressing is made in cold. The oil obtained is of very good quality and high trade. It has a light color and agreeable taste and odor. The second pressing IS made of the heated residue which is subjected to a high pressure. The oil obtained is colored and is refined before using for edible purposes. The residue IS used for the third extraction under the same conditions as for the second. The oil obtained from the third extraction is of inferior quality, not suitable for human consumption and is mostly used for the manufacture of soaps. The recovery of oil from expeller (screw pressing or hydraulic pressing) extraction is not complete. In Europe, a combination of prepressing and solvent extraction is used to obtain maximum recovery of oil. The direct solvent extraction is not suited for sesame seeds because of high oil content.

Purification

Sesame oil does not require extensive purification. The crude oil usually contains suspended meal ("foot") which is removed by settling, screening and filtering. The filtered crude oil from the extraction plant contains impurities like phosphatides, resins, free fatty acids and coloring matters. Alkali-refining removes gums, free fatty acids and some of the coloring matters. The oil is bleached with a relatively less quantity of bleaching earth than for other vegetable oils. Bleaching produces light-colored oil. Deodorization is necessary to produce bland oil. It is usually done by treating refined oil in vacuum with steam at 200-250°C. For use as a base of salad dressing, the oil must be stable under refrigeration. For this, winterization treatment is given to the oil. It consists of cooling the oil to remove components with high melting points that may settle out at low temperatures. Sesame oil, however, requires little or no winterization. The hydrogenation process brings about a considerable increase in the stability of the oil.

Cake and Meal

Sesame cake is obtained as a byproduct after extraction of oil. When powdered, the cake is converted into a meal or flour. As such there is no change in the chemical composition of cake when it is converted into meal. Four types of meals can be obtained from sesame seeds, namely whole seed meal, dehulled seed meal, defatted whole seed meal and dehulled-defatted meal. Of these, the dehulled-defatted meal is the most common and unless otherwise specified, the term sesame meal refers to the dehulled-defatted meal. A flow diagram for the production of sesame flour is given in Figure 5.6. The chemical composition of sesame meal varies significantly due to dehulling and the method of oil extraction. The meals or flours obtained from dehulled seeds contain more proteins, phosphorus, and less ash, crude fiber, calcium and oxalic acid than those obtained from whole seeds.

Sastry *et al.* (1974) studied the effect of heat treatment on the protein quality of sesame meal. Heat treatment did not affect the amounts of total protein and total lysine. However, autoclaving for a prolonged period (60 min) caused a significant decrease in the available lysine and dispersibility of proteins in water and NaOH solutions. Heat treatment of sesame flour did not affect the amino acid composition of its proteins except a slight decrease in basic amino acids.

Corn/Maize Processing for Oil

Corn oil is a byproduct of commercial processing of corn. Only about 8 to 10 per cent of the corn produce is industrially processed through dry-milling and wet-milling processes, where the germ can be separated and utilized for commercial oil extraction. Among the two important milling techniques, wet-milling process predominates over dry milling, due to a continuous expansion in the variety and utility of products, and a growing demand for them. A description of the technology of both wet and dry milling, the products and byproducts obtained, their physicochemical properties, nutritive values and utilization is beyond the scope of this chapter. The processes have been described with greater details in a few recent publications.

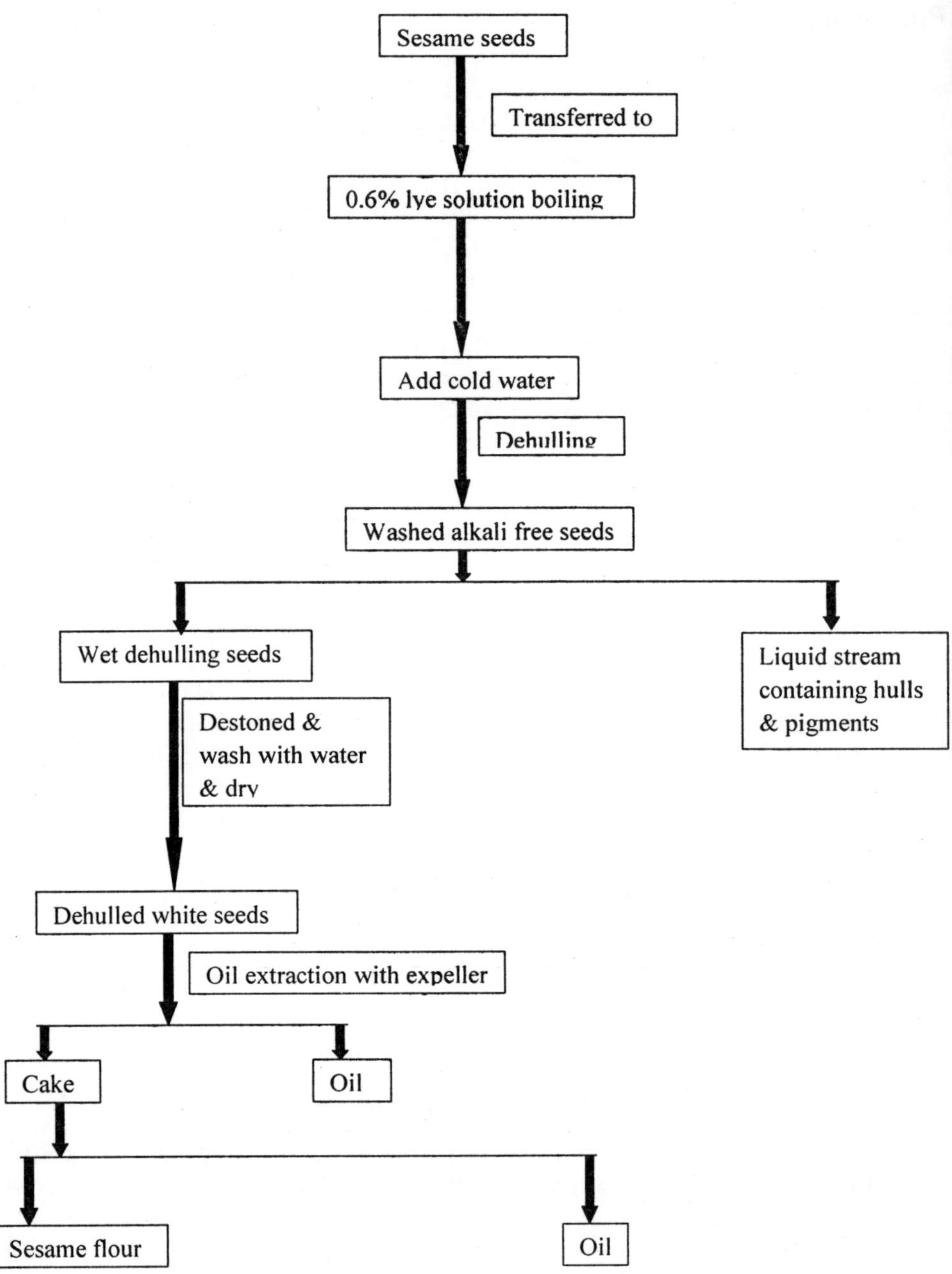

Figure 5.6: Processing of sesame for oil and flour

Dry Degerming

The main purpose of this system is to separate the germ, and produce a low-fat corn meal with a better shelf life. The degerminated corn dry-milling process, also called the tempering degerminating (TD) system, is operated to make a complete separation of pericarp, tip cap, germ, horny endosperm and floury endosperm, as is economically possible: retaining the maximum amount of the horny endosperm portion as discrete pieces, removing as completely as possible the germ and the pericarp to give a low-fat, low-fiber product, and recovering the maximum percentage of the germ as large clean pieces.

Process

Different steps in the separation process include cleaning, tempering by water addition, degerming, cooling, grading, aspirating, grinding, sifting and packaging. The corn used is always inspected for freedom from aflatoxin, insects, molds, and breakage susceptibility. The commercial mills, however, differ markedly in the types of equipment used and in preference for material routing. The initial cleaning is very important to remove tramp metals by magnetic separation, adhering dust, glumes and other light materials by scourers and aspirators, large and small foreign material by screens, cob pieces by washing with water, and stone and nonferrous metal by settlers. The electrostatic separators are often employed to remove rat pellets. The corn is moistened with water to bring the moisture content up to 20 to 22 per cent and held for 1 to 3 h to allow the water to equilibrate through the kernel. The degermination is accomplished in a variety of ways, but the most popular is with the Beall degerminator. However, some companies use impact mills or granulators for degermination for different results with and without tempering. The material discharged from the degerminator is dried to 15 to 18 per cent moisture and cooled. It is sized over stacked coarse sifters, + 3.5 to + 10 mesh, or reels, with the largest material passing through the aspirators to remove the free pericarps, and then to specific gravity tables, where germs are separated from the endosperm due to density differences. The endosperm fractions and the impure germ -endosperm mixture are passed through corrugated counter-rotating roller mills to remove the adhering pericarp and/or the germ pieces for better screening. Each roller mill set is preceded by an aspirator to remove the light material and is followed by a sizing device, usually stacked screens of appropriate sizes for the stock being handled. The roller mills are set with differential speeds and a gap appropriate for the material it receives, in order to minimize grit breakage. The endosperms fractions are finished in purifiers designed to remove fine pieces of fiber and are packaged according to the grade at 12 to 14 per cent moisture.

Products

The products of the TD system include prime endosperm products, gelatinized products, hominy feed and corn germ meal. The prime endosperm products include corn meal, bolted corn meal, degerminated corn meal, corn grits and corn flour. The Food and Drug Administration (FDA) has established standards to identify the food uses of these products. The dry-milled endosperm fractions can be cooked or gelatinized to varying degrees on steam-heated or gas-fired flaking rolls or through the extruder cooker to prepare a variety of foods such as ready-to-eat cereals, blended foods, and expanded snack foods. The nonfood uses are pet foods, adhesives, foundry core binders, and in oil well drilling mud. The byproducts include germ, hominy feed and corn germ meal. The finished germ, containing 18 to 20 per cent oil depending on the mill configuration, is dried to 3 per cent moisture and is used for oil extraction. The germ residue is also blended with pericarp fraction, inseparable mixtures of endosperm, pericarp and germ, and corn cleanings to give a commodity feed, hominy. This product is guaranteed to have a maximum level of 10 per cent protein and 5 per cent oil. It is an ingredient of swine, ruminant and poultry feed, as a source of energy and good-quality proteins. Solvent-extracted germ has been processed to germ flour containing 25 per cent protein using a process developed by the USDA. This material has become a useful, nutritious food ingredient. Typical yields of products from dry milling are: hominy feed 34 per cent, corn oil 2 per cent, grits, meal and flour 60 per cent and shrinkage 4 per cent.

Wet Degerming

This system is most widely used for corn processing to obtain starch, sweeteners, and corn oil. Its share is rapidly increasing due to the development of a process for enzymatically converting glucose to fructose. Sweetness equal to that of sucrose was achieved with the introduction in the late 1970s, of a high fructose corn syrup (HFCS) containing 55 per cent fructose and 45 per cent glucose. The product is currently widely used for soft drinks. The consumption of corn sweeteners has exceeded the sucrose consumption in U.S.A. In 1986-87, over 10.3 per cent of the total corn produce in U.S.A. was processed through wet milling. The current wet-milling plants are fully automated and computer-controlled.

Process

The cleaning operations are applied as described for the TD system. The cleaned corn is steeped for 30 to 35 h using process water recycled from the milling operations. Sulfur dioxide is added to solubilize the proteins, and a temperature of 35°C to 47 °C is maintained in the steeps to promote the growth of indigenous lactic acid bacteria. The steeping is usually conducted in countercurrent batteries with continuous or intermittent removal of extracted soluble, which are evaporated to thick syrup, often referred to as corn steep water or steep liquor. It contains 50 to 55 per cent dry substance composed of protein, peptides, amino acids, lactic acid and minerals. This fraction represents about 7.5 per cent of the original corn weight. About half of the steep water dry matter comes from the germ and half from the endosperm. The milling and separation steps have been described. The fully soaked corn (45 per cent water) is passed through an attrition mill to release about half of the endosperm starch and the germ, with minimum germ breakage. The germ fraction is separated from the denser components by flotation in continuously operated liquid cyclones. The germ is washed and dried. The underflow from liquid cyclones is further milled in vertical plate attrition impact mills or in impact pin mills. The slurry is passed over a series of screen-bend devices to separate the starch and gluten in continuous centrifugal machines. The gluten is concentrated and filtered. The filter cake is dried to produce a 70 per cent protein corn gluten meal at a yield of 5.5 to 6 per cent of corn. The starch slurry obtained through underflow of the centrifugal machine is further purified to about 0.3 per cent protein by diluting the slurry with fresh water and passing it through a series of liquid cyclones and finally dried with flash dryers as corn starch.

Products

The primary product of wet milling is starch. The yield is about 67 to 69 per cent of the corn dry weight with a recovery of 93 to 96 per cent of the original starch of the kernel. About 25 per cent of the starch produced is sold as starch products and over 75 per cent as hydrolyzed products, corn syrups, and dextrose. The general properties of corn starch and its application as unmodified and modified starch (acid-modified, maltodextrin, oxidized, pre gelatinized), waxy starch, high-amylose starch; sweetener products (corn syrup, dextrose, high-fructose corn syrup), corn oil, and feed products from wet milling of corn have been recently reviewed (Sprague and Dudley 1988).

Corn Germ Production and Processing

Production and Storage

Over 10 per cent of the total corn produce is industrially processed through wet and dry milling, particularly in the developed countries. In the U.S., about 21.5 MMT of corn was processed by wet milling and about 4 MMT by dry milling during 1986-87. The total corn oil production was 0.23 MMT in 1972, which increased up to 0.61 MMT in 1986 with increase in the proportion of corn processing through wet milling. The world production of corn oil has been estimated at 1.13 MMT in 1987-1988. Of the 500 MMT of world corn production, about 50 MMT is utilized for industrial processing. With an average of 10 per cent germ in corn, the total germ production would be around 5 MMT. With 50 per cent oil in germ, the potential for corn oil production is about 2.5 MMT. This can be further increased if a greater proportion of corn is diverted to wet milling.

The TD system does not separate the germ efficiently. The germ fraction contains only about 18 to 27 per cent oil depending on the mill configuration. This results in only about 2 per cent oil yield from the corn. The TD system needs to be improved to recover the complete and clean germ to increase the oil recovery. In the wet-milling system, the clean germ recovery is over 7.5 to 8 per cent. This germ contains about 50 per cent oil and the oil yield is about 4 per cent of the corn. The corn germ, being rich in oil, is a highly unstable material and undergoes deterioration due to the action of lipases and lipoxidases. Hence, the germ obtained in dry milling is dried to about 3 per cent moisture and immediately used for extraction. Alternatively, it is cooked to inactivate the deteriorative enzymes. Cooking by means of heated rollers at 124°C yields a product with acceptable flavor and stability. The cooked product exhibits little change in lysine or other amino acids compared to the original germ. Blessin *et al.* (1974) have developed a process for purifying the germ obtained by standard dry milling using additional aspiration, grinding and sieving steps. The germ obtained in wet milling is more clean and complete. It is washed to remove the bran and endosperm particles and dried before extraction of oil.

Oil Extraction and Refining

Corn oil is commercially produced only from corn germ isolated by wet milling or dry milling. The high-oil corn with about 20 per cent oil can be directly crushed for oil extraction. However, this has not yet become a commercial reality. Wet-milled germ is preferably expelled from an oil content of 50 to 60 per cent down to 20 to 25 per cent and finally extracted with hexane to a residual oil content of 1 to 2 per cent in the spent corn germ flakes. Most dry-corn mills recover the oil by tempering and flaking, followed by solvent extraction of germ, or the germ is sold to extracting companies. The solvent-oil mixture from the extractor is filtered to remove any solid material, and then the solvent is recovered from the oil in an evaporator. The germ flake from the extractor contains a certain amount of solvent, which is recovered by a steaming and a heating process.

Corn oil needs to be refined to remove insoluble materials, free fatty acids, phosphatides, and other gummy or mucilaginous materials. The insoluble and gummy substances are removed by filtration and settling process, while the free fatty acids are removed by treating the oil with an alkali solution in a continuous-type method of caustic soda refining. The free fatty acid content of alkali-treated oil is reduced to 0.01 to 0.03 per cent. To remove dark pigments, a bleaching treatment of the refined oil is carried out. The oil and adsorbent (Fuller's earth or clay) are vigorously mixed together at 220 to 240°F under atmospheric pressure, or at slightly lower temperatures if a vacuum process is employed. The treated oil is pumped through a filter press to remove the adsorbent with pigments. The corn oil contains a small portion of waxes which can cause turbidity when it is used as a salad oil. To remove the high melting point glycerides and waxes, the oil is winterized or chilled. The oil is cooled slowly with cold water, brine coolant or in a cold room to a temperature few degrees below that at which the oil is expected to remain clean, until the higher melting point components form definite well-built crystals which are then filtered off. Finally, the oil is deodorized by steam treatment to remove the relatively volatile constituents responsible for flavors and odors in the crude or partially refined oil. Modern deodorization equipments operate in the temperature range of about 425 to 475°F and at very high vacuum levels. The deodorization improves the stability of oil by removing peroxides and aldehydes. Corn oil, following deodorization, is usually cooled and finally packaged. A method for wet degumming, bleaching, deacidification and deodorization of corn oil has recently been presented.

Both mechanical pressing and solvent extraction processes leave significant amounts of triglycerides and bitter compounds in the germ flour, which lower its shelf life. Besides, considerable denaturation of proteins occurs due to heating during screw-pressing or sample preparation for hexane extraction. This lowers the functional properties of germ proteins. Recently, a new method, called high-pressure supercritical CO_2 extraction (SC-CO_2 extraction), has been tried for removing oil from corn germ with improved quality of defatted germ flour. The process produces good-quality oil from wet-milled corn germ which requires only mild refining. The oil obtained from wet- as well as dry-milled germ using 80001b/in^2 pressure and 50°C was lighter in color, lower in free fatty acid with a low refining loss. Many of the nutritional and functional qualities and storage characteristics of defatted germ flour obtained by hexane extraction were improved by SC-CO_2 extraction. The SC-CO_2 extracts almost all the triglycerides and bitter constituents, and also inactivates peroxidase enzymes in the germ meal. This results in the production of an excellent initial-quality defatted germ that has a better shelf life. The defatted, debittered, peroxidase activity free germ meal is a highly dispersible material that can be used as a protein supplement in food formulations. The commercial application of SC-CO_2 extraction method for oil extraction from oilseeds may be more economical and useful as it yields good-quality oil as well as a food-grade defatted protein-rich meal.

Rice Bran Processing

Milling Systems

The rough rice or paddy is milled basically to remove the outer husk and part of the bran. The brown rice and the partially or completely milled white rice are obtained as the main products, while the husk, bran and polish as byproducts. Rice milling is still carried out with the traditional and conventional methods in the less developed countries of Asia, while in the developed world modern rice milling systems are employed. Hand pounding of paddy is the traditional method that produces brown rice and husk mixed with little bran. It is still practiced by farmers for milling paddy for domestic consumption in Asia. The conventional mills can be categorized into three main types: huller mills, sheller-huller mills and sheller-cone polisher mills. In the U.S., most of the rice is milled in highly sophisticated and automatic modern rice mills. The design, operational efficiency and rice outturn from paddy in these rice milling systems have been extensively described.

The huller mill combines dehusking and polishing in the same machine in one or more steps. It causes friction, heats up the grain and causes high breakage. The total outturn ranges from 63 to 65 per cent. The husk and the bran are mixed together and cannot be profitably utilized for extracting oil. An improvement in their operational efficiency by adding a paddy separator, introducing a separate shelling device and using a huller machine only for polishing and the recovery of brokens from the bran may be useful to increase the head rice recovery and for a clean separation of bran for commercial oil extraction. A mini mill with improvements in the capacity to mill 50 to 200 kg of paddy in one lot may be very useful for small farmers of the developing countries.

The sheller-huller type mill uses an emery-coated disc sheller for shelling and a horizontal Engleberg-type huller for polishing the rice grains. As compared to huller mills, they give a higher outturn of rice, deliver purer bran and have better facilities for controlling the degree of polishing. The sheller-cone polisher-type mills have all the milling components such as a cleaner, a disc sheller, a paddy separator, one or more cone polishers and a grader. These mills can give total rice out turn of 68-70 per cent and are the best among the old-type indigenous mills. However, the disc sheller employed in such mills causes 20 to 30 per cent breakage with normal paddy and a loss of 1.5 to 2 per cent of rice as brokens in the bran.

Modern rice mills are in operation mainly in Japan, Germany and U.S.A. Few such mills have been introduced also in India. These mills are equipped with cleaners, rubber roller shellers, efficient paddy separators and precision type horizontal or vertical-type polishers and separators for bran, brokens and germ. The mills are also provided with facilities for bulk storage of paddy, with mechanical handling, drying and parboiling units, and few have also a solvent extraction unit to produce rice bran oil. Such a modern-mill complex can yield the maximum rice recovery and also help to utilize efficiently the rice bran and other byproducts.

A relatively new rice processing technology known as solvent extraction rice milling (X-M milling) is in commercial use in U.S.A. Paddy rice is cleaned and dehulled as in conventional milling but bran removal is conducted under more gentle conditions than in the usual pressure machinery method. Brown rice is spray-coated with rice oil and held for several hours, after which it is mixed with butane and rice oil as it goes to the milling machines. Passage through milling machines under these conditions causes much less kernel damage than doe's dry milling, and yields are consequently higher. The products obtained by the solvent process are white rice, defatted bran and rice oil. The white rice, in uncooked form, has an improved color but no difference in the eating quality. Defatted rice bran is free-flowing, light-colored, possesses a bland flavor, a pleasant odor and does not need further stabilization. The rice oil obtained is low in free fatty acids and can be refined to edible-grade oil with minimum refining losses.

Both modern rice mills and X-M milling systems, although efficient by all means, cannot be adapted as they are in the developing countries of Asia due to their higher costs of installation and a very large turnover capacity. Unless the procurement operation and distribution of paddy practices are stream-lined through cooperative societies or suitable government agencies, such mills are difficult to install and operate profitably in these countries. Attempts have, however, been made to design mini modern rice mills of equivalent capacity to that of the single huller or disc sheller. A replacement of the disc sheller unit of an existing mill by a rubber roller sheller has been attempted and such modernized mills are currently in operation in India. There is a need to install modern machinery in the rural rice-growing areas on a cooperative basis, where the paddy can be collected in relatively larger quantities, the rice bran obtained in a sizeable quantity can be suitably stabilized and transported to nearby extraction units.

The rough rice is often subjected to parboiling treatment before milling. Over 25 per cent of the world paddy is parboiled. In India, nearly 60 per cent of the total rice produced is parboiled before milling. Various methods of parboiling, effects of parboiling on processing characteristics and physicochemical properties of milled rice, rice bran and bran oil have been reviewed. It involves a hydro-thermal treatment by soaking, steaming and drying of paddy before milling. Basically, it is done to gelatinize the starch, remove air voids from the kernel, and heal the cracks. This process reduces milling breakage by imparting hardness to the grains and to make them more resistant to the pests. Although parboiling may not have much advantage in case of properly handled paddy, it is beneficial in poorly handled paddy, and when the conventional equipment is employed for milling. The parboiled paddy is difficult to polish as the external layers become stuck to the endosperm and have partially penetrated inside it. The fatty substances from the germ get spread outwards to the bran, making the grain slippery under the rubbing action of the polishers. Hence, the parboiled rice must be passed through more polishers than are needed for ordinary rice, and they must be kept cool by means of strong air suction. The yield of bran, however, remains low as compared to yields obtained from polishing of raw brown rice. The parboiled bran is richer in fat and ash and low in starch and B-group vitamins than the bran obtained from raw rice. Since the rice is subjected to heat treatment which destroys lipase, the fat hydrolysis and accumulation of fatty acids in parboiled rice bran are retarded. The output and quality of oil obtained by extraction of parboiled rice bran are, therefore, higher than those obtained from ordinary bran. The bran of otherwise poorly handled and infested paddy can be made available for commercial oil extraction through parboiling treatment before dehusking and polishing.

Bran Oil

Oil Extraction

The oil is extracted from raw or stabilized bran by a solvent extraction process. Hexane or petroleum ether is the most preferred solvents, although ethyl or isopropyl alcohols are also used to reduce the fire hazard. Details regarding solvent extraction of rice bran have been summarized in several publications. Bran is soaked in a solvent, which removes the oil by percolation, and is filtered from the bran solids. The solvent is removed from the miscella (oil plus solvent) by stripping, condensed and recovered; the separated oil is finally subjected to a refining process.

The fines (particles passing through 100 mesh sieve) occurring in rice bran tend to both clog and blind the filters, and pass through to produce turbid oils. The fines also increase the extraction/soaking time markedly and cause difficulties in settling out the non-oil constituents-tank settlings or foots. The proportion of fines in commercial bran can be up to 40 to 80 per cent. Attempts to treat fines for improving the extraction efficiency include an increase in the moisture content of bran to 15.5-16 per cent, a mild precooking of bran followed by cooling before extraction, hot-air drying, steam-cooking and extrusion. The fines in raw bran can be greatly reduced by steam-cooking and extrusion treatment. The extruded and pelletized bran exhibits a bulk density of 1.5 times higher than the other products. The extraction time to reach 1 per cent residual oil was found to decrease in the order of 116, 67 and 10 min for the hot air dried, steam-cooked and the extruded rice brans, respectively. The percolation rate of the extruded bran was reported to be 2 times higher than that of the steamed bran and 9 times higher than that of the hot air dried bran. Besides, the solvent to bran ratio for oil extraction was decreased from 3.18 for the hot air stabilized bran to 3.12 for the steamed bran and to 1.77 for the extruded bran. The size of pellets is reported to affect the extraction rate. Thus, extrusion stabilization appears to be the most effective method for economical extraction of good-quality oil from rice bran.

Refining of Bran Oil

The crude bran oil may contain 3 to 20 per cent free fatty acids (FFA) depending upon the quality of bran, 4.8 per cent wax, 5 to 8 per cent unsaponifiables and pigments, which need to be removed to convert it into an edible oil. The crude oil with more than 20 per cent FF A is suitable only for making soap or other technical products. The crude oil from high-acidity bran suffers a higher refining loss. For example, and oil containing 5 per cent free fatty acids may have a refining loss of over 40 per cent (Hogan 1967). Hence, the bran must be properly stabilized in order to obtain a crude oil of low acidity and to minimize the refining losses.

Several investigators have described the techniques to refine bran oil. Refining steps include degumming, dewaxing, removal of free fatty acids, bleaching and deodorization. Degumming and dew axing need to be carried out before alkali-refining to minimize the refining losses. Refining losses of up to 42 per cent are reported for undewaxed oil as against 31 per cent for dewaxed oil. The crude oil can be first degummed and dewaxed by the lipofrac process. Degumming involves the mixing of oil with 0.5 to 1 per cent phosphoric acid in 2 to 3 per cent water on oil-weight basis at 80°C for 15 min. The gums (phosphatides) are removed by settling. The degummed oil is then chilled to 10-20°C in the presence of 0.5 to 1 per cent (on the oil-in-water weight basis) lipofrac agents (surface-active compounds such as sodium lauryl sulfate and sodium stearate) for 2 to 4 h along with stirring. The chilled mass is then centrifuged to remove wax. The process includes degumming by phosphoric acid in oil phase, dewaxing in hexane phase with calcium chloride and lipofrac agents, followed by alkali-refining. The oil can, however, be first dewaxed by settling in tanks at warm conditions, followed by a deliberate cooling and settling to remove the wax completely. This oil can then be subjected to a degumming treatment.

The free fatty acids can be removed by alkali treatment. A calculated amount of alkali is mixed with oil at 50 to 60°C (2.77 M NaOH and 0.05 per cent NaOH excess), stirred for 10 to 15 min, followed by centrifugation and washing of oil with hot water to remove traces of soap. Bran oil with high FF A contents can be deacidified in a 2-stage process consisting of extraction with alcohol or reesterification with glycerol, followed by regular alkali treatment. The superheating of oil fixes the color; hence, this technique needs to be used only after bleaching of the oil. Acid-activated clay is the most effective in removing the color from the refined oil. The oil can then be winterized if it is to be used as a salad oil. After degumming and dewaxing, the oil is treated with 1, 3-specific lipase to convert free fatty acids into neutral glycerides. Residual free fatty acids are then removed by usual alkali neutralization. The process appears to be quite promising, since it can lower the refining losses and improve the yield of edible oil.

General Oil Processing

Refining

Crude oils and fats contain varying amounts of substances that may impart undesirable flavor, color, or keeping quality. These substances include free fatty acids, phospholipids, carbohydrates, proteins and their degradation products, water, pigments (mainly caratenoids and chlorophyll), and fat oxidation products. Crude oils are subjected to a number of commercial refining processes designed to remove these materials.

Settling and Degumming

Settling involves heating the fat and allowing it to stand until the aqueous phase separates and can be withdrawn. This rids the fat of water, proteinaceous material, phospholipids, and carbohydrates. In some cases, particularly with oils containing substantial amounts of phospholipids (*e.g.*, soybean oil), a preliminary treatment known as *degumming* is applied by adding 2-3 per cent water, agitating the mixture at about 50°C, and separating the hydrated phospholipids by settling or centrifugation.

Neutralization

To remove free fatty acids, caustic soda in the appropriate amounts and strength is mixed with the heated fat and the mixture is allowed to stand until the aqueous phase settles. The resulting aqueous solution, called foots or soap stock, is separated and used for making soap. Residual soap stock is removed from the neutral oil by washing it with hot water, followed by settling or centrifugation. Although free fatty acid removal is the main purpose of the alkali treatment, this process also results in a significant reduction of phospholipids and coloring matter.

Bleaching

An almost complete removal of coloring materials can be accomplished by heating the oil to about 85°C and treating it with adsorbents, such as Fuller's earth or activated carbons. Precautions should be taken to avoid oxidation during bleaching. Other materials, such as phospholipids, soaps, and some oxidation products, are also adsorbed along with the pigments. The bleaching earth is then removed by filtration.

Deodorization

Volatile compounds with undesirable flavors, mostly arising from oxidation of the oil, are removed by steam distillation under reduced pressure. Citric acid is often added to sequester traces of pro-oxidant metals. It is believed that this treatment also results in thermal destruction of nonvolatile off-flavor substances, and that the resulting volatiles are distilled away. Although the oxidative stability of oils is generally improved by refining, this is not always the case. Crude cottonseed oil, for example, has a greater resistance to oxidation than its refined counterpart, due to the greater amounts of gossypol and tocopherols in the crude oil. On the other hand, there can be little doubt as to the remarkable quality benefits that accrue from refining edible oils. An impressive example is the upgrading of palm oil quality that has occurred in the last decade. Furthermore, in addition to the obvious improvements in color, flavor, and stability, powerful toxicants (e.g., aflatoxins in peanut oil and gossypol in cottonseed oil) are effectively eliminated during the refining process.

Future Prospects

Malnutrition and under nutrition are prevalent in several parts of the developing countries of the world. These problems have been implicated in the inadequate supply of energy and proteins in many parts of the world. With continued increase in the world population and pressure on land use, man has been concerned not only about quantity but also about the quality of his food. Both plant and animal sources need to be improved to meet the nutritional requirements of the growing population. Animal foods, although excellent in nutritional quality, are of limited value for feeding mankind because their production is expensive, they possess a short shelf life, require sophisticated and costly processing technology for storage and distribution, and these are beyond the reach of the majority of the world population, particularly in the developing countries, due to their higher costs. Animal foods such as meat, milk, fish and their products will, therefore, continue to play only a supplementary role in human nutrition. The production and distribution of plant foods, on the other hand, is comparatively more economical because they exhibit a better shelf life and can be stored and processed with less expensive methods. Hence, greater emphasis has been placed throughout the world on increasing the production of plant foods, improving their nutritional quality, and developing simple and economical methods for their storage and processing.

Among the plant foods, cereals are grown over 73.5 per cent of the world harvested area while oilseeds (10.8 per cent), pulses and nuts (6.3 per cent), roots and tubers (5.2 per cent), sugar crops (2 per cent), and fruits and vegetables (2.2 per cent) occupy the remaining 26.5 per cent of the planted area. The oilseeds are important sources of edible and nonedible oils. In addition, some leguminous oilseeds are excellent sources of dietary proteins. The cake and meal obtained during oil extraction are mostly used as cattle feed and soil manure. However, if properly processed, it can serve as an important source of dietary proteins, energy and minerals.

There is a growing awareness of the importance of oils and fats in human nutrition. Vegetable oil sources account for about 68 per cent of world's edible fat production, the rest coming from animal fats (30 per cent) and marine oils (2 per cent). In most of the developed countries, the average energy intake from dietary fat currently ranges from 35 to 45 energy per cent. In many developing countries, intakes of 10 to 20 energy per cent of fat or less are common. Thus, on a global basis, there could be approximately six fold difference in the fraction of energy that is derived from fat by various populations. The available quantities of edible oils are still insufficient to adequately meet the requirement of the world's present population. In the developing countries with dietary fat comprising about 10 energy per cent, there is evidence that an increase to 15 to 20 energy per cent of fat would have beneficial effects. Such an increase will raise the energy density of the diet and help to satisfy energy needs. The increase in the production of oilseeds can be achieved by increasing the oilseed yield and by increasing the per cent of oil in new varieties. For most oilseed crops, there has been a continuing increase in the yield achieved, without a decrease in the oil content. A part of the increase has been achieved through improved production practices but much can be attributed to the use of better cultivars. In order to improve the yield of oilseeds and increase the production of edible oils, efforts at developing better cultivars with improved potential for yield, and suitable production practices need to be continued.

Vegetable oils contain large amounts of unsaturated fatty acids with 18 carbon atoms. The monounsaturated fatty acids consist almost entirely of oleic acid and polyunsaturated fatty acids consist of linoleic and linolenic acids. The importance of a specific group of polyunsaturated fatty acids, called essential fatty acids (EFA), has been well established. Young rats fed a diet deficient in EFA developed a variety of symptoms ranging from reduced growth rate and skin lesions to more recently recognized symptoms of reduced prostaglandin synthesis and abnormal thrombocytes. All these symptoms disappear or can be prevented by feeding essential fatty acids. Only the linoleic acid and fatty acids derived from it can cure or prevent all the symptoms. The essential fatty acids have been shown to be indispensable for many other species, including man. According to FAO/WHO, the minimum intake of linoleic acid required to prevent the EFA deficiency syndrome is 3 energy percent. The diversity of the symptoms of the essential fatty acid deficiency syndrome suggests a very important role for linoleic acid in most cells and organs. This fatty acid serves as a substrate for the biosynthesis of prostaglandins, hydroxy fatty acid and leukotrienes.

Various edible oils differ in their fatty acid composition. The major unsaturated fatty acids in edible oils are oleic, linoleic and linolenic. The optimum concentration of linoleic acid, being an essential fatty acid, is important from the view points of nutritional quality and storage stability. The committee on dietary allowances has advised an upper limit of 10 per cent dietary energy derived from polyunsaturated fatty acids. The ratio of the sum of all polyunsaturated fatty acids to the sum of all saturated acids (P/S) and the ratio of the content of linoleic acid to the sum of palmitic and stearic acids [L/(P + S)] are often taken as criteria for evaluating the dietary value of edible oils. On this basis, safflower oil exhibits higher P/S and L/(P + S) ratios than other common edible oils whereas palm oil has lower P/S and L/(P + S) ratio. Various oil blends, based on the existing edible oils which can exhibit optimum P/S and L/(P + S) ratios, can be developed to improve the nutritional quality of commonly consumed oils.

Oilseeds are generally processed to extract oil by employing various extraction techniques. In the developing countries the oil is mostly extracted by using traditional expellers. This results in significant losses of edible oil in the cake. The oil content of the cake ranges from 4 to 6 per cent or even up to 10 per cent, depending upon the oilseed and the processing equipments used. Modern technology for the processing of oilseeds to separate oil from meal has been evolved from the development and utilization of continuous expeller. Currently, three types of commercial processing systems are employed for oil seeds in the developed countries. These include (a) expeller pressing, (b) pre-pressing followed by solvent extraction and (c) direct solvent extraction. Regardless of the method used, caution must be exercised to maintain the quality of protein in the meal as it is used as a food or feed supplement. During processing, a significant amount of protein denaturation can take place. For feeding purposes, this denaturation is generally considered desirable because of the resultant improvement in digestibility. However, excessive heating during processing can result in extensive losses of certain amino acids, particularly the essential amino acid, lysine. The extent of such damage will depend on the processing time, temperature, moisture and the reducing sugar content of the meal. Research developments need to be directed towards (1) maximization of the yield of the oil extracted from oilseed, (2) minimization of the damage to the oil and meal, (3) production of oil as free of impurities as possible and (4) production of a residual cake or meal of the highest possible quality. The pace of these developments is increasing due to the application of new scientific practices such as biotechnology. For each step forward, an efficient manufacturing procedure will need optimization for an efficient production of quality products from oilseeds.

Edible oils are mostly used as cooking oil in developing countries whereas they are consumed in the form of margarine in developed countries. These oils are also used in the manufacture of cooking fats and oils, salad dressing and confectionery fats. The content of polyunsaturated fatty acids in margarine varies from 5 to 65 per cent depending on oils and fats used and their technological processing. To obtain spreadable product, liquid oils are mixed with higher melting fats. These can be obtained from the oil by a hardening process. By reducing the degree of unsaturation, the hardening process raises the melting point of the oil and decreases its susceptibility to oxidative deterioration. Interesterification in addition to blending, fractionation and hardening has also contributed to widen the range of uses of vegetable oils and fats, and making the various oils more interchangeable in their application. It is possible to modify the melting characteristics of fats or mixtures of fats without changing the fatty acid composition of the blend, thereby enabling a more efficient use of practically every commercial oil.

The partially defatted cake produced in an expeller press or the fully defatted meal obtained in the solvent extraction process are almost exclus-ively fed to animals or used as manures as these are unfit for human consumption due to the presence of toxic compounds; dark brown, black or grey pigmentation, bitter taste and high fiber content. Such meals being rich in proteins are potential sources of dietary proteins for the future. The postharvest technologies need to be developed to obtain cake/meal low in pigments and fiber by suitably modifying the oil extraction practices for removal of phenolics, bitter-tasting materials and other toxic compounds from the cake/meal to convert it into creamy white edible-grade products. Such preparations can be directly utilized as protein supplements in bakery products, snack foods and a variety of cereal- and legume-based traditional foods. Besides being rich in proteins, the processed meals are also rich sources of dietary minerals.

Vegetable oils have applications also in several nonfood products. The oils that are commonly used for nonfood uses are coconut, soybean, linseed, castor and tall oil. The use of coconut oil varies around 50: 50 for edible and nonedible purposes depending upon the market conditions, with about 60 per cent currently going into industrial products. A relatively low cost and a dependable supply make soybean oil as one of the most important sources of industrial products. Many latex paints contain linseed oil because the natural product imparts a superior adhesion to substrate surfaces. Epoxidized linseed oil is also used commercially in plasticizers or stabilizers for vinyl plastics. Castor oil is used in lubricants, plasticizers, coating, surfactants, pharmaceuticals, cosmetics, flavorings, fungistats and greases. Large-scale industrial applications of high-performance polymers and polymer composites can be foreseen in the realm of autos, aircrafts, recreational equipments, heavy machinery and construction. Thermo-plastic polymers with good moisture stability and electrical properties are expected to find an increased use in the electronic industry. The increased production of plastics will generate the need for additives, notably plasticizers and stabilizers that can be derived from vegetable oils. Strong solid adhesives and coatings that can reduce the fabrication and maintenance costs will also be in demand as will· be liquid surfactants designed to fulfill the precise requirements of the specialized emulsion polymerization process and textile processing.

Vegetable oils have been considered as fuels for engines since the beginning of this century. Certain vegetable oils such as sunflower and soybean have been found to be promising candidates for diesel fuel substitution. Indirect injected (IDI) diesel engines have multi fuel capabilities and run very well on vegetable oils. In an emergency, vegetable oils could be used as a fuel for short-term operation of direct injected (DI) diesel engines. Oilseed fuels have low sulfur content. They are safe to store and handle. However, fuel consumption is higher due to their lower heat energy. The glycerides of vegetable oils cause excess carbon, which could be reduced by chemical modifications. The sunflower methyl ester was found to have fuel characteristics comparable to diesel fuel. The price is, however, a major constraint in the use of vegetable oils as fuels. However, these oils and their derivatives may be considered as an alternative to diesel in case of fossil fuel crisis in the future.

Several plants are known to produce a significant quantity of oil which is nonedible due to the presence of certain toxic or undesirable flavor compounds. These oils, if properly processed to remove such factors and converted into processed fats, can serve as an important source of dietary fats/fatty acids. The minor oilseeds are potential sources of edible oils for the future. Efforts are needed to develop systems to increase the plantation of minor and nonedible oilseeds, their efficient collection and processing into edible-grade oils and processed fats. This is important to meet the requirements of dietary fats/fatty acids of world's growing population in the years to come.

6

Fruits and Vegetables in Human Nutrition

Introduction

The world's hunger is widespread and growing. Billions of people suffer from severe under nutrition in Asian, African, and Latin American countries, and from malnutrition in North American and Western European countries. There is nearly a 20-year gap in life expectancy between the rich and poor nations. Rapid population growth throughout the world necessitates a matching increase in for the food supplies to provide nutritionally adequate diet to everyone. Eradication of hunger and malnutrition depends not only on rapid increase in the agricultural productivity, but also on efficient use of farm products, especially in developing countries. Development of appropriate processing, transportation, and marketing technologies are important to make efficient use of farm products.

In the present day society, more and more people live in urban areas far away from the sources of fresh or natural foods. This has resulted in increased dependence on processed food products. This has forced the governments in these countries to explore the possibilities of processing of fruits and vegetables into more durable products, to maintain their supply during when, and in places where, the fresh fruits and vegetables would normally be unavailable. Modern food processing technologies have also expanded the farm produce market for growers in countries where the production greatly exceeds demand.

Fruits and vegetables are highly perishable food products. Water loss and postharvest decay account for most of the losses. These have been estimated to be more than 20–30 per cent in the tropics and subtropics. The problems of postharvest technology of fruits are critical in the battle to minimize wastage and to extract the maximum potential from the harvested crops. Wastage of fruits is so high in some instances that between the field and the consumer, bountiful amounts of highly nutritious crops are reduced to heaps of refuse. Reduction of high postharvest losses of fruits and vegetables includes proper harvesting, handling, low temperature storage and environmental control (controlled/modified and hypobaric storage), irradiation and use of chemicals and fungicides as well as packaging techniques, and processing of fruits and vegetables into suitable products.

Nutritional Significance

Fruits and vegetables play an important role in human nutrition particularly in supplying certain constituents in which other food materials are deficient. Fruits and vegetables contribute vitamins and minerals and the bulk of the dietary fibre to the diets of human beings. They neutralize the acid substances produced in the course of digestion of meat, cheese, and other high-energy foods. The value of fruits and vegetables as dietary fiber has increasingly been realized in recent years. The possible beneficial effects of fiber, vitamin C, β-carotene, and the vitamin B-complex derived from fruits on human health are being reexamined with the objective of minimizing certain important human diseases related to lack of fiber and antioxidant properties of some vitamins and minerals in diets. The nutritional composition of fruits and vegetables is presented in Tables 6.1 to 6.6. Ascorbic acid (vitamin C) is the principal vitamin supplied by fruits and vegetables in the diet. About 90 per cent of a person's dietary vitamin C requirement is obtained from fruits and vegetables (Figure 6.1). In addition, fruits and vegetables are rich source of β-carotene, thiamine, niacin, folic acid and Vitamin A.

Table 6.1: Proximate composition of major fruits grown in India

Fruit	*Water (per cent)*	*Energy (Kcal)*	*Proteins (per cent)*	*Fats (per cent)*	*Carbohydrates (per cent)*	*Minerals (per cent)*
Apple	84.4	58	0.2	0.6	14.5	0.3
Apricot	85.3	51	1.0	0.2	12.8	0.7
Avocado	74.4	80	1.8	20.6		1.2
Banana	75.7	85	1.1	0.2	12.6	0.6
Bar	85.9	55	0.8	0.3	12.8	
Grape	81.6	67	1.3	1.0	15.7	0.4
Guava	83.0		1.1	0.4	12.7	0.7
Mango	83.4	59	0.5	0.2	15.4	0.4
Orange	86.0	49	1.0	0.2	12.2	0.6
Papaya	90.7	32	0.5	0.1	8.3	0.4
Peach	89.1	38	0.6	0.1	9.7	0.5
Pear	83.2	61	0.7	0.4	15.3	0.4
Pineapple	85.4	52	0.4	0.2	13.7	0.3
Plum	81.1	66	0.5	0.2	17.8	0.4
Pomegranate	78.0	58	1.6	0.1	14.6	0.7
Sapota	74.0	98	0.7	1.1	21.4	0.5
Strawberry	89.9	37	0.7	0.5	8.4	0.5

Per 100g. edible portion. — Values not reported

Table 6.2: Mineral contents of major fruits grown in India

Fruit	*Calcium (mg)*	*Phosphorus (mg)*	*Iron (mg)*	*Magnesium (mg)*
Apple	7	10	0.3	8
Apricot	17	23	0.5	12
Avocado	14	27	0.7	23
Banana	8	26	0.7	33
Ber	30	30	0.8	–
Grape	16	12	0.4	13
Guava	17	28	1.8	
Mango	12	12	0.8	
Orange	41	20	0.4	11
Papaya	20	13	0.4	
Peach	9	19	0.5	10
Pear	8	11	0.3	7
Pineapple	18	8	05	
Plum	18	17	0.5	9
Pomegranate	10	70	0.3	12
Sapota	28	27	2.0	
Strawberry	21	21	1.0	12

Per 100g. edible portion, — Values not reported.

Table 6.3: Vitamin contents of major fruits grown in India

Fruit	*Vitamin A (IU)*	*Thiamine (mg)*	*Riboflavin (mg)*	*Niacin (mg)*	*Ascorbic acid (mg)*
Apple	90	0.03	0.02	0.1	4
Apricot	2700	0.03	0.04	0.6	10
Avocado	–	0.07	0.12	1.9	11
Banana	190	0.05	0.06	0.7	10
Ber	70	0.03	0.04	0.5	80
Grape	100	0.05	0.03	0.3	4
Guava	250	0.05	0.03	1.2	337
Mango	630	0.05	0.06	0.4	53
Orange	200	0.10	0.04	0.4	50
Papaya	110	0.03	0.04	0.3	46
Peach	1330	0.02	0.05	1.0	7
Pear	20	0.02	0.04	0.1	4
Pineapple	15	0.08	0.04	0.2	61
Plum	300	0.08	0.03	0.5	5
Pomegranate	–	0.06	0.10	0.3	16
Sapota	–	0.02	0.03	0.2	6
Strawberry	60	0.03	0.07	0.6	59

Per 100g. edible portion, — Values not reported.

Table 6.4: Proximate composition (per 100 g edible portion) of some of the important vegetables grown in India

Common Name	Energy (KCal)	Moisture (g)	Proteins (g)	Fats (g)	Carbohydrates (g)
Bitter gourd	.25	92.4	1.6	0.2	4.2
Brinjal	24	92.7	1.4	0.3	4.0
Cabbage	24	92.4	1.3	0.2	5.4
Capsicum	22	93.4	1.2	0.2	4.0
Carrot	42	82.2	1.1	0.2	9.7
Cassava	157	59.4	0.7	0.2	38.1
Cauliflower	27	91.0	2.7	0.2	5.2
Cucumber	18	96.3	0.4	0.1	2.5
Frenchbean	32	90.1	1.9	02	7.1
Garlic	30	'62.0	6.3	0.1	29.8
Lettuce	14	95.1	1.2	0.2	2.5
Muskmelon	17	95.2	0.3	0.2	3.5
Okra	35	89.6	1.9	0.2	6.4
Onion	50	86.8	1.2	0.1	11.1
Peas	84	78.0	6.3	0.4	14.4
Potato	97	74.7	1.6	0.1	22.6
Spinach	26	90.7	3.2	0.3	4.3
Sweet potato	114	59.4	0.7	0.2	38.1
Tomato	22	93.5	1.1	0.2	4.7
Watermelon	26	92.6	0.5	0.2	6.4
Yam	102	74.0	1.5	0.2	24.0

Table 6.5: Vitamin contents (per 100g edible portion) in some of the important vegetables grown in India

Common Name	Vitamin A (IU)	Thiamine (mg)	Riboflavin (mg)	Niacin (mg)	Ascorbic Acid (mg)
Bitter gourd	416	0.07	0.09	0.5	88
Brinjal	244	0.04	0.11	0.9	12
Cabbage	130	0.05	0.35	0.3	47
Capsicum	900	0.06	0.06	0.5	128
Carrot	11000	0.06	0.05	0.6	8
Cassava	0	0.05	0.10	0.3	25
Cauliflower	60	0.11	0.10	0.7	78
Cucumber	0	0.03	0	0.2	7
French bean	600	0.08	0.11	0.5	19

Contd...

Table 6.5–*Contd...*

Common Name	*Vitamin A (IU)*	*Thiamine (mg)*	*Riboflavin (mg)*	*Niacin (mg)*	*Ascorbic Acid (mg)*
Garlic	Trace	0.06	0.23	0.4	13
Lettuce	900	0.06	0.06	0.3	8
Muskmelon	558	0.11	0.08	0.3	26
Okra	172	0.07	0.10	0.6	13
Onion	Trace	0.08	0.01	0.4	11
Peas	640	0.35	0.14	2.9	27
Potato	2	0.10	0.01	1.2	17
Spinach	8100	0.10	0.20	0.6	51
Sweet potato	8800	0.10	0.06	0.6	21
Tomato	900	0.06	0.04	0.7	23
Watermelon	590	0.03	0.03	0.2	7
Yam	–	0.1	15	0.8	15

Table 6.6: Mineral contents (per 100g edible portion) in some of the important vegetables grown in India

Common Name	*Calcium (mg)*	*Phosphorus (mg)*	*Iron (mg)*
Bitter gourd	20	70	1.8
Brinjal	18	47	0.9
Cabbage	49	29	0.4
Capsicum	9	22	0.7
Carrot	37	36	6.7
Cassava	50	40	0.9
Cauliflower	25	56	1.1
Cucumber	10	25	1.5
Frenchbean	56	44	0.8
Garlic	30	310	1.3
Lettuce	35	26	2.0
Muskmelon	32	14	1.4
Okra	66	56	1.5
Onion	47	50	07
Peas	26	116	1.9
Potato	10	40	0.7
Spinach	93	51	3.1
Sweet potato	32	47	0.7
Tomato	13	27	0.5
Watermelon	7	10	0.5
Yam	12	35	0.8

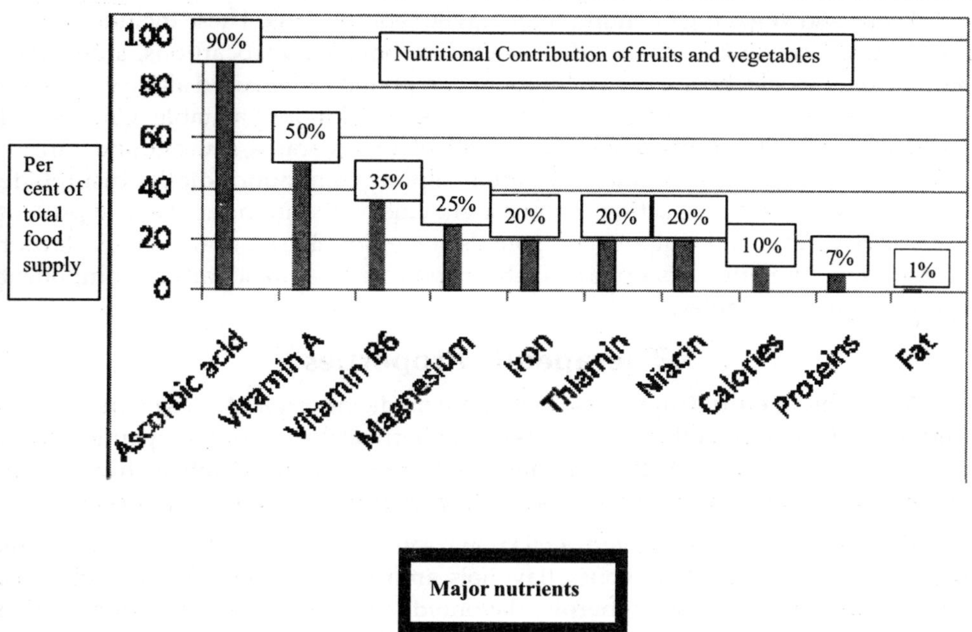

Figure 6.1: Nutritional contribution of fruits and vegetables compared to per cent of total food supply

(β-carotene) is essential structural component of the eye, and a prolonged deficiency of this vitamin may lead to blindness. These are also rich in calcium, magnesium, phosphorus, and iron, which are not present in many other food materials in quantities sufficient for the needs of the body.

Fruits and vegetables are particularly rich in pectin and gums. Pectin has been shown to delay gastric emptying, which may lead to favourable changes in glycaemic responses to foods. Pectin may also induce a satiety effect. It acts as a general intestinal regulator and detoxifying agent. It is used as a standard additive in commercial baby food formulations. Pectin has been shown to have antibacterial, antiviral, wound-healing, and metal-binding properties. It is shown to be effective in reducing cholesterol levels.

A substantial proportion of carbohydrates in fruits and vegetables are present as dietary fibre in the form of cellulose, hemicelluloses, pectin substances, and lignin. It is not digested and passed through the human intestinal system, as people are not capable of secreting the digestive enzymes necessary to break down these polymer complexes into simple units in the forms absorbed by the intestinal tract. Human body lacks enzymes such as cellulase, hemicellulase, and pectinase. Fibre was once considered to be an unnecessary component in the human diet although it was thought to relieve constipation. The epidemiological evidence obtained so far shows that dietary fibre can be the panacea to cure several human diseases, especially in the affluent societies.

Fruits and vegetables contain some flavour compounds. These include sugars, amino acids, organic acids, volatiles such as aromatic hydrocarbons, aldehydes, acetals, ketones, alcohols, ester and sulphur compounds when used along with other food items. These flavouring compounds make food more palatable. Onions and garlic are used as flavouring agents in variety of dishes, soups, sausage and curries. The pungency of garlic is due to volatile sulphur compound, diallyl disulphide which is produced by the action of allinase enzyme on the amino acid allicin present in garlic. Several volatile sulphur compounds are responsible for characteristic flavour of cole crops. Dimethyl trisulphide has been indicated as a major aroma component in cooked cole vegetables.

Therapeutic Properties

Many therapeutic drugs in use in modern medicine originated as plant extracts. Certain fruit and vegetable components exert pharmacological or therapeutic effects. Limonin and nomilin (and other limonoids) are present in citrus fruits such as orange, lemon, β-carotene may also play a role in the prevention of some forms of cancer.

Fruits and vegetables are very well known sources of different types of flavonoids which include flavonones, flavones, flavonols, anthocyanins, catechins and biflavans. Apart from fat soluble tocopherols, flavonoids are commonest and most active antioxidant compounds in human food, being active in hydrophillic as well as in lipophillic systems. This has considerable practical impact on nutrition. It protects food stuff from oxidative deterioration, prolongs their shelf life and keeping quality and improves taste, acceptability and wholesomeness of mixed dishes by inhibiting the oxidation of accompanying lipids. Flavonoids increase or stabilize biological activity of ascorbic acid. All fruits and vegetables rich in flavonoids are marked by the unusual stability of their vitamin C. The ascorbic acid stabilizing effect of flavonoids results in increased production of ascorbic acid-2-sulfate which improves membrane stability by synthesis of sulfated mucopolysaccharides and enhances cholesterol excretion. Ellagic acid is a naturally occurring phenolic constituent present in fruits, especially strawberries and other berries and has been shown to be effective as an anti mutagen and anticarcinogen and a potential inhibitor of common inducers of cancer.

Onion and garlic are known to reduce low-density lipoprotein and induce the formation of high-density-lipoprotein. The increase in high-density lipoprotein is more evident from consumption of white and yellow onion and is reduced with cooking. The allicin which is produced in garlic when tissue is injured by cutting or crushing has been reported to have hypocholesterolaemic action.

Cole vegetables contain a group of compounds known as indoles which have been recently linked with the prevention of cancers of colon, rectum and breast. Some of the species of *Cucumis, Luffa, Coccinia* and *Momordica* exhibit varying degrees of bitterness in fruits, leaves and twigs. This is caused by terpenes called momordicins. Owing to its bitter principle, the bitter gourd is used in Ayurvedic system of medicine. A hypoglycemic ingredient cheratin isolated from bitter gourd lowers blood sugars. Sapogenin compound called diosgenin present in some species of yam is used for production of cortisome and contraceptive drugs.

Processed Products

Fruits and vegetables are subjected to minimal processing such as sorting, grading, packaging and premarketing treatments such as precooling, and waxing to provide a convenient product with few or no preservatives. The processed products exhibit as near as the fresh like properties of original produce. Innovative techniques, being investigated include the use of high temperature short time treatment or low dose of irradiation to produce a product with fresh like quality and greater shelf-stability than fresh item.

Fruits and vegetables are processed into several products such as juices, jams, jellies, wines, dried fruits, canned products by employing different methods of food preservation. There are some losses of nutrients during processing. Once processed, the nutrient content of the product remains relatively stable. Hence, processing conditions should be such that processed products will suffer from minimum loss of nutrients compared to raw material. Considering the nutritional and pharmacological significance of fruits and vegetables and their products, they have a major role in nutrition and health of human population.

7

Principles of Preservation of Fruits and Vegetables

Introduction

The basic preservation techniques such as canning, freezing, dehydration, salting, pickling, and freeze-drying are designed to change perishable foods into more usable form and to prevent undesirable changes caused by microorganisms, or by chemical, physical and biochemical reactions in the food itself. Equally important is the retention of texture, colour, nutritive value and palatability of foods. Hence, an ideal processing technology would involve prevention of growth of microorganisms and at the same time, retention of flavour, texture, nutritive value and palatability of the product.

Nature of Spoilage

Microbial Spoilage

One of the major objectives of any processing technology is to prevent microbial spoilage of foods by destroying microorganisms that are present and then eliminating recontamination by outside microorganisms. Heat is still most commonly used to destroy microorganisms. Recontamination by outside microorganisms in processed food is prevented or delayed by altering the pH of the medium and removal of oxygen etc. However, foods cannot be stored indefinitely because of other changes induced by the processing.

Physical, Chemical and Biochemical Changes

The most common deteriorative reactions occurring in food is oxidation of stored fats, vitamins, and flavouring compounds. Another enzymatic change that reduces the shelf-life of stored product is breakdown of pectins and other cell wall materials

(Figure 7.1). As these changes are catalyzed by one or more enzymes, (existing and de novo synthesis) they can be retarded by heat destruction. Removal of molecular oxygen from containers can also reduce the rate of oxidation.

Fruits and Vegetables

Cutting and slicing

Enzymes, Oxygen, Metals, Substrates, Polyphenols, Tannins

Browning

Figure 7.1: Enzymatic browning in fruits and vegetables

Certain enzymatic and nonenzymatic changes occurring in food also shorten their shelf-life and cause browning, discolouration, and may impart off-flavour to the product (Figure 7.1). Recent developments in preservation techniques have made it possible to retain most of the desirable chemical and physical properties of the product even after extended storage.

Nutritive Losses

In addition to qualitative changes, reduction in nutritive value occurs during processing and storage of processed fruits and vegetables. The best example is the reduction in ascorbic acid content in processed fruits and vegetables. An ideal processing technology, therefore, must prevent such nutritive changes in the product.

Principles of Preservation

The food is preserved for a short or long duration. The short term preservation can be achieved by taking care in handling the produce to low temperature storage and removing air from the atmosphere surrounding the produce. It is possible that in long term preservation, a product may keep well for a pretty long period and even for an indefinite period as far as microbial spoilage is concerned but slow chemical changes however, may render the food inedible if kept for an unduly long period. Long term preservation can be achieved by canning, drying and dehydration, fermentation and by using chemical preservatives. The basic six principles, on which various methods of preservation are based, are as follows:

Moisture Removal

- ☆ Drying
- ☆ Dehydration
- ☆ Concentration
- ☆ Intermediate moisture processing
- ☆ Freeze-drying

The metabolic activities of microorganisms require plenty of free water. Removal of the biologically active water through drying or dehydration stops the growth of microorganisms. It also reduces the rate of enzymatic activity and biochemical reactions. The effect of water removal on nutrients is relatively small if dehydration temperature is moderate and food is subsequently packaged. Freeze dehydration offers decisive advantages in nutrient preservation over dehydration at elevated temperature.

Heat Treatment

- ☆ Blanching
- ☆ Pasteurization
- ☆ Sterilization

Heat treatment results in denaturation of proteins *i.e.* inactivation of the microbial and native food enzymes. Pasteurization frees the food from human pathogens and most vegetative microorganisms whereas sterilization means the destruction of all viable microorganisms. Heat sterilization severely affects labile vitamins and reduces the nutritional quality of proteins mainly through Maillard reactions.

Low-temperature Treatment

- ☆ Cold storage
- ☆ Refrigeration
- ☆ Freezing
- ☆ Dehydro-freezing

Low temperature treatment particularly freezing is a relatively harmless method of food preservation. Low temperature inhibits microbial growth and slows down the rate of chemical and enzymatic reactions. Many vegetables are blanched before freezing in order to avoid undesirable quality changes due to enzymatic activity at freezer temperature. Vitamin losses are minimal compared to other methods of food preservation. Most of the fruits and vegetables are preserved at low temperature.

Acidity Control

- ☆ Fermentation
- ☆ Acidic additive

Spoilage of low acid foods is relatively rapid. The growth of food spoilage organism is significantly inhibited in an acidic environment. Acidic fermentation lowers the pH of carbohydrate containing foods by producing lactic acid. The acidity of some foods may be increased by acidic preservatives such as vinegar or citric acids, producing some inhibitory effect on spoilage. Juices, pickles, sauerkraut are some of the examples of preservation of food by acidity control.

Use of Preservatives

- ☆ SO_2
- ☆ Benzoic acid

- Sorbic acid
- Sugars and salts

Chemical additives can substantially contribute to the preservation of foods by providing inhibitory environment for microbial growth and for enzymatic and chemical reactions. Several preservatives such as sodium benzoate, potassium metabisulfite, sorbic acid etc. are commonly used for preservation of foods. The use of sugars and salt creates unfavourable environment for growth of microorganisms by reducing water activity of the foods. Sugars are utilized mainly in preparation of jams, jellies, marmalades and preserves. The salts are commonly used for pickling purpose. Cucumber, cabbage and bamboo shoots or fruits like unripe mangoes, and lime etc. are commonly preserved by salting process.

Irradiation

Irradiation is the recent method of food preservation. The free radical mechanism of irradiation destroys microorganisms but is also detrimental to nutrients particularly vitamins. Gamma rays are commonly used for preservation of vegetables. Cobalt-60 is the radiation source generally preferred for irradiation of foods including vegetables. Potatoes, onions are commonly preserved by radiation treatment.

8

Raw Material for Processing

Introduction

The processor is primarily interested in colour, aroma, flavour and texture whereas the grower is more concerned with yield, disease resistance and ease of cultivation and harvesting. A processed product of excellent quality cannot be made from poor quality raw material. Quality characteristics of product may be divided into three categories; sensory, hidden and quantitative. The sensory characteristics of quality include colour, gloss, size, shape, defects, odour, and taste which the consumer can evaluate with his senses. Those characters which consumers cannot evaluate with his senses are nutritive value, presence of harmful adulterants and presence of toxic substances. The quality of finished product depends upon the quality of raw material used for processing.

Colour

Colour increases the attractiveness of the product. Colour is also associated with flavour, texture, nutritive value, maturity and wholesomeness. Coloured fruits when harvested at firm ripe stage should be fully and uniformly coloured. Three major classes of pigments occur in fruits and vegetables; the chlorophyll, carotenoids and xanthophyll and anthocyanins. The characteristic yellow colour of mango, citrus and pineapple is mainly due to carotene and xanthophyll esters. There are great demands for juice with deep orange colour especially in concentrates and bases. Hence, varieties yielding deep orange colour are preferred for processing. Red colour in fruits and vegetables is mainly due to anthocyanins. This colour gradually passes out into syrup or brine used in Canning. Hence, colour retention in canned fruits or vegetables is important in processing. Guava, lychee, banana and broad beans may turn brown or pink during processing due to the presence of leucoanthocyanin, and this appears to differ with variety. The varieties which retain their original colour during processing are preferred.

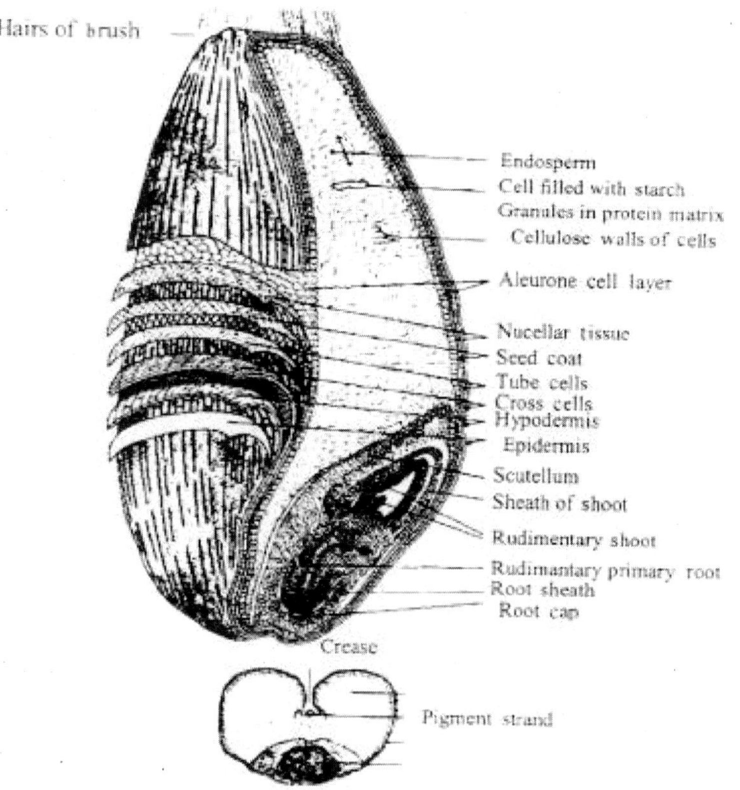

Wheat kernel section

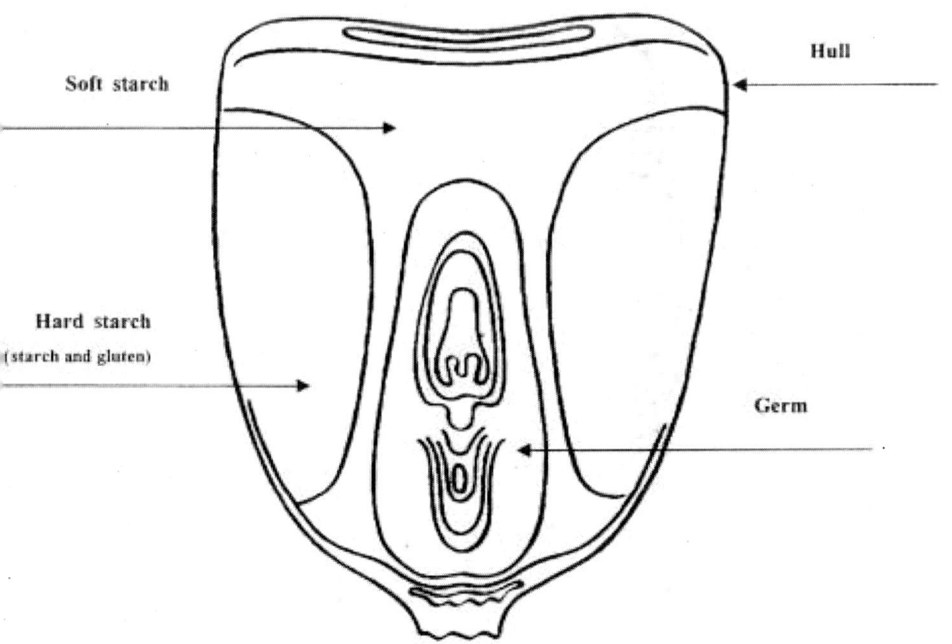

Maize kernel

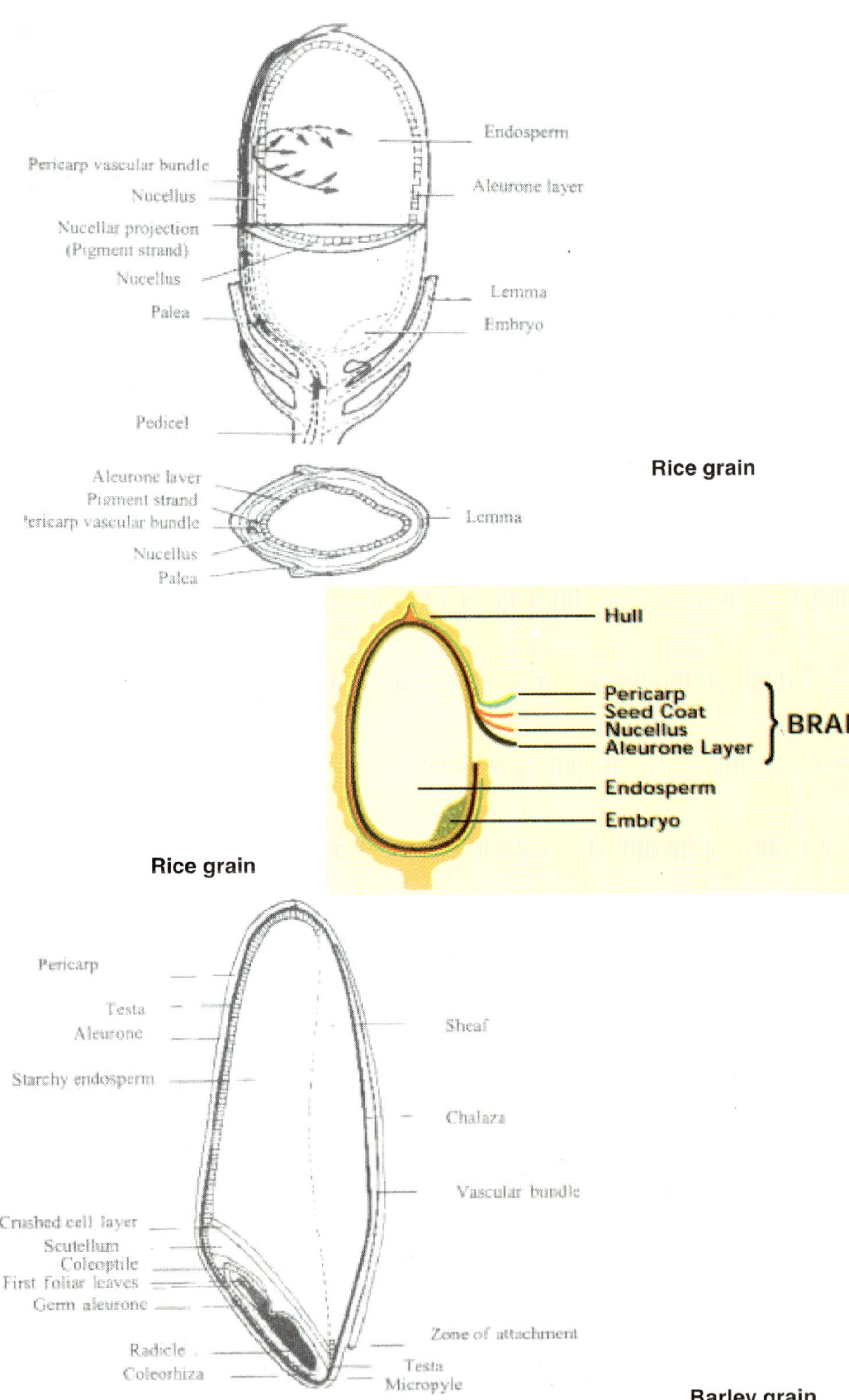

Rice grain

Rice grain

Barley grain

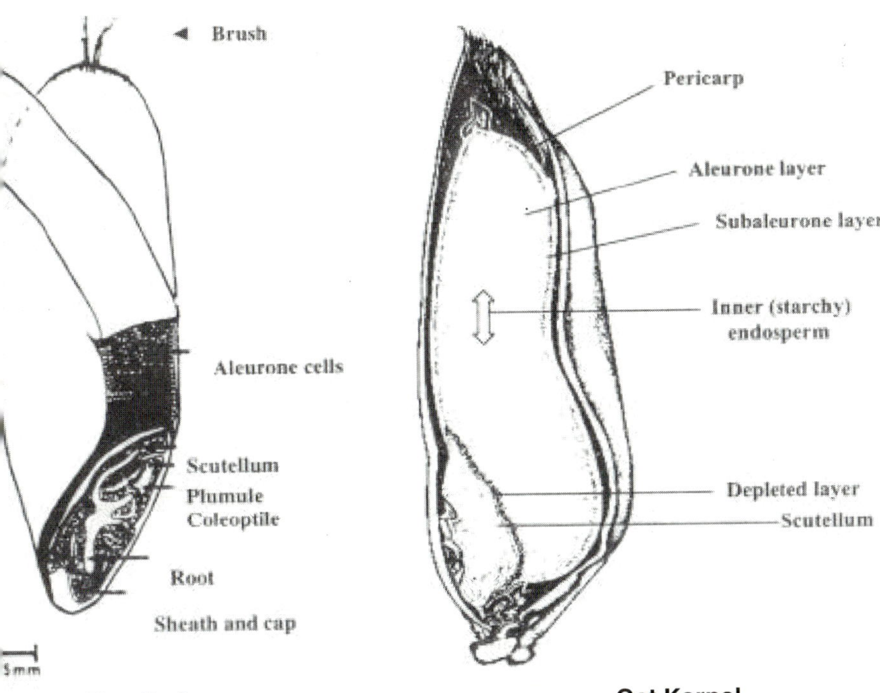

Rye Grain **Oat Kernel**

Oil mill

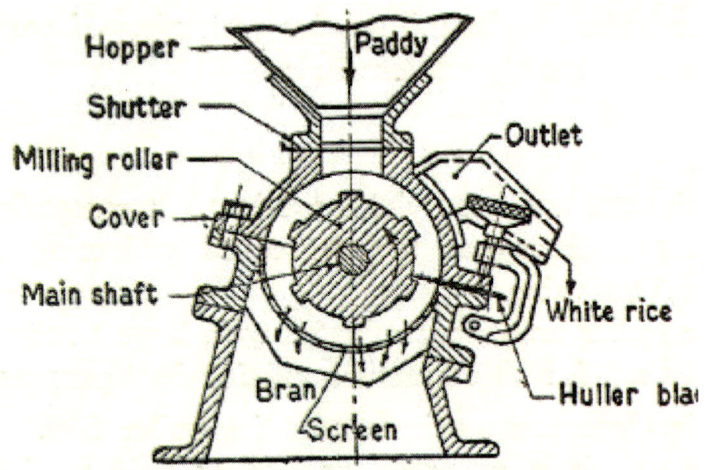

Engelburg Type Huller

Oil mill

Mini dhal mill

Dhal polisher

Sheller

Mini dhal mill

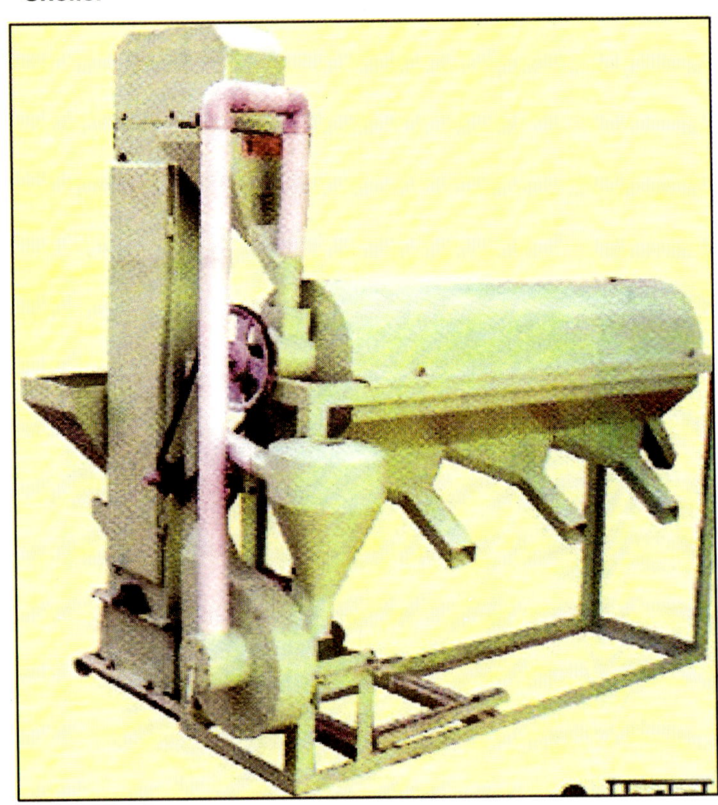

Dhal grader cum sorter

Paddy huller

Electric rotary huller

Rubber rolls huller

Corn/maize sheller

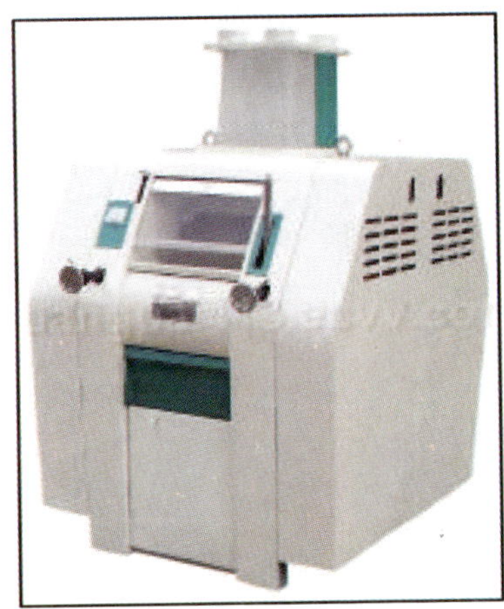

Roller mill

Hammer mill

Disc huller

Dryer

Screw type juicer

Fruit pulper

Kettle

Steam jacketed kettle movable

Retoard

Autoclave

Fermentor

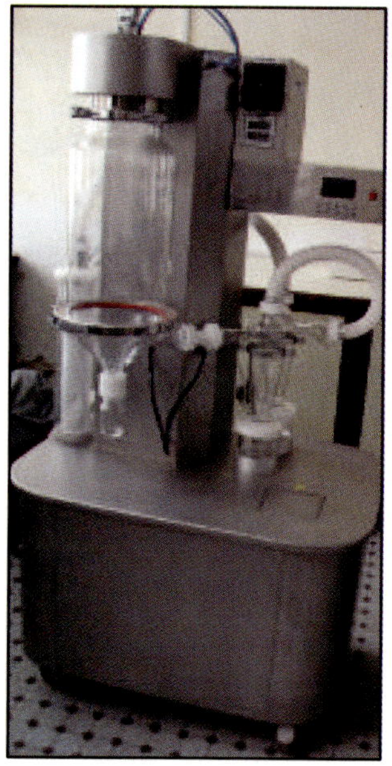

Freeze drier

Ultra grinding mill

Potato slicer

Potato peeler

Roasing machine

Screw type juice extractor

Cap sealer

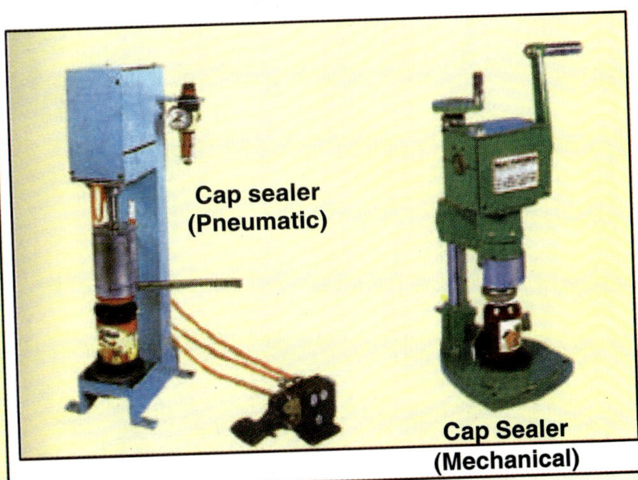

Cap sealer (Pneumatic)

Cap Sealer (Mechanical)

Fruit pulper

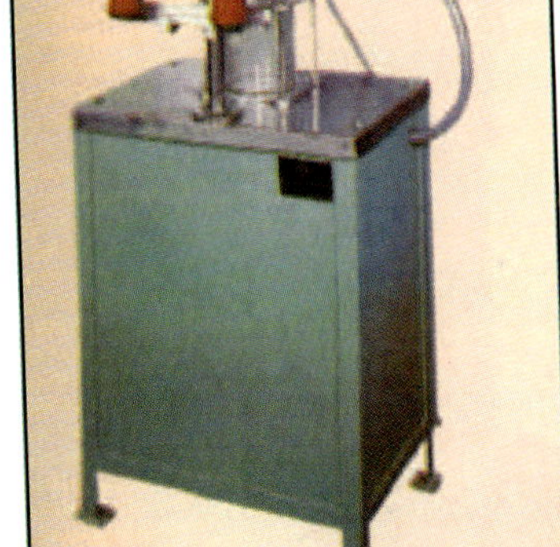

Vacuum filling

Fruit mill machine

Canning machine for fruits and vegetables industry

Basket press

Cone-type juice extractor

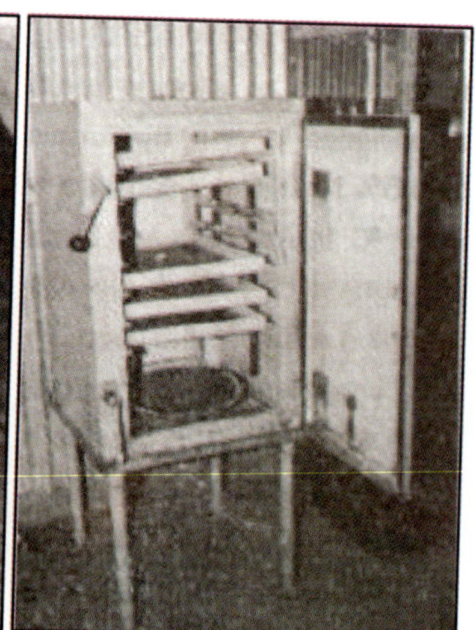

Sulphur fumigation box

Pulses and their products

Black or kidney beans

Navy bean

Navy bean/great northern bean

Chickpea

Pigeon pea

Winged bean

Black bean

Broad bean

Common bean

Black eyed bean

Red kidney bean

Horse gram

White bean

Lupine seed

Peas

Mung bean

Lupine seed

Soybean

Oilseeds Processing

Maize/corn cobs

Maize/corn oil

Soybean seeds

Peanut pods

Soybean oil

Peanut oil

Groundnut seeds

Sunflower seeds

Safflower oil

Sunflower oil

Safflower seeds

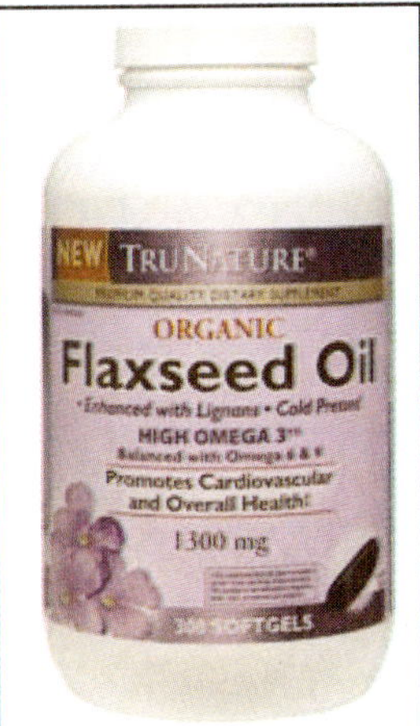

Flaxseed oil

Flaxseeds

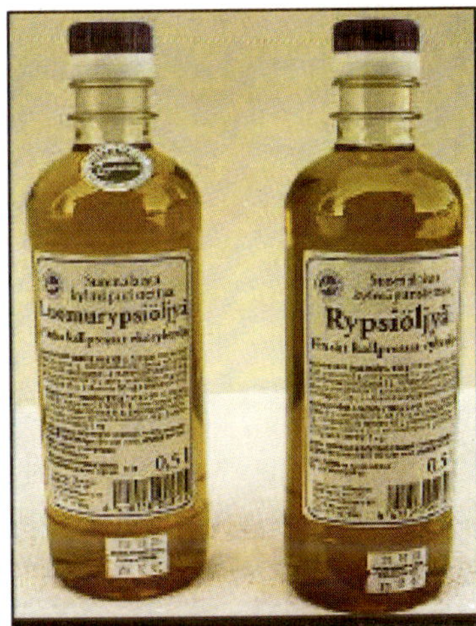

Rapeseed oil

Palm seeds

Rapeseeds

Palm seed oil

Cotton seeds and oil

Sesame oil

Sesame seeds

Niger seeds

Coconut and oil

Rice bran oil

Mango kernels

Mango kernel oil

Poppy seed oil

Cocoa bean fruit

Olive oil

Hemp seed nut

Cocoa bean seeds

Shea seeds

Hemp seeds

Poppy seeds

Apricot seeds/nuts

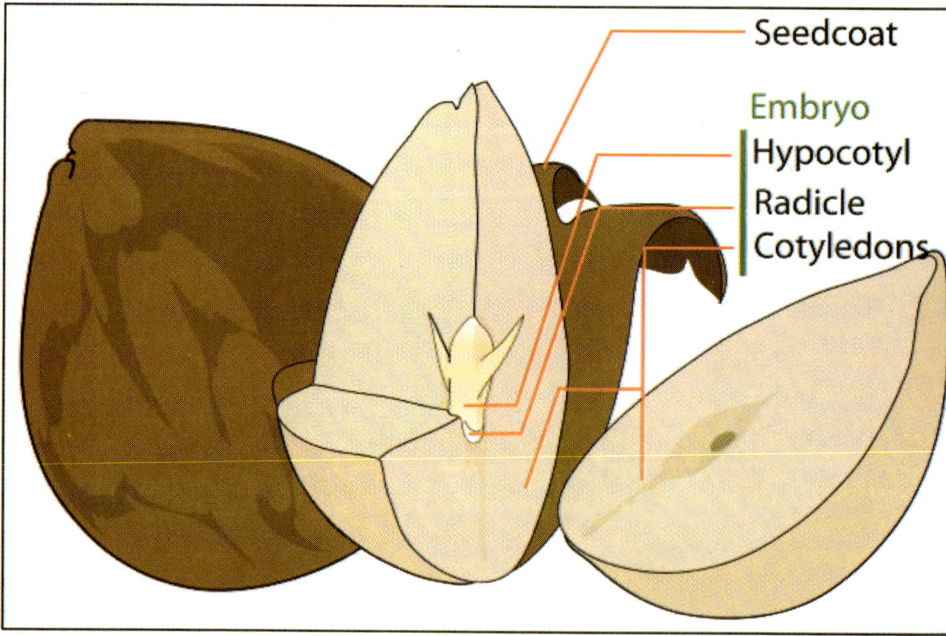

Avocado seed

Apricot kernel oil

Winged bean

Crambe seeds

Lupine seeds

Almond seeds

Jojoba seeds

Castor oil

Almond seed oil

Jojoba seed oil

Linseed oil

Neem seeds

Castor seeds

Linseeds

Mahua seeds

Karanj seeds

Kokum fruits

Sal seeds

Dhupa seeds

Neem seed oil

Size and Shape

Grading of the raw material into various size and shape categories is usually one of the first steps in food processing operations. Size grading is done mainly to facilitate succeeding operations such as cutting, peeling, to obtain uniformity in the product, and to provide consumers with the, preferred size. It may also be an indirect means of grading for other quality characteristics; for example small peas, okra, and cucumbers for pickling are usually less mature and consequently more tender and desirable. Size is of major interest to the grower as it is directly proportional in certain crops, *e.g.*, pineapple, to the yield per hectare. Certain sizes are in greater demand than others for specific purposes; for example, the desirable long shreds/cuts can be made only from large onions. Evenly-sized materials are easier to handle in large quantities, and promote less wastage, rapid production and high quality.

Shape of the raw produce sometimes determines the suitability for processing. To reduce losses during mechanical trimming and handling, the shape of a fruit or a vegetable should readily lend itself to such processes. Selection and breeding of raw produce primarily for a suitable shape is yet to be attempted in most crops.

Texture

Texture and consistency, the structural features of fruits and vegetables, are attributes of prime importance. Texture characteristics involve touch sensations, which determine the firmness, softness (yielding quality), juiciness, grittiness, fibrousness, and mealiness of a fruit or a vegetable. In general, these factors can be readily measured with such instruments as tenderometers, texture meters, puncture meters, succulometers, fibrometers and pressure testers. In addition, certain physico-chemical tests are employed successfully for measuring esthetic properties. Certain vegetable crops, such as corn, tend to accumulate solids as they approach maturity (become less desirable); moisture determination may be used to measure this quality characteristic. Determination of alcohol-insoluble solids is particularly suitable for vegetables such as peas, sweet corn, and limabeans. Fruit specific gravity, which is related to the alcohol-insoluble solids content of fruits, can be used to identify this stage.

Fruits and vegetables used for processing should be firm enough to withstand necessary heat treatment. The firmer canned product given by "pear" tomatoes as compared with "round" tomatoes has been attributed to differences in pectin substances. Consistency or viscosity is of importance in such products as juices, jams, jellies, and preserves. Some juices (e.g., orange juice) are consumed in a naturally cloudy state, and juices of apple, grape, and other berries have traditionally been consumed in a clear condition and the problem has been to clarify them effectively and to maintain them in a brilliant condition throughout their storage life. However, in recent years, unclarified or opalescent juices have also come into production. The juices of some fruits lack much of the flavour and colour of the fruit itself, as these are associated with insoluble cellular material. Fruits rich in carotenoids, such as mango, papaya and tomato, are usually converted into pulpy products by disintegrating the whole fruit to give purees or nectars.

Flavour

Flavour distinguishes one food from another. It is difficult to evaluate instrumentally, and is still measured largely by subjective methods such as taste panels or the profile method. Though many deciduous tropical and subtropical fruits have attractive flavours, the aroma components tend to be heat-sensitive. In passion fruits, the distinctive aroma and flavour are very elusive, and are lost during heat treatment. Various methods are adapted during processing of citrus to enhance the flavour. Incorporation of fine pulp and small quantities of peel oil in juices and addition of fresh juice to the concentrate before packing are used.

Defects

Most defects are still largely evaluated by the consumer's eye, though in some cases instruments may be used. The presence of defects frequently lowers the grade of products which are otherwise of very high quality. Defects may be caused by heredity or unfavourable environmental conditions. The green colouration in ripe tomatoes, the tendency of some types of sweet potatoes to disintegrate and peas to split on canning is some of the examples of defects caused by genetic factors. Insects and microorganisms are major sources of defects in fruits and vegetables. Microorganisms not only reduce the crop yield but also deform the produce and cause internal defects such as off-colouring, corky tissue, mould and rot. The wholesomeness of the product decreases with such infections. Mechanical defects may occur during handling. Bruises upset normal biochemical reactions, causing discoloration, off-flavour and more rapid deterioration.

Nutritive Value

Succulent fruits and tubers high in total solids generally contain more total stored food values. They contain more nutrients as they approach maturity, although the proportion of each changes. On the other hand, many vegetables become fibrous and woody as their total dry matter increases, and there is a decline in the digestible carbohydrates, proteins, minerals and vitamins.

Toxicity

Various chemical compounds are used extensively in fruit and vegetable production. Commercial canning and juice extraction operations may remove the residues in tomatoes, but sometimes these persist. Persistent insecticides belonging to the chlorinated hydrocarbon group such as aldrin, dieldrin, and DDT, cause problems due to their preferential absorption by crops such as carrots, radish and potato. Edible tissues may accumulate amounts even beyond permissible limits. These residues may lead to bitter or musty flavour in the processed products, and present a health hazard.

Yield of Finished Product

The yield of finished product mostly depends upon the raw material used for processing. The good quality raw material gives maximum and acceptable final finished product. Good quality raw material eliminates much wastage.

Postharvest Physiology of Fresh Fruits and Vegetables

Harvested fruits and vegetables continue to maintain physiological systems and sustain metabolic processes that were present before harvest. Losses of water and substrates used in respiration can no longer be replaced and deterioration of the product begins.

Maturation, ripening and senescence induce many changes in fruits and vegetables. Although a strict physiological distinction between fruit ripening and senescence is unclear, ripening hastens the onset of senescence and the probability of cell injury and death. Fruit ripening involves many complex changes, including seed maturation, colour changes, abscission from the parent plant, tissue softening, volatile production, wax development on skin, and changes in respiration rate, ethylene production, tissue permeability, carbohydrate composition, organic acids and proteins. Product respiration, transpiration and ethylene production are major factors contributing to the deterioration of fresh fruits and vegetables.

Gas Exchange

Respiration

The energy is generated by respiration, which is the oxidative catabolism of carbohydrates. Respiration can be considered a series of enzymatic reactions, involving three pathways: i. glycolysis (glucose to pyruvate), ii. the tricarboxylic acid cycle (pyruvate to CO_2), and iii. Oxidative phosphorylation [reduced nicotinamide adenine dinucleotide (NADH) + reduced flavin adenine dinucleotide (FADH) to adenosine triphosphate (ATP)].

$$C_6H_{12}O_6 + 6O_2 + 36\text{ ADP} \rightarrow 6CO_2 + 6H_2O + 36\text{ ATP}.$$

Measuring Gas Exchange

1. Static method: A specific gas composition is generated around an object and the gas flow is closed for a specific period of time. Gas composition is measured at the beginning and end of the period.
2. Flow through method: A specific gas composition is generated around an object and the gas composition of the inward and outward is measured.
3. Modified atmosphere method (MA method): Using a package with film of known O_2 and CO_2 permeability, the equilibrium concentrations that develop are measured. Gas exchange rates can then be calculated.

Respiration Quotient

The ratio of moles of CO_2 produced pre mole of O_2 consumed is called the respiratory quotient (RQ).

Fermentation

The main source of CO_2 production by plant tissue is respiration. At low O_2 or high CO_2 atmospheres, however, fermentation becomes increasingly important. The main fermentative metabolites found in plant tissues are ethanol, acetaldehyde and lactic acid. Ethanol fermentation is the catabolism of pyruvate to ethanol; the combination of glycolysis and fermentation can be expressed as:

$$C_6H_{12}O_6 + 2ADP \rightarrow 2CO_2 + 2C_2H_5OH + 2ATP$$

Transpiration

Transpiration is a mass-transfer process in which water vapour moves from the surface of fruit or vegetable to the surrounding air. This process of moisture loss induces wilting, shrinkage, and loss of firmness and crispness of fruits and vegetables and thus adversely affects the appearance, texture, flavour and mass of produce. Most fruits and vegetables loss their freshness after 3 to 10 per cent mass loss. Transpiration is considered to be the primary cause of postharvest loss and poor quality in leafy vegetables, such as lettuce, chard, spinach, cabbage and green onion and is considered the major cause of commercial and physiological deterioration in citrus fruits.

Factors Affecting Transpiration

1. Water vapour pressure deficit
2. Temperature
3. Storage environment
4. Product respiration
5. Size, shape and surface of a commodity
6. Maturity and ripening

Controlling Transpiration

Transpiration of fresh fruits and vegetables can be reduced by minimizing the water vapour pressure difference between the produce and the air and/or by increasing the resistance of the product by storage conditions, packaging (modified humidity packages, MHP) and surface coatings (commercial carnauba waxes).

Ethylene [C_2H_4]

Ethylene can induce plant responses at very low concentrations (0.1µl/L or less) and has a wide variety of effects on virtually all stages of a plant life cycle (e.g., germination, vegetative growth, flowering, fruiting, abscission, ripening, senescence and dormancy). Ethylene is sometimes called the "ripening gas" because its most important postharvest effect is the acceleration of ripening and senescence. However, there are other postharvest effects of ethylene, including induction or suppression of potato sprout growth, loss of chlorophyll (degreening) and induction or suppression of disease resistance. Some of these effects are desirable and some are undesirable, depending on the intended use of the stored plant product.

Production of Ethylene

Biological ethylene production: In flowering plants, the formation of ethylene from methionine major source of ethylene, via the ACC pathway is now recognized as the major source of ethylene, whereas non-flowering plants do not possess ACC oxidase and produce ethylene from other unknown pathway when they are stressed or damaged. The rate of ethylene synthesis is controlled by the activity of ACC synthase, but it can sometimes be limited by ACC oxidase activity.

Non-biological ethylene production:

1. Incomplete combustion of organic fuels
2. Smoke production
3. Kerosene and illuminating gases

Ethylene, Respiration and Fruit Ripening

Respiration rate highest in an immature vegetables and fruits. One group has a pattern similar to vegetables and is called non-climacteric fruits. The other group called climacteric fruits exhibits a temporary respiratory increase, which occurs when the fruit reaches full maturation and size and is entering the ripening stage. In climacteric fruits, the respiration climacteric and subsequent ripening and senescence are associated with a coincident increase in endogenous ethylene production. Exogenous ethylene application; no increase in respiration in climacteric fruits; but increase in respiration in non-climacteric fruits.

Ethylene Addition

Ethylene gas is explosive at concentration between 2.7 – 36 Kpa. Ethylene (2-chloroethane phosphoric acid) commercially available under various trade names, ethyrel, florel, cerone, prep and CEPA in liquid form. If pH increase above 5 ethylene gas produce from above solutions.

Ethylene Removal

Potassium permanganate ($KMnO_4$) oxidizes ethylene to CO_2 and H_2O. Ultraviolet (UV) lamps or Ozone (O_3) generators oxidize ethylene. Ethylene adsorbed by activated or brominated charcoal. Cold storage reduces ethylene production. Soil bacteria that use ethylene as a biochemical substrate. Hollow fiber technology and pressure swing adsorption as well as a modified pressure swing adsorption called vacuum swing adsorption which keeps ethylene level below 1µl/L.

9

Minimal Processing and Storage

Introduction

Recent developments in nutrition have created greater demands for fresh fruits and vegetables because of their greater contribution to texture, colour, flavour and nutritive value than the processed products. Various postharvest handling operations help in retention of quality of fresh produce. This has increased the availability of fresh fruits and vegetables during most of the year thus increasing per capita consumption. Hence, postharvest handling operations including grading, packaging, precooling is important in influencing the quality of fresh fruits and vegetables. In recent years, attempts have been made to process the produce to increase their functionality without changing their fresh-like properties. This type of processing which include unit operations such as washing, sorting, peeling, slicing etc. is termed as "minimal processing". The fruits and vegetables remain in raw form having cells alive and respiring. The main requirement however would be that the products have as near as possible the fresh-like properties of original produce.

Fruits and vegetables can lose their typical fresh appearance and characteristic in only a short time after harvesting. However, if they are quickly cooled, this change can be retarded. This rate of change is increased if the cells are bruised or they are ruptured in any way which occurs if the product is peeled, sliced etc. This causes an increase in respiration rate which can lead to physical decay through oxidative and other chemical reactions. It is, therefore, essential that processing of fresh produce should be done to improve the functional properties of fresh produce.

Postharvest storage life of fresh produce or processed products depends upon preharvest cultural/environmental conditions. They are important determinants of quality and nutritive value. After fruits and vegetables are harvested, they must be cleaned, sorted for size, colour and maturity and packed for selling in the fresh market. Additional steps must be undertaken such as washing with disinfectants and application of waxes or other protective agents. These operations are usually carried out in packing shed. However, certain fruits and vegetables are directly packed in fields. Postharvest technology employed in California, U.S.A. for handling of grapes, tomato and banana is presented in Figures 9.1–9.3 respectively.

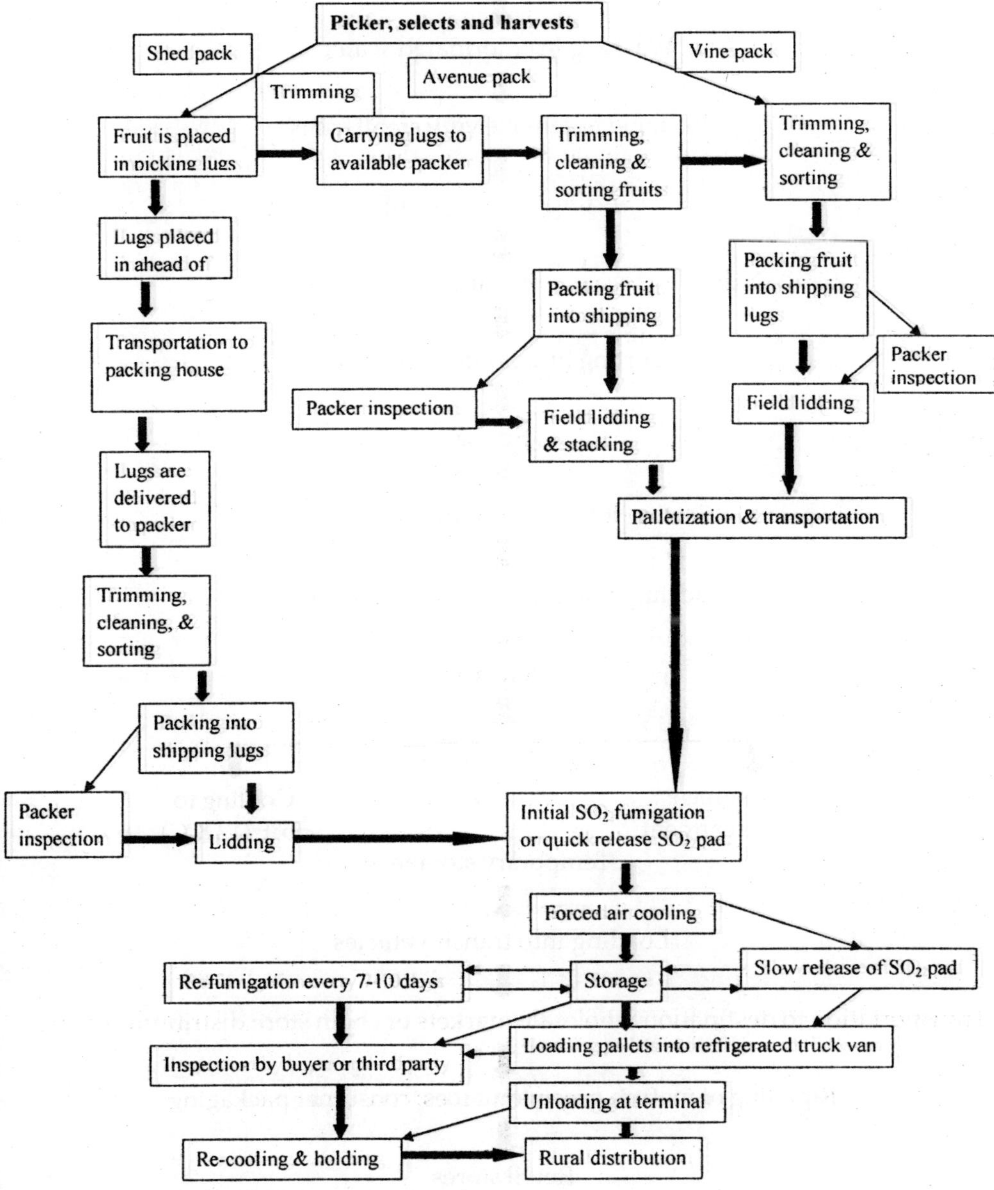

Figure 9.1: Handling system for table grapes

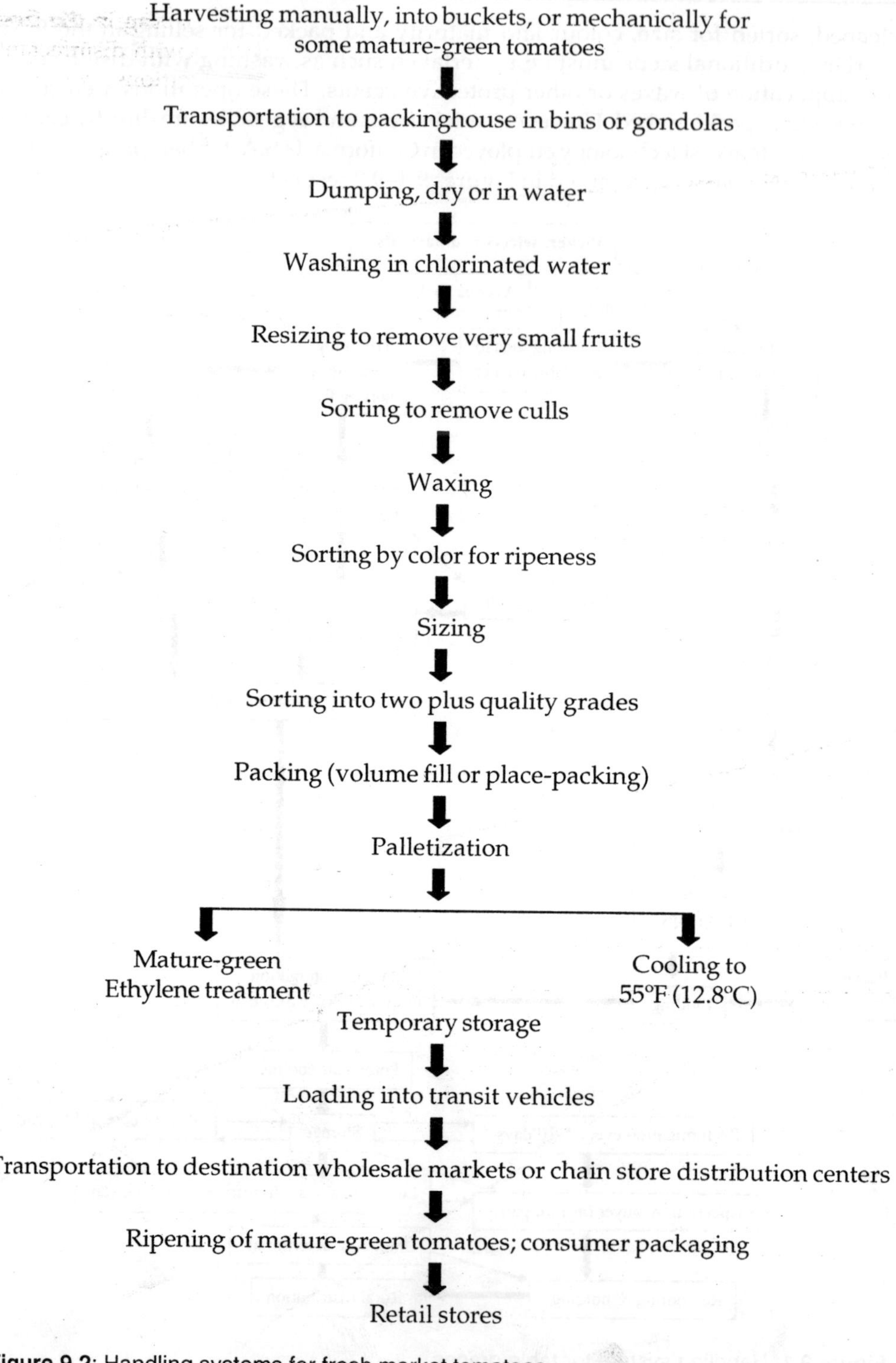

Figure 9.2: Handling systems for fresh market tomatoes

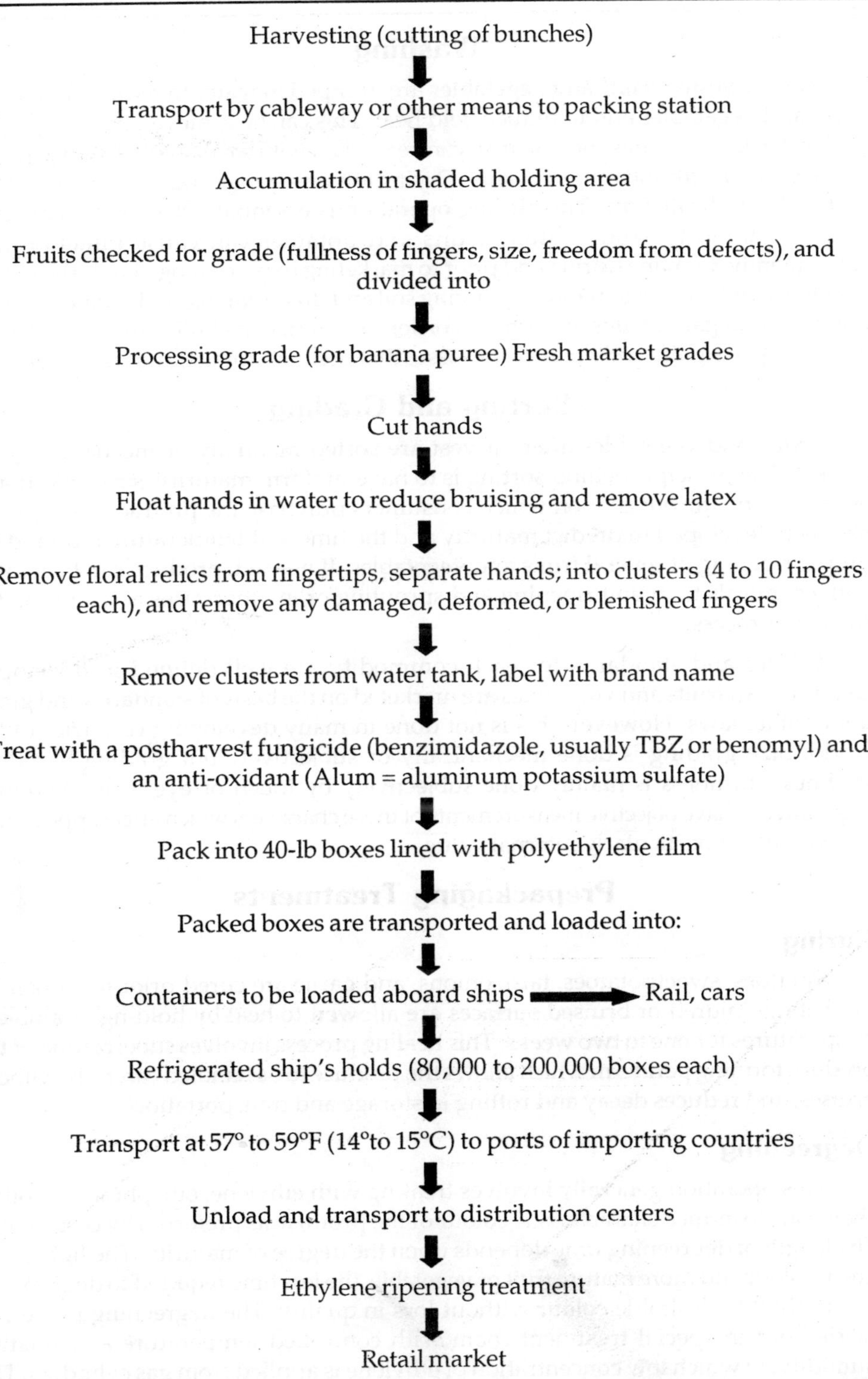

Figure 9.3: A flow diagram of a postharvest handling system for bananas

Washing

The harvested fruits and vegetables are dumped in water to reduce damage to fruits and vegetables and to reduce field heat. This can be achieved by dumping in cold water. Certain fruits such as citrus can be washed with weak solution of detergents and scrubbed with soft nylon brushes. Drying after washing can be accomplished by high velocity heated air. This drying operation is essential. Otherwise, increased relative humidity around stone fruits and potatoes can cause infection and subsequently decaying during and prior to marketing or processing. Similarly melons are often dry-brushed to remove adhering soil and dust. Onions, and potatoes, sweet potatoes, and garlic bulbs are cured in order to suberize the bulbs, tubers and roots which help to significantly reduce the losses that occur in storage, and transportation.

Sorting and Grading

Fruits and vegetables after harvest are sorted manually or mechanically for market. The principle behind sorting is to have uniform maturity, size, colour and other quality attributes upon which consumers purchase the produce. Equipment has been developed to predict maturity and the time and temperature required for proper ripening of several fruits and vegetables. If needed, produce can be treated with chemicals to inhibit ripening and sprouting or in other cases accelerating the ripening process.

Grades and standards for each commodity are well defined in developed countries. All fruits and vegetables are marketed on the basis of standards and grade under State laws. However, this is not done in many developing countries of the world. Size grading is done mechanically or subjectively but grading based on freshness, firmness is mainly done subjectively by touch or eye. It is, therefore, imperative to have objective measurements of those characters which affect appearance, sale value and particularly internal value.

Prepackaging Treatments

Curing

Potatoes, sweetpotatoes, taro, onions, and garlic are cured prior to storage or marketing. Injured or bruised surfaces are allowed to heal by holding at ambient temperatures for one to two weeks. This healing process involves suberization of the produce forming periderm which allows the produce to be handled carefully without bruises, and reduces decay and rotting in storage and transportation.

Degreening

This operation generally involves treating with ethylene, ethephon, and other chemicals to induce characteristic colour of the produce as preferred by consumers. The length of degreening time depends upon the degree of maturity. The lighter the green colour and more mature fruit or vegetable, the less time required to degrade the chlorophyll to desirable colour without loss in quality. The degreening process is carried out in special treatment room with controlled temperature and relative humidity, in which low concentration of ethylene is applied from gas cylinders. The best degreening temperature without loss in quality is 27°C and 80-90 per cent relative humidity. Ethylene accelerates the decomposition of chlorophyll without affecting the synthesis of carotenoid pigments.

Precooling

High temperatures are harmful to the keeping quality of fresh fruits and vegetables. Harvesting of produce at high temperatures is sometimes inevitable, especially when harvesting is carried out during summer season. Precooling is a means of removing field heat. The aim is to quickly slow down respiration of produce, minimize microbial growth, reduce transpiration rate and ease the load on cooling system. Various methods used for precooling are as follows:

Air Cooling

Cooling may be accomplished by forced air over iced water containing fungicide to control microbial contamination. Prompt cooling and storage is highly essential to achieve the benefits of precooling. Forced-air (1.7°C) cooling may last about 2-4 hours.

Hydrocooling

Hydrocooling lowers field heat from the produce particularly fruits such as peaches, apricots, cherries and leafy vegetables, resulting in better retention of turgidity, texture and freshness. For some fruits such as apricots, cherries and citrus fruits, a fungicide in cooling (containing ice) water may be added to control decay.

Vacuum cooling

This is the most rapid method of precooling used for leafy vegetables. The vacuum cooler consists of a large autoclave with steam injectors. The vacuum is achieved in three stages: 15 in., 0.2 in, and finally 0.016 in level. The principle employed is evaporative cooling. At 29.9 in. pressure, water evaporates at 100°C. At 0.018 in., water vaporizes at O°C. Vacuum cooling takes heat from head lettuce in 3 to 4 min with drop in temperature from 22 to 0°C.

Treatments to Control Browning

The discolouration of fresh or minimally processed fruits and vegetables occurs due to enzymatic reaction. The main enzyme associated with fruit and vegetable browning is polyphenol oxidase. The most commonly used chemical for control of enzymatic browning IS ascorbic acid since sulfites are no longer permitted to be used on fresh produce. It functions as a reducing agent, antioxidant and as a metal-sequestering material. The ascorbic acid solution is more stable at low concentration and at lower pH levels. For this reason, it is frequently used in conjunction with citric acid. Polyphenol oxidase can be inactivated by thiourea and L-cysteine.

Treatments to Retard Textural Loss

Softening of fruits is a common feature in minimally processed product. This is mainly due to degradation of pectin. Calcium chloride has been the compound of choice for use in maintaining fruit and vegetable texture in minimally processed product. Calcium chloride imparts a taste to the product if too much is absorbed into fruit. Natural cytokinins, zeatin and dihydrozeatin are effective in preserving banana, broccoli and other products.

Waxing

Waxing of certain fruits and vegetables is accomplished to reduce the rate of respiration and to enhance product gloss, improving merchandising and marketing. Waxing is applied by foaming, dipping, or brushing. Wax emulsion may contain some fungicides to protect the produce and also reduces transpirational and respirational losses.

Packaging

The important step in preparation of fruits and vegetables for fresh produce market involves placement in containers. Mechanical injuries such as cuts, punctures, pressure bruises, abrasion and friction must be avoided prior to packaging. Produce must be immobilized within the container. Immobilization can be accomplished by wrapping and packaging, by placing in various types of trays or by certain volume fill techniques. Proper and uniform design of container is necessary. These are packed in lug boxes, field pack or carton boxes. Padding is essential to achieve immobilization of the produce. Fruits and vegetables must be cushioned against impact. Various types of cushioning pads are used as a bottom pad and/or between layers of fruits and vegetables.

Storage

Fruits and vegetables often require some storage to balance day-to-day fluctuations between harvest and sale or for long term storage to extend marketing beyond the end of harvest season. Storage of fresh fruits and vegetables prolongs their usefulness and in some cases improves their quality. It also controls market glut. The principal goal of storage is to control the rate of transpiration, respiration, diseases and insect infestation and to preserve the commodity in its most usable form for the consumers.

Refrigerated Storage

Storage of food at temperature above freezing and below 15°C is known as refrigerated storage. The storage of fruits at refrigerated temperature controls growth of microorganisms reduces metabolic activities of intact tissue and inhibits deteriorative chemical reactions and reduces moisture loss. The most important characteristics of harvested fruits and vegetables is the persistence of aerobic respiration throughout the storage period. The respiration involves metabolism of carbohydrates and organic acids in the presence of atmospheric oxygen with the ultimate production of carbondioxide, water, heat and small amounts of organic volatiles and other substances. For maximum storage life of fruits and vegetables at refrigerated temperature, it is desirable that (1) aerobic respiration be allowed to continue at a slow rate so that maintenance processes associated with life continue to function and (2) temperature be suitably low so that major biochemical deteriorative reactions are slowed as much as possible. The fruits and vegetables which normally respire rapidly often have short storage lives. Okra (Bhendi) and peas have high rate of respiration. When temperature reduction is used to slow the rate of aerobic respiration, this usually delays senescence and decay of the fruits and vegetables and many times enables some fruits to be ripened at controlled rates.

There are large differences in the rates at which different fruits and vegetables respire, a particular fruit does not respire at constant rate when held at constant temperature. This irregularity in rate of respiration is most apparent in climacteric fruits. The climacteric maximum frequently but not always occurs at optimum ripeness of the fruit and sharp decline in respiration rate following maximum is often associated with over ripeness, a stage referred to as senescence. The climacteric phenomenon is not exhibited by vegetables. It does not occur in some fruits (Table 9.1). Non-climacteric fruits generally exhibit a decline (Figure 9.4) in respiratory activity during storage.

Table 9.1: Respiratory characteristics of fruits

Climacteric Fruits	*Non-Climacteric Fruits*
Apple, Apricot, Asian pear	Bell pepper
Avocado, Banana	Carambola
Blackberry	Cherry
Blue berry	Citrus
Cranberry	Grape
Cherimoya	Lychee
Eggplant	Pineapple
Feijoa	Pumkin
Fig	Rambutan
Guava	Strawberry
Jujube	Tamarillo
Kiwifruit	Watermelon
Mango	Cucumber
Muskmelon	Olive
Papaya	
Passion fruit	
Peach	
Pear	
Persimmon	
Plum	
Raspberry	
Sapota	
Tomato	

Temperature

Temperature in storage facility is normally maintained at the desired level for commodities being stored. Temperature below the optimum range *tor* a given fruits or vegetables causes chilling injury, temperature above the optimum reduces storage life. The storage temperature *tor* different fruits and vegetables recommended by USDA are presented in Table 9.2. Growth of pathogenic microorganisms occurs rapidly in the range 10-37°C but only slowly in the range 3.3 to 10°C. Below 3.3°C pathogenic microorganisms can no longer grow. The storage temperature has profound effect on growth of microorganisms, metabolic activity and other chemical reactions.

Sometimes small changes in temperature results in marked changes in rate of quality loss. It has been reported that William pears have 12 week storage life at -1.5°C as compared to 10 weeks at -10, 9 weeks at -0.5°, 7.5 weeks at 0° and 6 weeks at 1°. Hence, careful control of handling and storage temperature is necessary in order to achieve maximum storage life and minimum loss of nutrients.

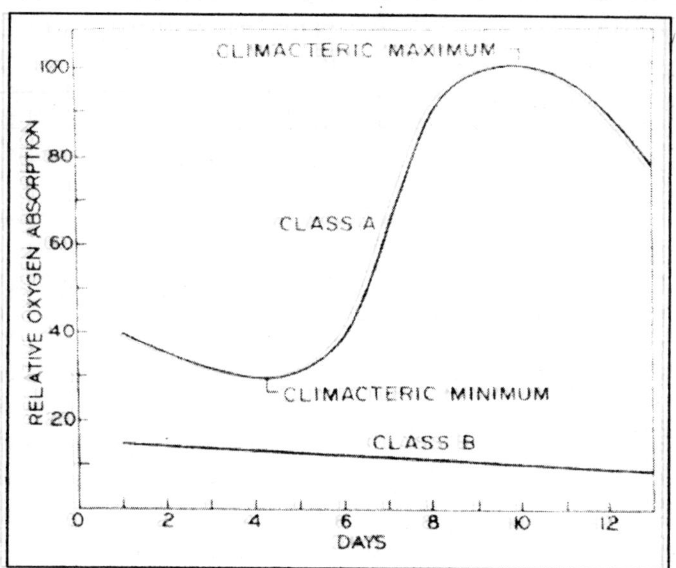

Figure 9.4: Respiratory behaviour of climacteric and non-climacteric fruits. Class A represents climacteric fruits and class B represents non-climacteric fruits

A substantial number of fruits and vegetables of tropical and subtropical origin develop physiological disorders when exposed to temperature below their optimum storage temperature but above their freezing points. This type of disorder is known as "chilling injury", low temperature injury or cold injury. The common symptoms of chilling injury include browning, pitting or skin blemishes, excessive rotting and failure to ripen.

Relative Humidity

Control of relative humidity in storage facility is essential if maximum storage life of fruits and vegetables is desired. The recommended level of relative humidity for storage of fruits and vegetables are presented in Table 9.3. Relative humidities which are less than optimum result in wilting or shriveling of fruits and vegetables and unnecessary economic losses due to reduction in

Table 9.2: Optimum storage temperature for some fruits and vegetables

Fruit/Vegetable	*Safe Storage Temperature (°C)*
Fruits	
Banana	12-13
Lime	7-9
Grape	0-1
Mango	9-10
Melons	4-5
Oranges	3-4
Papaya	7-9
Pomegranate	4-5
Vegetables	
Brinjal	7
Cucumbers	7
Tomato ripe	7-10
Tomato green	13
Sweet potato	13
Okra	7
Potato	3

product weight. Moisture losses of 3-6 per cent results in marked loss in quality of many vegetables. Relative humidity higher than optimum may cause excessive growth of microorganisms. While using high relative humidity during storage, care must be taken to prevent the growth of surface microorganisms.

Table 9.3: Relative humidities recommended for various foods stored at optimum storage temperature

Relative Humidity	*Fruits and Vegetables*
Less than 85 per cent	Dried fruits, garlic, dry onions
85- 90 per cent	Banana, melons, pineapple, plums, potatoes, tomatoes
90-095 per cent	Banana (green), sweet corn, cucumbers, leafy vegetables, peas

Atmospheric Composition

The atmospheric composition in storage facility is controlled by addition of gases allowing commodity to produce or consume gases or by physically or chemically removing undesirable gases from the storage room. Gases such as carbonmonoxide (CO), carbondioxide, ethylene and nitrogen can be added to a storage facility from bottled supply or produced by on-site-generators. As fruits and vegetables undergo respiration, they consume O_2 and release CO_2. This effect can be successfully used to control the desired concentration of these gases in storage. High concentration of undesirable gases is removed by scrubbing devices. Carbon dioxide can be absorbed in water or lime. Ethylene can be removed by potassium permanganate or UV light. In certain cases, external concentrations of gases are desirable and accumulated gases can be adjusted by ventilation.

Controlled Atmosphere (CA) Storage

The term CA storage implies addition or removal of gases resulting in atmospheric composition substantially different from that of normal air. Thus CO_2, O_2, CO, C_2H_4, acetylene or N_2 may be manipulated to attain various gas combinations. However, by common usage, CA is the term used for increased CO_2, decreased O_2 and high N_2 levels as compared to normal atmosphere CA storage is the most important innovation in fruit and vegetable storage since the introduction of mechanical refrigeration. This method if combined with refrigeration, markedly retards respiratory activity and may delay softening, yellowing, spoilage and other breakdown processes by maintaining an astrosphere with more CO_2 and less O_2 than normal air. The recommended levels of O_2, CO_2 and optimum storage temperatures are presented in Table 9.4.

Modified Atmosphere (MA) Storage

Modified atmosphere (MA) storage is often used interchangeably with CA. MA storage (*e.g.* packaging in film bags) also requires decrease in O_2 and an increase in CO_2 or N_2. There is no attempt to control atmosphere at specific concentration. MA storage differs from CA only in degree of control of atmospheric composition. The modified atmosphere packaging (MAP) of fresh, refrigerated produce to extend shelf-life has already found wide acceptance in many countries of the world. Shorter distribution chains are more conducive for marketing these types of product because closer control of refrigerated temperature is required. In this case, the product is sealed in a pouch atmospherically and provides heads pace for gas released during respiration. The product continues to respire absorbing oxygen and giving off CO_2. In MAP, starting with atmospheric conditions, the product is exposed to larger concentration of oxygen for a considerable time before equilibrium conditions occur. However, if there is initial flushing of head space with low oxygen or high CO_2, there is a significant extension of shelf-life of product. In this case, the film used for packaging is mainly instrumental in retarding moisture loss from product.

Table 9.4: Controlled and modified atmosphere conditions for storage of some fruits vegetables

Fruit/Vegetable	*Temperature °C*	*Atmospheric per cent O_2*	*Composition per cent CO_2*
Fruits			
Apple	3	3	5
Banana	13	2	5
Lime	10	5	7
Mango	13	5	5
Grapes	0	5	2-5
Papaya	13	1-2	0
Pineapple	7	2-5	0
Strawberries	0	4-10	0-20
Vegetables			
Cabbage	0	3-5	5-7
Carrot	0-5	None	None
Cucumber	8-12	3-5	0
Okra	10	3-5	0
Peas	0	5-10	5-7
Tomato	13	3-5	0

Hypobaric Storage

It is a type of CA storage with emphasis on reducing the pressure exerted on storage material. This method not only reduces O_2 concentrations but also increases the diffusion of ethylene (C_2H_4) by evacuation from the tissues of fruits consequently extending storage life.

Cooling and Storage

Pre-cooling, refrigeration, proper relative humidity and optimal atmospheric composition in storage facilities and packages are essential to reducing postharvest losses of commodities that are destined to reach the consumer in fresh condition (Table 9.5).

Table 9.5: Recommended pre-cooling methods and storage conditions for various fruits and vegetables

Produce	*Pre-cooling*	*Storage Conditions*
Apples	RC,FA, HC	0–5°C, 1% -3% O_2, 1-5% CO_2
Asparagus	HC, PI	0-2°C, 95-100% RH
Apricots	RC, FA	0-5°C, 95% RH, 2-3% O_2, 2-3% CO_2
Artichokes	HC, FA, PI	0-5°C, 90-95% RH, 2-3% O_2, 2-3% CO_2
Beans	RC, FA, HC	8°C, 2-3% O_2, 4-7% CO_2
Beets	RC	0-4 °C, 95% RH
Black berries	FA, RC	-0.5- 0°C, 90-95% RH
Blue berry	FA	1°C (3-4 °C), 90% RH
Broccoli	FA, HC, PI, LI	0-5°C, 90-95% RH, 1-3% O_2, 5-10% CO_2
Brussels sprouts	FA, HC, PI	0°C, 92-100% RH
Cabbage	RC, FA	0°C, 92% RH
Cantaloupes	HC, FA, PI	2-5°C, 95% RH
Cauliflower	HC, VC	0°C, 95-98% RH
Carrots	RC, PI	0-2°C, 95% RH
Chinese cabbage	RC, FA, HC	0°C, 95-100% RH
Celery	FA, HC, VC	0-5 °C, 90-95% RH, 2-4% O_2, 3-5% CO_2
Cucumbers	RC, FA	10-13°C, 50-55% RH
Eggplant	RC, FA	8-12°C, 90-95% RH
Figs	RC, FA, HC	0-5°C, 5-10% O_2, 15-20% CO_2
Garlic	RC	0°C
Grapes	FA	-1 to 0°C, 85% RH
Kiwifruit	FA, RC, HC	-5 to 0°C, 90-95% RH, 1-2% O_2, 3-5% CO_2
Leeks	HC, PI	0°C, 95-100 RH
Lettuce	HC, PI, VC	0°C, 95% RH
Mushrooms	FA, VC	0-5°C, Normal O_2, 10-25% CO_2
Nectarines	FA, HC	-5 to 0°C, 90-95% RH
Okra	RC, FA	7-12°C, 90-95% RH, Normal O_2, 4-10% CO_2
Onions	No Pre-cooling	0°C, 75% RH
Peaches	FA, HC	-1 to 0°C, 85% RH
Pears	FA, RC, HC	-1.5 to -0.5°C, 90-95% RH
Peas green	FA, HC	0°C, 95-98% RH
Peas southern	FA, HC	4-5°C, 95% RH
Peppers chilli	RC, FA, VC	0-10°C, 32-50% RH
Peppers, sweet	RC, FA, VC	7-13°C, 45-55% RH
Plums	FA, HC	-0.5 to 0°C, 90-95% RH

Contd...

Table 9.5–*Contd...*

Produce	*Pre-cooling*	*Storage Conditions*
Potatoes	RC, FA	3-10°C, 90% RH
Pumpkins	No Pre-cooling	10-13°C, 70% RH
Radish	PI	0°C, 90-95% RH, 1-2% O_2, 2-3% CO_2
Raspberries	FA	0-0.5°C, 90-95% RH
Rutabagas	RC	0°C, 98-100% RH
Spinach	HC, VC, PI	0°C, 95-100% RH
Squash, summer	RC, FA	5-10°C, 95% RH
Squash, winter	No Pre-cooling	10°C, 50-70% RH
Straw berries	RC, FA	0°C, 95% RH, 5-10% O_2, 15-20% CO_2
Sweet cherry	RC, FA, HC	0-5°C, 3-10% O_2, 10-15% CO_2
Sweet corn	HC, VC, LI	0°C, 95% RH
Sweet potatoes	No Pre-cooling	10-15°C, 85% RH
Tamarillos	RC, FA	3-4°C, 85-95% RH
Tomatoes	RC, FA	12-20°C, 3-5% O_2, 0-3% CO_2
Turnip	RC, HC, VC, PI	0°C, 95% RH
Watermelons	No Pre-cooling	4-10°C, 80-85 per cent RH

RC: Room cooling; FA: Forced-air cooling; HC: Hydro-cooling; VC: Vacuum cooling; PI: Package icing; LI: Liquid icing.

The storage of produce has two main objectives:

1. To provide short-term storage to balance the daily fluctuations of supply and demand.
2. To provide long-term storage to extend the marketing season.

Background for Cooling and Storage Condition

Temperature

1. Ten degree reduction in temperature reduce respiration rate by a factor of 2 to 4.
2. Reduce pathogens and insects attacks by low temperature.
3. Low temperature presence the quality of fresh fruits and vegetables and extending their storage lives.
4. Some cases lower temperatures may cause chilling injury.
5. The temperature measurement units are best placed about 15m off the floor for easy reading.
6. Sufficient air circulation to keep the product at a uniform temperature and to prevent condensation is also important.
7. Flow rates of 0.001 to 0.002 $m^3.S^{-1}$. of air per 1000 kg of product are generally adequate.

8. The timing, degree, and type of cooling that can be used depend on the commodity to be stored and its end use since there are important differences in composition and physiological characteristics among the fruit and vegetables species used as foods.
9. Many commodities are susceptible to chilling injury (e.g., banana, cranberry, cucumber, green pepper, and tomato), whereas others are sensitive to high CO_2 levels (pear, lettuce).
10. Chilling injury can result in several sources of postharvest losses, including; surface lesions, water soaking of tissues, internal discolouration, break down tissue, failure to ripen normally, accelerated senescence, greater susceptibility to decay and compositional changes.
11. Tropical and subtropical fruits are more sensitive than temperature fruits to the chilling injury.

Types of Cooling and Refrigeration

1. Room cooling
2. Mechanical cooling
3. Evaporation cooling
4. Carbon dioxide cooling
5. Alternative methods of cooling
   (*a*) Night time cooling
   (*b*) Well water cooling
   (*c*) High altitude cooling
   (*d*) Underground cooling
   (*e*) Thermoelectric cooling

Calculating Refrigeration Requirements

The amount of refrigeration that is required for cooling or storage is often expressed in kilowatts or tons of refrigeration. By definition, a ton of refrigeration is the amount of heat absorbed in 24 hr by a ton of ice melting at 0°C. One ton of refrigeration is equivalent to 12660 KJ. h^{-1} (12000 Btu. h^{-1}). The gas concentrations of ambient air are 78.08 per cent N_2, 20.95 per cent O_2, and 0.03 per cent CO_2.

Modification of atmospheric composition: i. Structural design of the storage room, and ii. controlled atmosphere storage.

Oxygen Control Systems

Gas separator systems may also use for O_2 control. The different gas separator systems are as follows:

1. Pressure swing absorption (PSA): Systems are used to generate a stream of air that is very high in N_2 and low in O_2. PSA systems can provide rates of 105 to 385 M^3. h^{-1} at 98 per cent purity with compressors ranging from 30 to 112 KW respectively.

2. Hollow fiber membrane separators (HFMS): It works on the principle that some gases can diffuse through membranes at higher rates than others. In the case of air CO_2 and O_2 have much higher permeation rates than N_2.
3. High-temperature ammonia cracking (HTAC).

Carbon Dioxide Control System

There are five commercially available scrubbing systems for the removal of excess carbon dioxide; these are as follows:

1. Caustic soda (NaOH)
2. Hydrated lime [Ca $(OH)_2$]
3. Water
4. Activated charcoal and
5. Molecular sieves.

Ethylene Control Systems

Ethylene (C_2H_4) induces ripening in many fruits and can also cause some physiological disorders in vegetables. The amount of ethylene produced by the commodity can be reduced by decreasing the surrounding O_2 and increasing CO_2 level.

Modified Atmosphere Storage

Modified atmosphere differs from controlled atmosphere storage in that the atmospheric composition is not actively controlled. Most modified atmosphere (MA) systems use semipermeable membranes to regulate gas exchange between the MA and the ambient air. The composition of the air changes as a result of the respiratory action of the produce and the permeability characteristics of the membrane used (Table 9.6).

Table 9.6: Difference between modified and controlled atmosphere storage

Sl.No.	*Modified Atmosphere Storage (MA)*	*Controlled Atmosphere Storage (CA)*
1.	In MA atmospheric composition is not actively controlled.	In CA atmospheric composition is actively controlled.
2.	Semipermeable membranes are used to regulate gas exchange between the MA and the ambient air *e.g.*, Silicone membrane.	Membranes are not used.
3.	MA does not achieve the same degree of atmospheric control as does CA approach.	CA achieves the atmospheric control.
4.	It is less expensive.	It is more expensive.
5.	Membrane can be made of polymeric films or wax or may be edible coatings of individual fruits.	It cannot be prepared from polymeric films or wax or any coating material.
6.	MA is better approach for short term storage.	CA is better approach for long term storage.

Contd...

Table 9.6–*Contd...*

Sl.No.	*Modified Atmosphere Storage (MA)*	*Controlled Atmosphere Storage (CA)*
7.	Small quantities of produce used in MA storage.	Large quantities of produce used in CA storage.
8.	MA is often used in association with packaging.	CA is not used in association with packaging.
9.	The time required for the gas composition to stabilize to quasi steady state is an important factor in MA storage.	Not necessary to stabilize gas composition in CA storage.
10.	MA widely used in European countries.	Use other than European countries.
11.	Membrane used in MA increase CO_2 concentration in the storage room.	CA required very low level of O_2 in the storage room.
12.	A pallet MA package system is used successfully for fruits and vegetables.	No such type of pallet in CA.

Packaging Materials

Properties of the packing materials are given in Table 9.7 and recommended modified atmosphere conditions for fruits and vegetables are given in Table 9.8.

Table 9.7: Properties of packaging films

Material	*Properties*
Paper	Strength, rigidity, opacity, printability
Aluminium foil	Negligible permeability to water vapour, gases and odours, grease proof, opacity and brilliant appearance, dimensional stability, dead folding characteristics.
Cellulose film (coated)	Strength, attractive appearance, low permeability to water vapour, gases, odours and greases; printability.
Polythene	Durability, heat-sealability, low permeability to water-vapour, good chemical resistance, good low-temperature performance.
Rubber hydrochloride	Heat-sealability, low permeability to water vapour, gases, odours and greases, chemical resistance.
Cellulose acetate	Strength, rigidity, glossy appearance, printability, dimensional stability.
Vinylidene chloride copolymer	Low permeability to water vapour, gases, odours, and greases, chemical resistance, heat sealability.
Polyvinyl chloride	Resistance to chemicals, oil and greases; heat sealability.
Polyethylene terephtalate	Strength, durability, dimensional stability, low permeability to gases odours and greases.

Requirements and Functions of Food Packaging Materials

1. They must be non toxic and compatible with the specific foods.
2. Sanitary protection
3. Moisture and fat protection

4. Gas and odour protection
5. Light protection
6. Resistance to impact
7. Transparency
8. Tamper proofness
9. Ease of opening
10. Pouring features
11. Reseal features
12. Ease of disposal
13. Size, shape, weight limitations
14. Appearance, printability
15. Low cost
16. Special features etc.

Table 9.8: Recommended modified atmosphere conditions for fruits and vegetables

Commodity	Storage Temperature (°C)	Atmosphere (per cent)	
		O_2	CO_2
Apple	0–5	2–3	1–2
Apricot	0–5	2–3	1–2
Avocado	5–13	2–5	3 – 10
Banana	12–15	2–5	2 – 5
Cherry (sweet)	0–5	3–10	10–12
Grapefruit	10–15	3–10	5–10
Kiwifruit	0–5	2	5
Mango	10–15	5	5
Papaya	10–15	5	10
Peach	0–5	1–2	5
Pear	0–5	2–3	0–1
Pineapple	10–15	5	10
Strawberry	0–5	10	15–20
Asparagus	0–5	20	5–10
Beans, snap	5–10	2–3	5–10
Broccoli	0–5	1–2	5–10
Brussels sprouts	0–5	1–2	5–7
Cabbage	0–5	3–5	5–7
Cantaloupe	3–7	3–5	10–15
Cauliflower	0–5	2–5	2–5

Contd...

Table 9.8–*Contd...*

Commodity	*Storage Temperature (°C)*	*Atmosphere (per cent)*	
		O_2	CO_2
Corn, sweet	0–5	2–4	10–20
Cucumber	8–12	3–5	0
Honeydew melon	10–12	3–5	0
Lettuce	0–5	2–5	0
Mushrooms	0–5	Air	10–15
Bell peppers	8–12	3–5	0
Spinach	0–5	Air	10–12
Tomatoes (mature)	12–20	3–5	0
Tomatoes (immature)	8–12	3–5	0

10

Canning

Introduction

Canning as a means of food preservation is one of the important developments as far as the annals of history of food processing are concerned. It was first developed for inhibiting microbial spoilage of a food a little over 150 years ago by a Frenchman, Nicholas Appert. Canning is based on the principle of destroying microorganisms by heat and then preventing recontamination from outside.

Containers

Containers used in commercial canning include enamel coated steel, tin-coated steel, and glass. Plastic containers resistant to high temperatures are being developed and their use on a commercial scale is steadily increasing.

Metal Containers

Advances in electroplating technique have made it possible to place a very thin, uniform tin coating on steel cans. The presence of tin serves two purposes, protects steel containers, and a small portion of tin leaching out from the can stabilizes the colour of canned products. Products- that are corrosive to tin are packed in containers having thin layer of a variety of lacquers or coatings. Aluminium containers are also commonly used and most of these containers are made with lids equipped with pull-out device.

Acid Resistant Cans

Tin cans are made of steel plate of low carbon content, lightly coated inner side of the tin metal of thickness of about 2.5 urns. The acid resistant lacquer is ordinary gold coloured enamel and the cans when treated with it are called R-enamel cans. Acid fruits of which the colouring matter is insoluble in water *e.g.* peach, pineapple, apricot, grape are packed in plain cans and acid fruits of which the colouring matter is soluble in water *e.g.* raspberry, strawberry, red plum and coloured grapes are packed in lacquered cans.

Sulphur Resistant Cans

These cans are also of a golden colour and are called C-enamel or SR cans. These cans are used for non-acid products (e.g. peas, corn, limabeans, red kidney beans).

Glass Containers

There has been a significant improvement both in the quality of material and handling of glass containers in canning industry. Superior glass containers equipped with improved 'twist-off' lids are now available for commercial canning operation.

Equipments

The equipments used for canning has changed considerably. The recently introduced sterilization equipments used in the canning operation have been developed (1) to overcome the problems of inefficient heat transfer into the can, and (2) to change sterilization phase of canning from a traditionally small scale batch method to a continuous process. The most commonly used sterilization equipments are rotary cooker, hydrostatic cooker, and hot air and gas flame sterilization equipments.

Process

Steps in Canning

Several fruits and vegetables are canned and preserved for long time. Various steps involved in canning process are outlined in Figure 10.1.

Washing

Washing of fruits and vegetables in water has numerous advantages. Water acts as coolant and helps in removing field heat from the harvested produce. Washing reduces the load of spoilage bacteria that is on the product thus increasing effectiveness of sterilization process. In addition the quality and appearance of the produce are improved. Washing is usually accomplished in special equipment designed to direct high pressure sprays effectively so as to accomplish the desired effect.

Sorting/Grading

Damaged and inferior products are removed and if required the produce is size sorted. Size sorting is usually done mechanically by passing the produce over screens containing different size holes or slits. Defect removal is largely done by passing the produce over an inspection belt where trained personnel look for defects and undesired material. Some products are sorted by their difference in optical properties.

Blanching

Blanching consists of heat treatment usually given by using steam or hot water prior to canning or dehydration. This is done to inactivate enzymes and remove air from fruit or vegetable tissues. This can be achieved by using hot air and also infrared

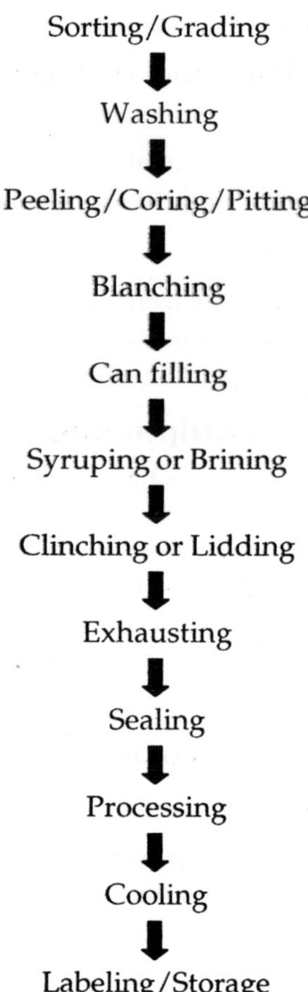

Figure 10.1: Steps in canning process

and microwave radiation. The degree of enzyme inactivation indicates the effectiveness of blanching treatment. The activity of polyphenol oxidase is followed in fruits and that of catalase in cabbage and peroxidase in other vegetables. Commercially, continuous blanchers are favoured over batch type which involves 2-10 min exposure to live steam series. Blanching in hot water is advocated in some countries with solid content of water maintained at equilibrium level to minimize leaching losses. Individual Quick Blanch (IQB) process is a method that produces less effluent than conventional steam blanching. This process is comprised of a heating step and holding step. Heating is done by a condensing steam with raw produce one layer deep and heat is applied to raise the temperature (87°C) of the produce to inactivate enzymes. In the holding step, after the pieces leave heating section, pieces achieve a uniform temperature and are allowed enough time to inactivate enzymes and to yield the desired texture.

Peeling/Coring/Slicing

Some fruits such as mango, apples, pears require peeling before canning. Several methods are used for peeling such as hot water, hot sodium hydroxide and mechanical or abrasive peelers.

Freezing/Exhausting/Sealing

The containers used for canning of fruits and vegetables are usually limited to metal cans and glass bottles. After the product is filled into container, it is usually passed through steam before sealing to reduce oxygen in the can head space. After filling the sterilized cans, hot sugar syrup (for fruits) or brine solution (for vegetables) is poured in the can. The purpose of adding syrup or brine is to improve the taste of canned product and to fill up the interspace between the fruits or vegetables in the can. The syrup or brine is added to can at 79-82°C leaving suitable head space in the can. Then cans are partially sealed by clinching process. Then containers are exhausted either by heat treatment or by mechanical means. Cans are passed through a tank of hot water at about 82-87°C for 15-25 min or on a moving belt through a covered steam box. After exhausting, the cans are sealed by special closing machines known as double seamers.

Retorting or Processing/Cooling

The sealed cans or glass containers are processed by immersing in hot water or boiling water or exposed to steam under pressure. The exposure needed to destroy spoilage microorganisms depends on the type of product, container size and type of retort. In general, fruit cans are processed at temperature of 100°e for 15-20 min whereas vegetables (except acidic) are processed at about 115-121°C for 20-30 min. Most fruit and vegetable processing is done in rotary agitating retorts. After the cans are heated sufficiently, they are then quickly immersed in cool water. These are labeled and stored in cool place.

Presently, continuous retorts, which involves feeding the cans through a lock similar to that used in torpedo tube in submarine, then rotating them through a heating chamber under pressure and then cooling continuously in a separate, continuous cold water cooler or spray cooler. Hydrostatic cooker under pressure is most commonly used in canning operation. Static retorts are used for sterilization of canned foods with a pH lower than 4.5. These static retorts are being replaced with agitating retorts involving movement of cans and food mechanically thereby reducing process time considerably.

Fruits

Mango

Fresh, firm, fully developed, evenly matured and sound fruits are selected. The fruits are thoroughly washed in water, preferably in running water to remove dust, spray residues etc. The fruits are handled carefully to avoid bruising. Sometimes gentle rubbing by hand may be necessary to remove adhering dust during washing. The ripe fruits are peeled with stainless steel knives. The peeled fruit is sliced longitudinally so as to obtain two symmetrical slices. The side slices can also be removed and packed as such. The slices are placed in plain cans of appropriate size and sugar syrup of 35-40 °Brix containing 0.25 per cent citric acid. The filled cans are exhausted before sealing to drive out air from the fruit tissues and to prevent discolouration as well as loss of flavour during storage. The filled cans are exhausted to a centre temperature of 76.7-79.5°C for 10 min. While still hot, cans are removed one by one from water and clean lid is placed on the filled can and sealed it by means of a can sealing machine. The sealed cans are placed in a boiling water bath to a centre temperature of 90°C (20-25 min for A 2 ½ tall cans, 15-20 min for Butter and No. 1 tall cans). The cans are cooled quickly preferably in running cold water, wiped dry, labeled and stored in cool and dry place.

Pineapple

Giank Kew and Queen and Kew are important varieties of pineapple used for canning. The pineapple fruits are graded, cored and sliced by using slicing machines and outer peel is removed by using curved knife. For the peeled slice, a circular ring is punched out using stainless steel puncher. The core is then removed with a corer. The rings thus prepared are graded and dipped in water to remove adhering broken and then filled into cans. Sometime broken rings are cut into segments of suitable size and canned as tit-bits of chunks. Both slices and tit-bits are highly popular among the consumers. Broken slices and smaller segments can be passed through screw-type crusher to get pineapple pulp known as crush. From the pulp, pineapple juice of high quality can be extracted by passing it through a hydraulic basket-type press or by centrifuging in hydro-extractor which is a basket-type centrifuge. After blending the juice to proper °Brix-acid ratio, it can be canned.

Vegetables

Peas

Peas for canning should be uniformly ripe and should retain their green colour even after processing. They should possess good texture and flavour. Large size peas are generally preferred for canning. Pods are shelled by means of a pea-podding machine. They are generally graded by size using sieves with mesh sizes ranging from 0.7 to 1.0 cm or more. Sometimes, they are graded on the basis of their density by floating them in a brine of 1.04–1.07 specific gravity. The graded peas are blanched in boiling water for 2–5 min. Steam can also be used for blanching the peas. The blanched peas are filled into cans manually or by means of a filling machine. Brine containing 1.1 kg of sugar and about 0.67 kg of common salt per 45 liters of water is used as the covering liquid. Sometimes, an edible green colour and mint flavour are also added to the brine. After the cans are filled with the peas, they are passed through the exhaust box, closed (sealed) and processed for 40–45 min (A 2 ½ tall cans) in a retort at 10 psi and cooled.

Okra

Tender green okras of uniform size are selected for canning as whole or slices. These are blanched for 1- 2 min in boiling water or steam and then cooled and filled in the can with 2 per cent brine. Slices can be canned in viscous tomato sauce with suitable spicing.

Spoilage of Canned Foods

Canned foods are subjected to spoilage in storage for various reasons. Spoilage occurs due to physical or chemical changes or by microorganisms.

Symptoms of Spoilage

1. *Swelling or bulging of cans:* The bulge is due to accumulation of CO_2 and other gases inside the can as a result of decomposition of the contents caused by microorganisms. Food gives offensive and sour odour and is discoloured. It may contain toxin due to presence of *Clostridium botulinum.*
2. *Hydrogen swells:* Accumulation of hydrogen gas due to action of acids on tin plate causing bulging of can. A food remains free from harmful microorganisms and fit for consumption.
3. *Ripper:* A can with a mild positive pressure is called a flipper. Flipper occurs due to over-filling or under-exhausting.
4. *Springer:* Mild swelling at one end may be due to hydrogen swell or insufficient exhausting or over filling of the can. Food remains safe for consumption.
5. *Rat sour:* It occurs mostly in non-acid foods like vegetables by microorganisms without gases. The product gives sour taste and odour. Such type of food is not fit for consumption.
6. *Leaker:* A very small leakage due to faulty seaming and faulty lock seaming. Pinholes result due to corrosion or rusting of the can.
7. *Bursting of cans:* Food can may burst due to excess pressure of gas inside produced due to decomposition of food by microorganisms or by hydrogen gas formed due to chemical reaction.
8. *Breather:* Tiny holes through which air can pass but not microorganisms. The food remains fit for consumption.

Discolouration of Products

1. *Chemical reactions:* Brown discolouration of fruit products is generally known as non-enzymatic browning. The colour changes may be caused by reaction between (1) nitrogenous matter and sugars, (2) nitrogenous matter and organic acids, (3) sugar and organic acids, (4) organic acids among themselves.
2. *Metallic contamination:* Discolouration of fruit product due to metallic contamination is generally caused by iron and copper salts.
   (*a*) *Ferric tannate:* Tannins from fruit react with iron tin plate to form ferric tannate which is black in colour.
   (*b*) *Iron sulphide:* Sulphurdioxide may react with hydrogen formed by fruit acid acting on the tin plate and forms H_2SO_3 which reacts with iron to form black iron sulphide.

(c) *Copper sulphide:* Copper of canning machinery gets oxidized to copper oxides and reacts with hydrogen sulphide formed in the can and the copper sulphide (black colour) is formed in the reaction. Copper sulphide discolours the product.

3 *Hydrogen:* Fruit acids react with tin of the can to produce hydrogen. When lacquering is not proper, hydrogen formed in can bleach red or purple pigments of fruits and discolour the fruit products.

7. *Corrosion due to oxygen:* Filling or exhausting of the cans 5-6 min at 82-88°C gives better results. Canned fruit and fruit products must be stored at low temperature. The products are quickly cooled to about 38°C after processing immediately.

Microbes/Spoilage

1. *Non-poisonous spoilage:* It is mainly due to under sterilization. It is caused by variety of yeast. Bacteria seldom occur in canned foods. Spoilage is evident by bulging of cans due to accumulation of carbondioxide.
2. *Poisonous spoilage:* This type of spoilage occurs mostly by thermophilic bacteria. In this case, gas formation is very rare and the cans retain its normal shape. Non-acid fruits are more susceptible to flat-souring. It usually occurs in pasty and viscous materials. If cans are not cooled properly, contents of can remain favourable for multiplication of bacteria. Thermophilic bacteria are more difficult to kill than the facultative ones. Some thermopiles produce hydrogen and other produce hydrogen sulphide.
3. *Spoilage by fungi: Byssochlamys fulva* fungi are responsible for spoilage of canned food. Remedy measure is adequate sterilization of the canned product.

Aseptic Canning Process

Aseptic canning reduces the problems associated with slow heat penetration in containers by an integrated approach in which product is sterilized and cooled separately from the container and then sterilized and cooled product is filled in presterilized containers and sealed in sterilized environment. The main prerequisites for aseptic canning are: 1. efficient and versatile equipments for sterilization and subsequent cooling of products both rapidly and uniformly 2. efficient method of sterilizing containers and lids both simultaneously and rapidly 3. facilities for filling and closing the containers under sterilized environment.

Continuous heat exchangers and efficient heat transfer into and throughout the product is necessary for achieving proper sterilization. This is accomplished by applying pressure and constantly delivering the product at a pressure. Wetting of heat transfer wall by a thin layer of product acts as the major deterent for efficient heat transfer from metal wall into the product. Thin film is difficult to remove unless it is continuously removed mechanically. The tenacity and resistance of this film for removal depends on the product (ingredients, viscosity, solids, and oil content) and the operating pressure. Similar problem is encountered in cooling of product also

Heat exchangers are used to achieve efficient heat transfer without any deleterious effects on the product.

Both mechanically assisted (scraper blades) and unassisted (tubular coil, or plate types) scrape-surface heat exchangers are commonly used to constantly remove thin product film. In the case of mechanically assisted scrape surface heat exchangers the spinning shaft with scraper blades constantly and rapidly cleans the heat transfer surface 500–1000 times a min thereby preventing scorching or over-cooking of products. Mechanically unassisted heat exchangers relay on product viscosity to create the turbulence required to remove the product film. The synchronization of product flow, agitation and maintenance of clean heating and cooling wall assures better quality of the product.

The advantage of aseptic canning process over the conventional one is based on the kinetic difference between thermal destruction of nutrients and thermal inactivation of microorganisms. Microorganisms are destroyed much faster than vitamins in high temperature short-time (HTST) process. Aseptic canning system carries out all the canning operations simultaneously in a closed interconnected system. Several operations are synchronized mechanically to move the product, containers, covers, and finished canned product continuously without interruption. Four types of heat exchangers (steam injection, swept-surface, tubular, or plate) can be used in this system. An increasing number of products such as tomato paste, fruit concentrates, and banana puree are packed by aseptic canning process.

Storage of Canned Foods

Canned fruits and vegetables retain their good flavour and nutrient properties for an extended time if they are held at moderately low temperatures. The effect of high temperature storage varies with the product and seems to be related to its pH. The ascorbic acid in canned orange juice, a typical acid product is lost faster during high temperature storage than the ascorbic acid in green beans, a high pH product. Storage duration and temperature also influence pigment stability.

11

Juices and their Products

Introduction

Fruit juices have an important place in human diet being rich in essential minerals, vitamins and other nutrients. Besides, they are delicious and have universal appeal. The nutritive value of fruit beverages is far greater than that of synthetic product. Fruit juices have their best taste, aroma and colour when they are freshly extracted or expressed. But all fruits are not suitable for making fruit juice, either because of difficulties in the extraction of juice or because the extracted juice is of poor quality. The best juice is, therefore, extracted from freshly picked, sound and suitable fruits, when they are at the optimum stage of ripening.

Equipment for Juice Extraction

Cone-type Juicer

Orange, *mosambi,* grapefruit, lemon and lime are cut by a special machine in which the fruit is placed in a conical cap in a wheel which brings the fruit against a stationary or revolving knife. The burrs (or rose cones) are generally made of stainless steel or non odorous hard wood. In this machine, excessive tearing of the tissues can be avoided.

Screw-type Juicer

Pieces or segments of the fruit are fed through a hopper at one end of a feeding screw revolving inside a conical jacket, which is perforated in sections or throughout. The diameter of the perforations depends upon the type of fruit. The juice and pomace comes out at the end of the conical jacket. Screw-type juice extractors are useful in the case of a tomatoes, grapes, pineapple, etc.

Plunger-type Press

Roller presses made of hard granite or wood are specially designed to extract juice from sour limes (Kagzi lime). Fruits like apples, grapes, berries, etc. are crushed in grater or crusher and the juice is extracted by means of a hydraulic press.

Apple Grater

The entire fruit is crushed and pressed. Apples are crushed and grated between the cylinder and the corrugated plate and then juice is extracted.

Crusher

A grape crusher consists of two fluted or grooved rollers made of wood or metal. The fruit fed to the hopper at the top falls between the rollers and gets crushed. Strawberries and some other berry fruits which contain gums are first heated and then crushed. Tomatoes can also be crushed in these crushers.

Basket Press

Basket press is operated manually or by hydraulic pressure. It has been found useful in the case of apple, grape, pomegranate, phalsa etc.

Rack and Cloth Press

In this type of press, the crushed fruit is spread as a layer in a coarse woven cloth of cotton or nylon and folded into a square suited to the side of the platform of the press. The built-up layers are subjected to hydraulic pressure by means of a pump. The juice is pressed out gradually due to increasing pressure in the pile. The released juice is collected at the base of the pile.

Extraction of Juice

Preparation of Fruits and Processing

The freshly harvested and well ripened fruits are used for juice extraction. The fruits are washed thoroughly with water and in some cases scrubbed also while washing to remove any adhering dust and other extraneous matter. Residues of sprays of arsenic and lead if present are removed with dilute HCl. All mouldy and decayed parts are removed as in the case of fruits prepared for canning.

The method of juice extraction differs with the structure and composition of the fruit. There are two types of extractions. In one case, the fruits are crushed and pressed continuously in one operation whereas in the other, the fruits are crushed or cut into small pieces or comminuted in a mill and these are subsequently pressed in a suitable press. Depending upon the type of fruit, suitable machine as described in earlier section should be used for extraction of juice.

Straining and Deaeration

Several types of equipment are used for straining juices. The juice flows out through the sieve into the jacket and is collected at the outlet below while the coarse residue passes out at the lower end of the sieve. The juice is allowed to stand to settle. The finely suspended particles in the juice are removed with filter press. Freshly extracted and screened juices contain appreciable quantity of oxygen which is removed (deaeration) before packing.

Clarification

Various fining agents used in clarification of juice includes enzymes proteolytic or pectin decomposing (pectinol or pectinase), hydrolytic and starch liquefying, chemical agents such as gelatin, albumen, casein and clays such as kaolin and bentonite. The clarification of juice can be achieved by precipitation by freezing at -10 to -170C. In some juices, colloidal material usually coagulates when the juice is heated and settles down readily. In this case, juice is heated to about 82°C for a min or less and then cooled down rapidly. Then juice is passed through filter press.

Fruit and Vegetable Juices

Apple Juice

The process for extraction of apple juice is outlined in Figure 11.1. Freshly prepared apple juice appears slightly cloudy than brilliantly clear. The juice clarification is achieved by using gelatin or pectinol (pectic enzyme mixture). The clear juice is passed through filter using filtered juice and is pasteurized at 80-82°C for 30 seconds in flash pasteurizer.

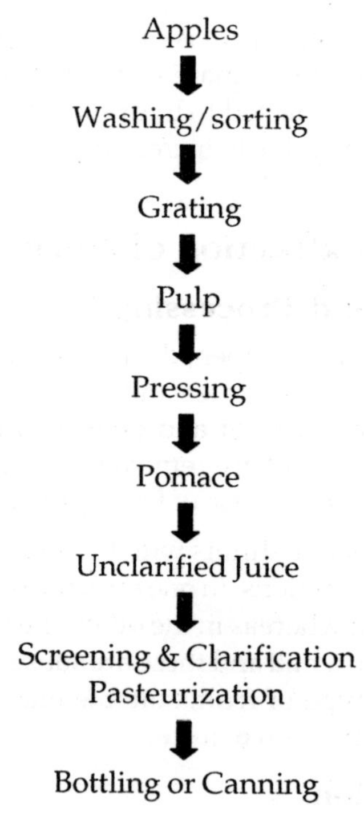

Figure 11.1: Flow diagram for extraction of juice from apple

Grape Juice

In production of juice from coloured varieties, the berries are heated to extract anthocyanins and other soluble substances which give colour to the juice. This treatment is not required in case of white grapes. The crushed grape mixture is heated to above pasteurization temperature and pressed in the same way used for apple. The hot pressed juice contains higher soluble solids, non-sugar solids, acidity, tannins and colouring substances compared to cold pressed juice. The pressed juice is screened, filtered and pasteurized. The pasteurized juice is stored at low temperature from 1-6 months to facilitate precipitation of proteins and separation of tarter. It is achieved by freezing and thawing followed by siphoning of cold juice or by adding calcium acid malate, lactate or phosphate to the juice or pectinol followed by filtration. The clear juice is pasteurized and bottled. The grape juice is occasionally preserved with sodium benzoate.

Lime Juice

Juice is extracted from limes (West Indies or Persian type) mainly for beverage purposes. The whole fruit is fed to mill in which the hopper has a roller having blunt projection. The angular space between the walls of the hopper and the roller is so adjusted that the fruit is pressed, after which it passes through two set of granite stone rollers which press the fruit further and expel the juice. The yield of juice is reported to be about 40 per cent. The peel oil content in juice is reduced by deoiling process when it is extracted by crushing whole fruit in a screw type juice extractor. After deoiling, it is clarified by filtering using a filter aid. The peel oil in contact with the high acidity of juice produces off-flavour due to conversion of terpene into pcymene. Commercially, lime juice is pasteurized at a temperature of 90°C for few seconds.

Mandarin Juice

Mandarins (tangerines) are loose skinned oranges grown in India. The juice contains 10-12.5 per cent T.S.S., 0.37 to 0.75 per cent acidity and a TSS/acid ratio from 17.2 to 28.1. Some varieties develop bitterness on processing which has hindered wide scale processing in India. The yield of juice is about half to that of flesh and juice can induce off-flavour development. The excessive amount of peel oil in the juice causes the taste to be too aggressive, but removal to an optimum level (0.0075 ml/100 g ml juice) by deoiling process gives carrot-like taste.

Orange Juice

Sweet oranges are mainly used for extraction of juice in different parts of world. The best quality juice is produced when °Brix/acid ratio is between 13 and 19. Oranges are first washed with food grade detergent to remove dirt and rinsed with water containing 20-30 ppm chlorine. The juice is then extracted by using rotary juice extractor or citromat juice extractor. The juice yield should be more than 55 per cent or more of the whole fruit weight. Juice extracted with less pressure yields juice containing high colour value and better flavour.

The juice is then screened by using cylindrical perforated screen with rotating paddles or reinforced perforated screen having a rotary stainless steel screw in the centre. During the screening, coarse particles are removed by passing through raw juice. It can be used for preparation of beverages. The screened juice is blended and frozen as single strength juice. This juice has a short storage life of a few days because of loss of cloud and rapid microbial spoilage. By using this juice, chilled orange juice, pasteurized canned orange juice and canned or bottled orange juice for infants can be prepared. Excessive oil in the screened and blended juice is removed by deoiling process by using vacuum evaporator. After deoiling, juice is deaerated which keeps 'the oxidation at low levels in the juice. The deoiled and deaerated juice is pasteurized at 92°C for few seconds and cooled and filled separately in pasteurized bottles.

Tomato Juice

In manufacturing tomato juice and puree and other products, the tomatoes are washed and sorted similar to preparation for canning. The fruits are chopped to 0.4-0.6 inch for crushing prior to juice extraction. Immediately following chopping, they may be subjected to hot-break procedures. In the hot-break method, the chopped or crushed tomatoes are rapidly heated to at least 82°C for 15 seconds to inactivate the pectic enzymes. Heat treatment is usually given in rotary coil tanks followed by a heat exchanger and holding tube to achieve 104°C to retain at least 90 per cent of the potential serum viscosity in the original fresh tomato. The hot-break produces a better-quality juice with respect to cooked tomato flavour and body. A heavier-bodied homogenous juice is obtained because of the inactivation of pectic enzymes and efficient extraction of pectin. In the cold-break procedure, tomatoes are scalded to loosen the skin before chopping. The fruits are chopped or crushed at temperatures less than 66°C and then transferred into a holding tank, where they remain static for periods ranging from few seconds to few min. During this holding period, the pectic enzymes liberated during procedure is claimed to give better coloured, flavoured juice, and a better retention of vitamin C. Quick processing of the extracted juice is necessary to produce high quality juice from cold-break procedure.

Following either hot break or cold break process, the chopped tomatoes are conveyed to a cyclone for juice extraction. About 70-80 per cent of the juice is extracted. The juice is deaerated immediately after extraction to prevent loss of vitamin C, acidified with citric acid to enhance the flavour, salted if necessary and filled in cans or bottles. The fill machines are adjusted to give maximum fill to exclude headspace as much as possible. The cans are closed at about 82-88°C followed by water cooling or the juice is pre-sterilized' at 121°C for about 40 seconds and the hot juice is poured into cans to ensure sterilization of the containers. A minimum closing temperature of 93°C is suggested, after which the cans are held inverted and conveyed for a minimum of 3 min at this temperature prior to water cooling.

Preservation of Juices

Fresh juices are highly attractive in appearance and possess good taste and aroma, but deteriorate rapidly, if kept for some time. The spoilage of juice occurs due to (1) fermentation by mould, yeast and bacteria, (2) enzymatic reactions affecting colour and flavour, (3) chemical reaction affecting taste and aroma of the juice, (4) oxidation of glucosidal material making the juice bitter and (5) contaminants from equipments spoiling the colour and aroma. The juices can be preserved by various methods.

Pasteurization

It is a heating of the juice to 100°C or slightly below for a sufficient time (10-15 min) to kill microorganisms which spoil the juice. Some spores and spores bearing bacteria like *B. subtilis* and *B. mesentericus* can survive after pasteurization; the juice is, therefore, made acidic to avoid the growth of these bacteria. For mould spores, heating at 79°C for 5-10 min is good enough. For destroying pectic enzymes, the juice is heated at 82°C for about 4 min or at 88°C for one min.

Preservation by Chemicals

Sometimes pasteurized squashes and cordials have a cooked flavour. To avoid this, chemical preservatives are used.

Sodium Benzoate

This is a salt of benzoic acid and is used in the preservation of juices and squashes. Chemically, pure sodium benzoate is practically tasteless and odourless. Generally, pH of the fruit juices is maintained at 3.5 to 4.0. Sodium benzoate at 0.06 to 0.1 per cent concentrations is most effective against *Bacillus subtilis,* yeasts and moulds. In European countries esters of parahydroxy benzoic acid like methyl, ethyl and propyl esters are used most effectively. Benzoic acid has been found to cause no deleterious effects when used in small amounts. It is readily eliminated from body primarily after conjugation with glycine to form hippuric acid (benzoyl glycine). This detoxification step precludes accumulation of benzoic acid in the body.

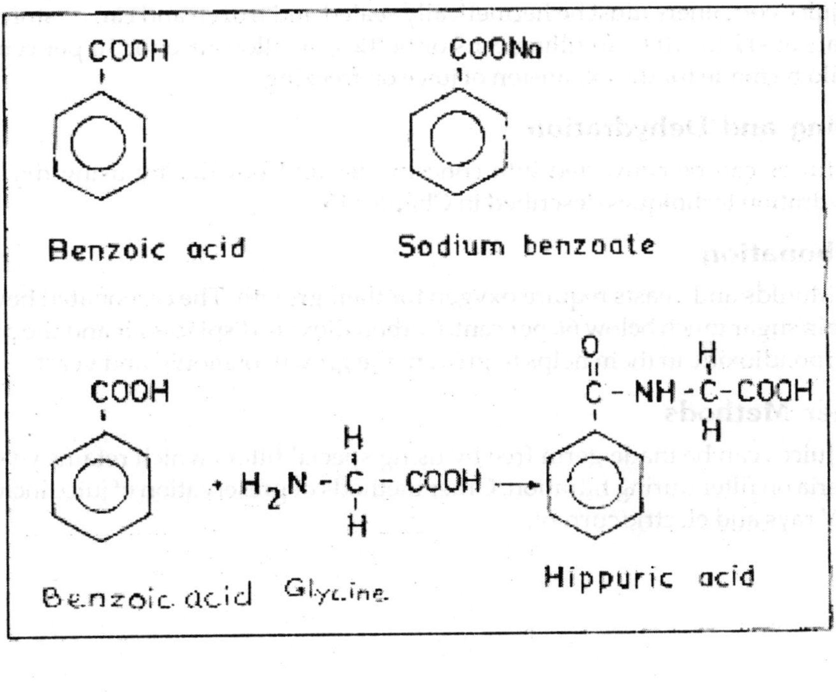

COOH — Benzoic acid

COONa — Sodium benzoate

$C_6H_5COOH + H_2N-CH_2-COOH \rightarrow C_6H_5-CO-NH-CH_2-COOH$

Benzoic acid + Glycine → Hippuric acid

Sulphurdioxide

Potassium metabisulphite $K_2S_2O_5$ (KMS) is decomposed in presence of weak acids like carbonic, citric, tartaric and malic acids which are present in juice and squash. It has been shown that H_2SO_3 molecules prevents the multiplication of yeast and HSO_3^- ions inhibit the growth of *E. coli* at concentration of 10 mg per 100 ml. The growth of yeast is not inhibited by HSO_3^- ions. According to the Indian Fruit Product order, the maximum amount of sulphurdioxide allowed in squashes and cordials is 350 ppm (1). It cannot also be used in tin containers because it creates

$$K_2S_2O_5 + H^+ \rightarrow SO_2$$

$$SO_2 + H_2O \rightarrow H_2SO_3$$

$$HO_2SO_3 \rightarrow H^+ + HSO_3^-$$

pinholes and forms hydrogen sulphide which has a disagreeable smell. It has better preserving action than sodium benzoate against bacteria. It helps to retain colour of beverages for longer time. This being gas helps in preserving the surface layer of the juice. It is highly soluble in juices and squashes.

Sugar

Fruit juices containing 66 per cent or more sugar do not ordinarily ferment. The sugar acts as a preservative by osmosis and not as a true poison for microorganisms.

Freezing

The properly frozen juice retains its freshness, colour and aroma for a long time. It is mostly useful in the case of juices whose flavour is adversely affected by heating. The juice containers must be hermetically sealed and frozen and can be stored up to 2 years at -12 to -10°C. In filling cans or bottles, an allowance for 10 per cent space should be made for the expansion of juice on freezing.

Drying and Dehydration

Juices can be converted into concentrates and powder by using drying and dehydration techniques described in Chapter 15.

Carbonation

Moulds and yeasts require oxygen for their growth. The carbonated beverages contain sugar much below 66 per cent. Carbon dioxide displaces air and the presence of carbondioxide in them helps to prevent the growth of mould and yeast.

Other Methods

Juices can be made germ free by using special filters which retains yeasts and bacteria on filter during filtration. Other methods of preservation of juice include use of UV rays and electric current.

Products

Beverage (RTS)

The dietary value of fruit and vegetable beverages far exceeds that of synthetic products. These beverages are rich sources of minerals, vitamins, and other nutrients. The most common natural beverages are made from mango, citrus, pineapple, apple, grape and guava. The demand for natural beverages has increased significantly during the past few decades. Due to the rapid advancement in juice industry, specialized equipments capable of handling a large quantity of raw materials are presently available in the industry, but simple household juice extractors are still used for small quantity. Pure juice is unfermented and is practically unaltered in its composition during processing. Fruit juice beverage is produced by altering the composition. The beverage normally contains 10-15 per cent juice, 12-15°Srix sugar and 0.2–0.3 per cent acidity (Figure 11.2).

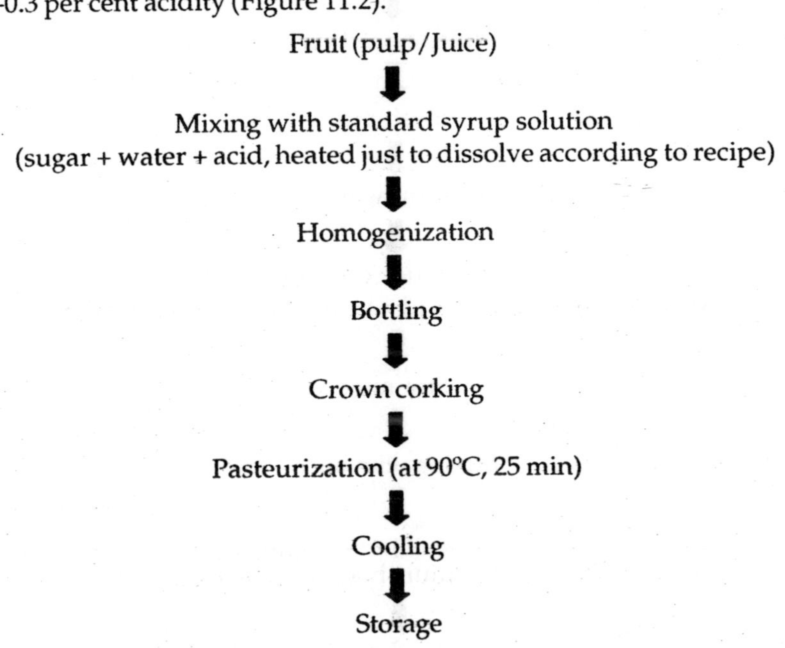

Figure 11.2: Process for preparation of RTS beverages

Squash

Sweet orange, mandarin (loose jacket orange), sour lime, lemon, grapefruit, apple, mango, pomegranate, passionfruit, pineapple, ber, phalsa, tomato, mulberry, guava, jamun, strawberry and litchi are used for preparation of squashes (Figure 11.3).

Extracted juice is filtered through muslin cloth, and seed and other coarse tissues are removed. This pure and clean juice is used for the preparation of squash. The squash from orange juice is prepared using 30 per cent juices. The °brix and acidity are adjusted to 450 and 0.75 per cent, respectively. Potassium metabisulpite (KMS)

610 mg/kg of finished product is used as a preservative. The mixture is strained through coarse muslin cloth and heating is given for few min and then filled in sterilized bottle and stored at cool dry place. For attractiveness and appearance, the essence and edible colour are added as per requirements.

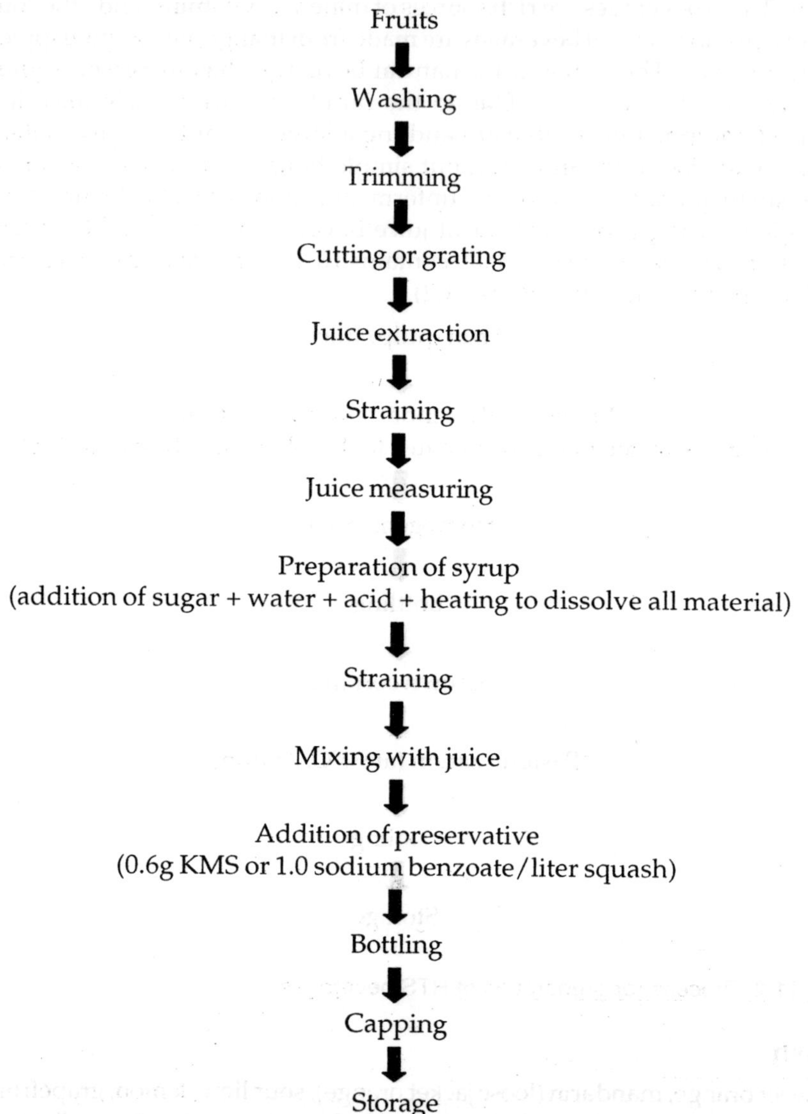

Figure 11.3: Process for preparation of squash from various fruit juices

Nectar

Fruit nectars are basically extended fruit juices containing added sugar (up to 20 per cent) and fruit acid. They are prepared from a variety of fruits. The nectars should contain minimum quantities of added substances. Nectar is obtained by blending the thin pulp of the fruit with sugar and citric acid. The pulp is gradually thinned with water, after which the sugar, citric acid and other ingredients are mixed. The finished product contains 15-20 °Brix and mild acid taste. The nectar is filtered and heated to 85°C. It is filled into plain or lacquered cans and cooled in running water to about 38°C. The cans are then left to cool at atmospheric conditions.

Cordial

Cordials are sparkling, clear, sweetened fruit juices. The cordial contains 25 per cent juice, 30 per cent total solids and not less than 3.5 per cent acidity. For preparation of a clear juice for cordial preparation, raw lime juice is clarified by storing in wooden barrels using 700 ppm sulfurdioxide for 10-12 weeks. The clear juice is siphoned off and filtered using a filter aid. The filtered clear juice is mixed with sugar syrup and refiltered again to get a brilliantly clear cordial. The required quantities of preservatives are added to the final product and then bottled (Figure 11.4).

Concentrate

Fruit and vegetable juices are mostly aqueous solutions and contain low total solids usually in the range of 8-17 per cent. It is expensive to package, store and ship single-strength juices, and in many cases it is desirable to remove a part or all of the water from such juices. The refrigeration, storage, transportation and distribution costs of single-strength fruit juices could be greatly reduced by concentration *i.e.* removing substantial water from the fruit juices. Further, concentration of fruit juices offer several advantages: (1) reduced requirements for storage space by virtue of smaller volumes, (2) reduced refrigeration loads, (3) reduced transportation costs, (4) cheaper packaging, (5) more profitable export marketing via efficient handling, and (6) improvement in product quality, stability and shelf-life.

Juice concentrate can be prepared from either freshly extracted or pasteurized single strength juice or from a stored and pasteurized single-strength juice. In many cases, concentrates are prepared by evaporation and distillation processes. However, evaporation causes heat damage to the flavour of the concentrate. The citrus juice processors use two modern citrus juice evaporators to minimize heat damage and retain quality. They are low temperature evaporation and high temperature short time evaporation. During evaporation, the volatile components are recovered for reincorporation into the concentrate so that the reconstituted juice has the characteristic aroma and flavour of the original juice. Methods of collecting the aromas include absorption and compression of volatile compounds in liquid sealed vacuum pumps, in cold traps, as well as condensation at temperatures around 0°C.

Commercially, three types of concentrate are manufactured; frozen, canned and chemically preserved. Frozen concentrate is sold primarily for direct consumer use. Canned concentrate is generally hot packed. Chemical preservation of concentrate is carried out usually with sulphurdioxide, and such concentrates are mostly used for the production of bottled soft drinks and confectionery.

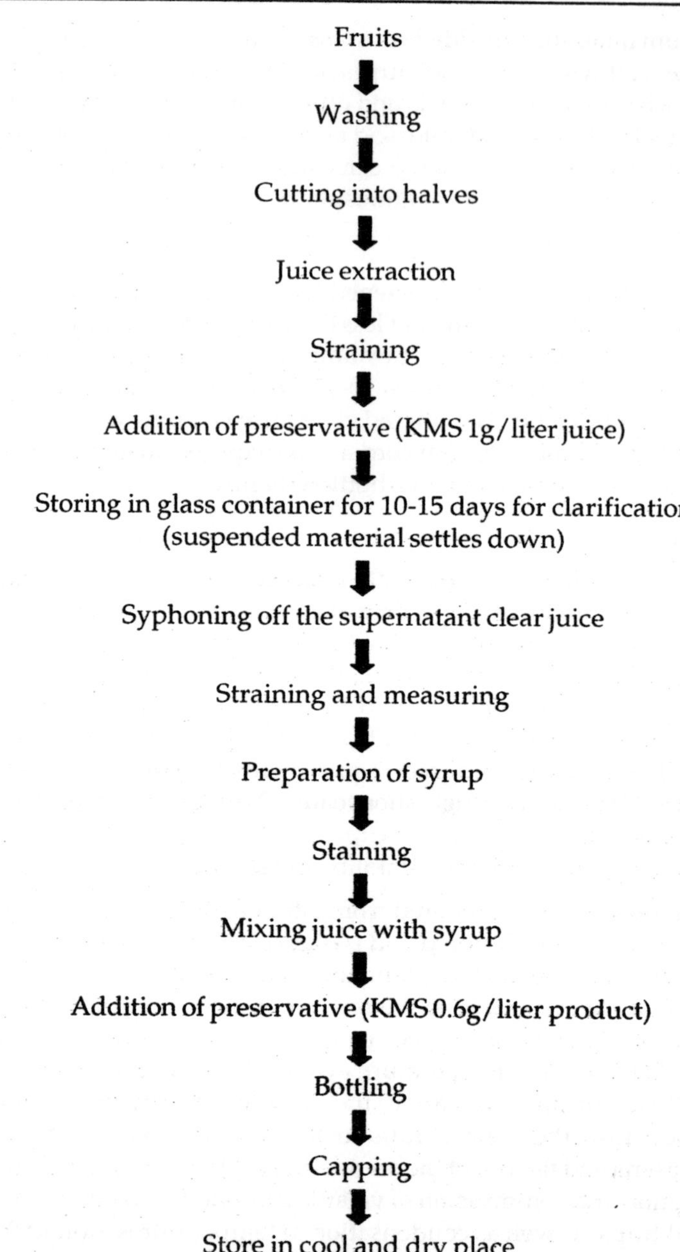

Figure 11.4: Process for preparation of cordial from various fruits

Wine

Wine is prepared from grape and other fruits and is named accordingly, for example, apple wine (or cider); pear wine (perry), berry wine, pineapple wine and

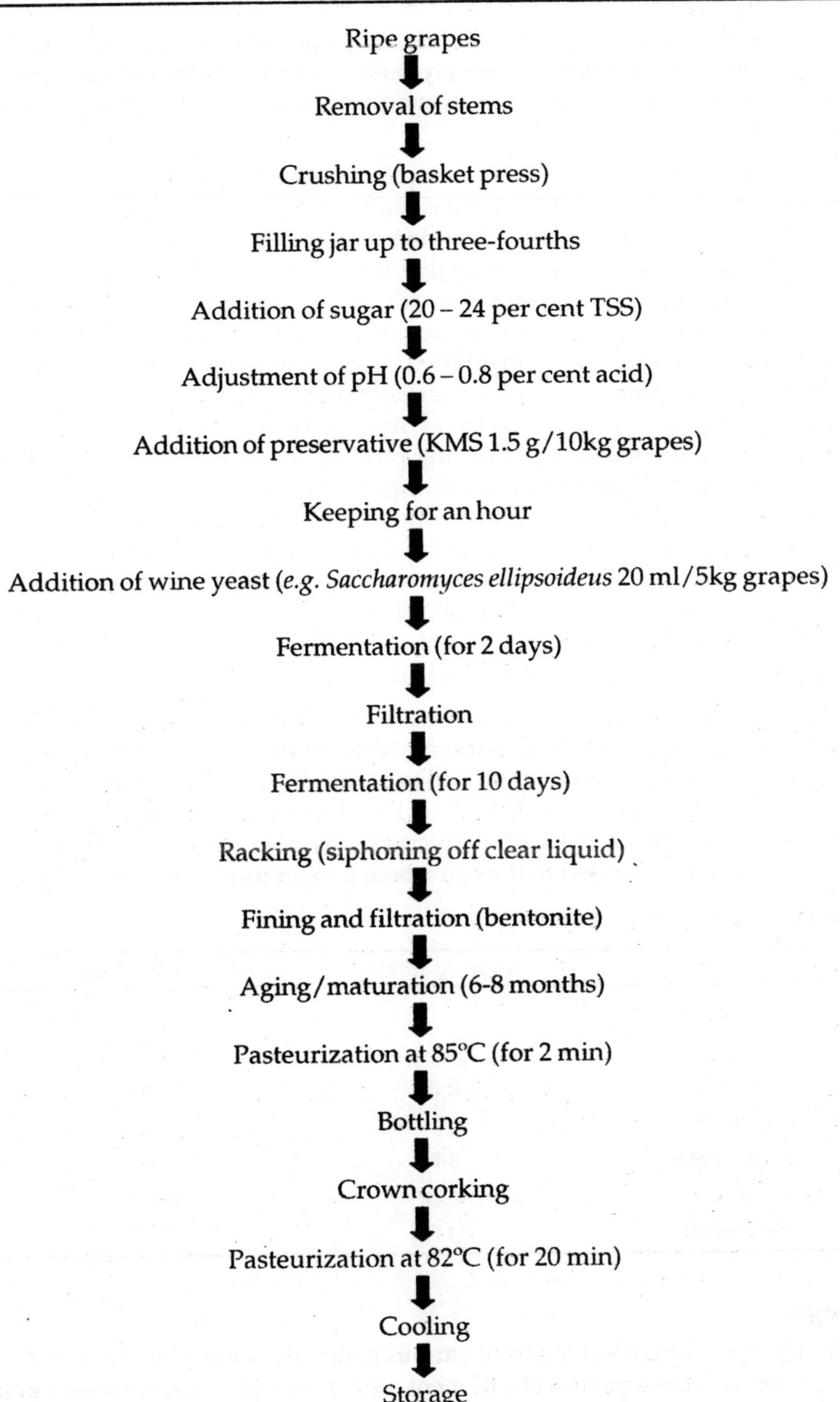

Figure 11.5: Processing of grapes for wine making

apricot wine. For preparation of fruit wine, fully ripened fruits are selected to extract the pulp and then the pulp is used for preparation of juice or direct fermentation. For fermentation of pulp or juice, *Saccharomyces cerevisiae* var *ellipsoideus* yeast is used (Figure 11.5).

The production of wine involves complex biochemical processes in transforming juice into wine. A pure culture of yeast is multiplied prior to adding it to the must in container, barrels or vats. To the culture in bottle, pasteurized juice is added and the yeast multiplies rapidly. This fermenting juice is mixed with a larger volume of pasteurized juice in a larger container by adding in the proportion of 1: 10. After this, a water seal is attached to the barrel in order to permit the release of any accumulated pressure during subsequent fermentation. The rate of fermentation will depend on the °Brix of fermenting must. When fermentation is complete, the' clear wine is syphoned, filled and sealed air-tight to exclude all air. In the course of time, the wine matures. During the maturing or ageing process, which takes 6 to 12 months, the wine loses its raw and harsh flavour and acquires a smooth flavour and characteristic aroma. During the maturation process, there is a natural clarification of the wine. Filter aids, egg albumen and bentonite can also be employed to bring about clarification. It is desirable to pasteurize wine to destroy spoilage microorganisms and coagulate colloidal material which causes cloudiness in the wine. Wines are generally pasteurized at 82-88°C for 1-2 min and then bottled. The bottles are closed with corks of good quality and then labeled.

The wines prepared from fruit juices are classified as table wine or dessert wine (Table 11.1). "Champagne" is a sparkling wine produced from certain varieties of grapes in which fermentation is allowed to proceed to completion in the bottle. The bottles are strong enough to withstand high pressure built up during fermentation. Sherry is a wine matured at high temperature by placing the filled barrel for 3-4 months in the open, exposed to the sun where temperature is as high as 54 to 60°C.

Table 11.1: Classification of wines

Wine	*Sugar (g/100ml)*	*Alcohol (ml/100ml)*
Table		
Dry	0-1	9-11
Semi dry	2-3	10-12
Dessert		
Slightly sweet	3-8	11-13
Sweet	8-11	12-14
Very sweet	11	13-18

Vinegar

Vinegar is a liquid derived from various materials, containing sugar and starch, by alcoholic and subsequent acetic fermentation. It should contain at least 4 grams of acetic acid per 100 ml and a corresponding quantity of the solids/salts of the material from which it is made.

Types

Vinegar is made from various fruits and also from sugar. Some important types of vinegar are listed in Table 11.2.

Table 11.2: Vinegars produced from various raw material

Type	*Raw Material*
Cider vinegar	Apple juice
Grape vinegar	Grape juice
Spirit Vinegar	Alcohol
Malt Vinegar	Malted Barley
Other Vinegars	Orange, Pineapple, Banana, Pear, Peach, Juices

Method of Preparation

Two different processes are involved in the preparation of vinegar *i.e.* (*i*) transformation of the sugary substances into alcohol by yeast, and (*ii*) conversion of the alcohol into vinegar by acetic acid bacteria.

$$\underset{\text{Glucose}}{C_6H_{12}O_6} \xrightarrow{\text{Yeast}} \underset{\text{Ethyl alcohol}}{2C_2H_5OH + 2\,CO_2}$$

$$2C_2H_5OH + 2\,O_2 \xrightarrow{\text{Bacteria}} 2CH_3COOH + H_2O$$

Ethyl alcohol

Acetic acid

Preparation of Vinegar Slow Process

In this process, the juice or sugary solution is filled into barrels and allowed to undergo alcoholic and acetic fermentations slowly. The barrel is placed in a damp, but warm place. In about 5 to 6 months, the sugar solution turns into vinegar. The main drawbacks of this method are (*i*) alcoholic fermentation is often incomplete, (*ii*) acetic fermentation is very slow, (*iii*) quality of the vinegar is inferior, and (*iv*) yield is low.

Orleans Slow Process

In this process, about three-fourth of the barrel is filled with the juice, inoculated with mother vinegar. The barrel is kept at 21-27°C, and fermentation is allowed to proceed till the acid reaches its maximum strength. Under favourable conditions, it usually takes about 3 months for the complete conversion of the liquid into vinegar. Vinegar produced by the "Orleans slow process" ages during the process of fermentation, and is clear and of superior quality.

Quick Process

To improve its flavour, vinegar is kept in plain oak barrels for about six months.

material settles down leaving a major portion of the liquid clear. This clear liquid can be syphoned out for further clarification. This can be accomplished either by using finings such as Spanish clay, bentonite, casein, gelatin or by filtering through pulp filters or aluminium plate and frame presses. The vinegar, after ageing and clarification, is pasteurized to check any spoilage. It is heated in an open vessel to about 66°C and then cooled to room temperature. Bottled vinegar is pasteurized by immersing the bottles in hot water till the vinegar inside attains a temperature of 60°C. Caramel colour which is the only permitted colour in the case of vinegar is employed for colouring vinegar.

12

Jam, Jelly and Marmalade

Jam

Jam, jelly and marmalade constitute a major class of processed products of fruits. Jams are prepared by boiling the fruit pulp with sufficient quantity of sugar to achieve a desired consistency and also to hold the fruit tissues together. An ideal jam must have at least 68.5 per cent soluble solids when cold. The method of preparing jam and jelly are similar except that pulp and pieces of fruit are used in jam. It may be prepared from a single fruit or from a combination of fruits. It differs from preserve in that it needs not contain whole fruit. Unlike preserve, jam preparation is a single step operation. A jam manufacturer can choose fresh, frozen, sulphited, dried fruit pulp or fruit pulp preserved by heat (Figure 12.1).

Commercial Process

Jam is prepared by crushing the fruit (or cutting into small pieces) and then cooking (Figure 12.1) with the addition of sugar to a uniform consistency. The proportion of sugar to be added in jam depends upon the kind and amount of sugar in fruit pulp, and acidity in fruits used for jam making. However, it is important to achieve proper ratio between pectin, acid, and sugar. The pectin in fruit imparts jam a desired set while the sugar facilitates its preservation. Most jams have a minimum sugar percentage of 68.5. Citric, tartaric or malic acid is sometimes used to supplement natural acids present in the fruit. Permitted colour or flavour may also be added at the end of the boiling process.

Home-scale Processing

Fruits such as mango, apple, sapota, cashewapple, pineapple, grapes, carambola, loquat, cherry and litchi are used for preparation of jam. Fully ripe fruits having good colour, flavour and aroma are selected and thoroughly washed in fresh water. The fruits are peeled and stones and cores are removed. Then small pieces of fruits are

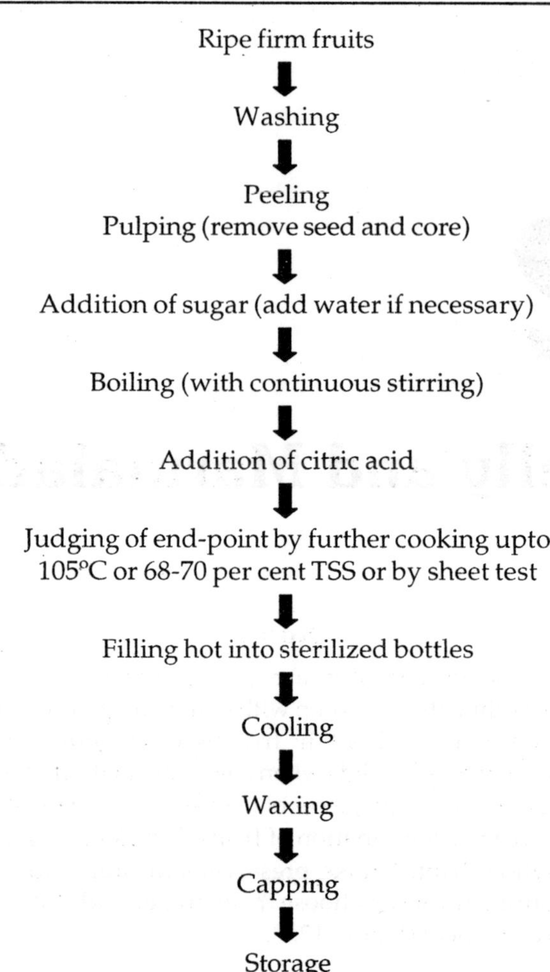

Figure 12.1: Process for preparation jam from various fruit pulps

made with the help of stainless steel knife. To the sour fruits, an equal quantity of sugar by weight is added. For sweeter pulp, three fourth quantity of sugar by weight is added and citric acid at the rate of 1.5-2 g/kg of fruit is also added. The ingredients are thoroughly mixed and mixture is allowed to stand for 1/2 to 1 hour. The mixture is cooked slowly with occasional stirring and crushing until the cooking mass approaches the desired consistency. When mass becomes sufficiently thick in consistency, a spoon is dipped into it and the run-off of the product is observed on the sides of the spoon after cooling. The product falls off in the form of a sheet instead of flowing in a single stream. This indicates that jam is ready for filling into container. The hot jam is filled into clean, sterilized dry glass bottles or other suitable containers. The filled container are then capped without ·delay. The containers are inverted for about 5 min to sterilize the lid also and then allowed to cool. The containers are stored in a cool and dry place.

Storage of Jam

Storage of jams in glass bottles which cannot be hermetically sealed poses problems as the surface of the jam is susceptible to mould growth. Another major difficulty in long term storage of jam is that, it loses moisture unless the jars are kept in cool place. Jams are rarely spoiled by yeast because of their jelly consistency. A relative humidity of approximately 80 per cent has been shown to be acceptable for long term storage of jams. Keeping the bottles unsealed after placing waxed tissue paper on the surface, can minimize mould growth. Jars are inspected periodically and are sealed at the time of shipping. If jam is packed in jars which can be hermetically sealed, it is filled while still hot and pasteurized at 80°C for 30 min.

Jelly

In jelly making, pectin is the most important constituent as it forms jelly when mixed with proper amounts of sugar and acid. It is possible to achieve a desired level of consistency by utilizing proper proportion of pectin, sugar, and acid (Figure 12.2).

Commercial Process

Jelly preparation involves extraction of pectin from fruits followed by clarification of pectin extract. The amount of pectin in the fruit extract determines the final consistency of jelly. On an average, 1 per cent pectin in extract is considered sufficient to produce a good jelly. The clarified pectin extract is then cooked by adding sugar until a desired setting is achieved. Thickening or gelling of pectin-sugar mixture depends upon pectin molecules cross-linking by means of hydrogen bonds. Hence, the mixture must contain sufficient pectin. It should have ideal pH and sugar or solid contents above 60 per cent. All these factors influence the gelling properties. If pectin is low, less than 1 per cent in jellies, the product should be fortified by using a high pectin fruit such as woodapple or adding commercial pectin. Acidity can be increased by adding lemon juice. The product pH should be between 3.0 to 3.5 for optimum gelling. Gels can be obtained in product with lower sugar levels by using a low methoxyl-pectin which will produce a gel in a product without requiring large quantities of sugar. The firming action is initiated by adding calcium salts which will cause cross linking of the pectin molecules, thus providing the desired gelling properties. The end point is determined by utilizing one of the standard tests (sheet test, weighing test or boiling point test). Hot jelly is packed in containers which are hermetically sealed and pasteurized at 80°C for 30 min.

Home-scale Process

Sound and tart fruits are selected and washed thoroughly in cold water. The washed fruits are cut with a stainless steel knife and pieces are covered with water. Citric acid is added at the rate of 1.5 to 2.0 g/kg of the fruit. The mass is boiled, crushing it occasionally with wooden ladle for about t half an hour until it exhibits stickiness. Then the mass is strained through a coarse muslin cloth to separate the extract. Another extract is prepared from the residue by following earlier procedure. The first extract is mixed with the second extract and mixed extract is stored in container for about 8- 10 hours. Subsequently, the clear extract is decanted in another stainless steel pot.

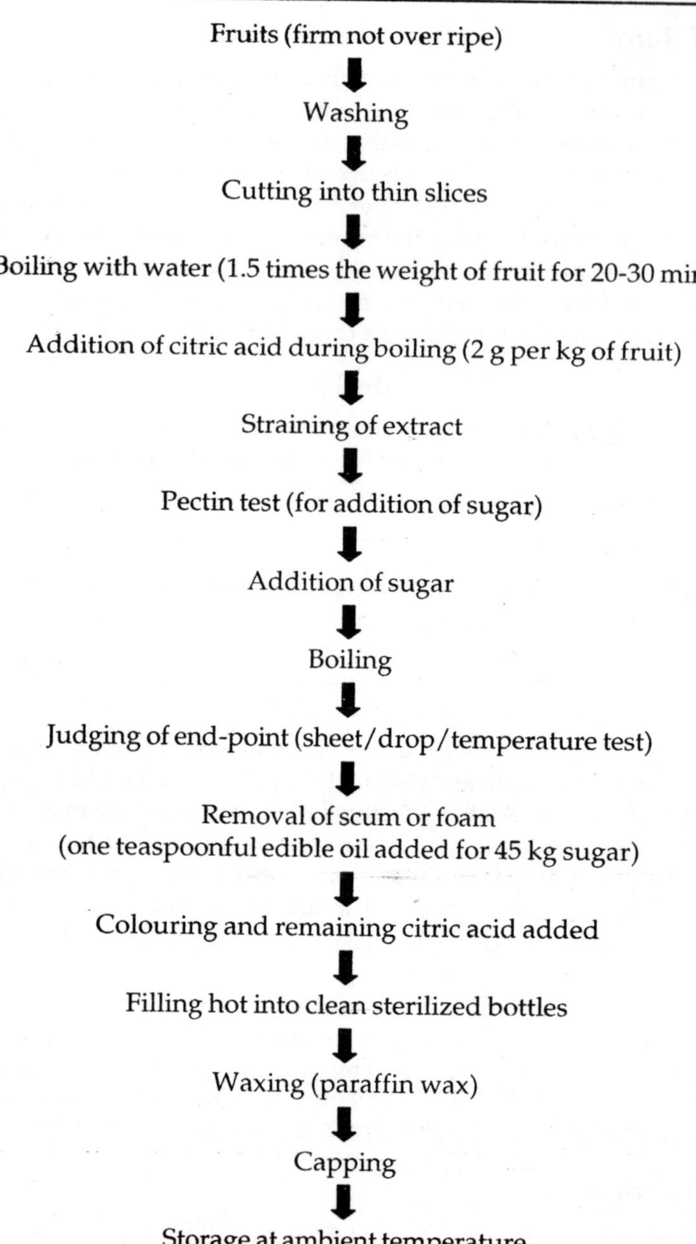

Figure 12.2: Processing of fruits for preparation of jelly

Test for Pectin

A teaspoonful of extract is taken in a stainless steel dish and two teaspoonful of methylated/rectified spirit is added. Formation of one big clot indicates high pectin in the extract. If formation of many clots takes place, it indicates medium pectin in the extract. If thin gelatinous precipitate is formed, it indicates poor pectin in the extract. The material poor in pectin are concentrated till it gives test for high pectin.

For every one kg of extract which contains high or medium quality of pectin, 3/4 or 1/2 kg of sugar is added, respectively. The mixture is cooked until it gives the sheeting test. Foaming is controlled by addition of small quantity of edible oil or by removing scum. Then the finished product is poured into clean, dry, already sterilized glass jars in hot condition kept on a wooden board for preventing breakage. The product is allowed to cool and the jars are sealed air-tight, which are stored in cool place. A good quality jelly is transparent, and keeps its shape like square box when cut.

Deformities in Jelly

Failure of Jelly to Set

- ☆ Lack of acid or pectin
- ☆ Addition of too much sugar
- ☆ Cooking below end point
- ☆ Cooking beyond end point
- ☆ Slow cooking for a long time

Cloudy Jelly

- ☆ Overcooking
- ☆ Faulty pouring
- ☆ Non-removal of scum
- ☆ Premature gelation
- ☆ Use of non-clarified juice
- ☆ Use of immature fruits
- ☆ Over-cooking

Synersis

- ☆ Excess of acid
- ☆ Too low concentration of sugar
- ☆ Insufficient pectin
- ☆ Premature gelation
- ☆ Fermented Jellies

Marmalade

Preparation of marmalade is similar to jelly, but the pectin and acid ratios are maintained relatively higher in marmalade. There are mainly two types of marmalade–jam marmalade and jelly marmalade. Extraction of pectin and boiling are similar to jelly. The peels are cut into 2.5 × 0.5 cm pieces and boiled in water with repeated changes for 15 min and are added at the end of cooking of pectin extract to supplement

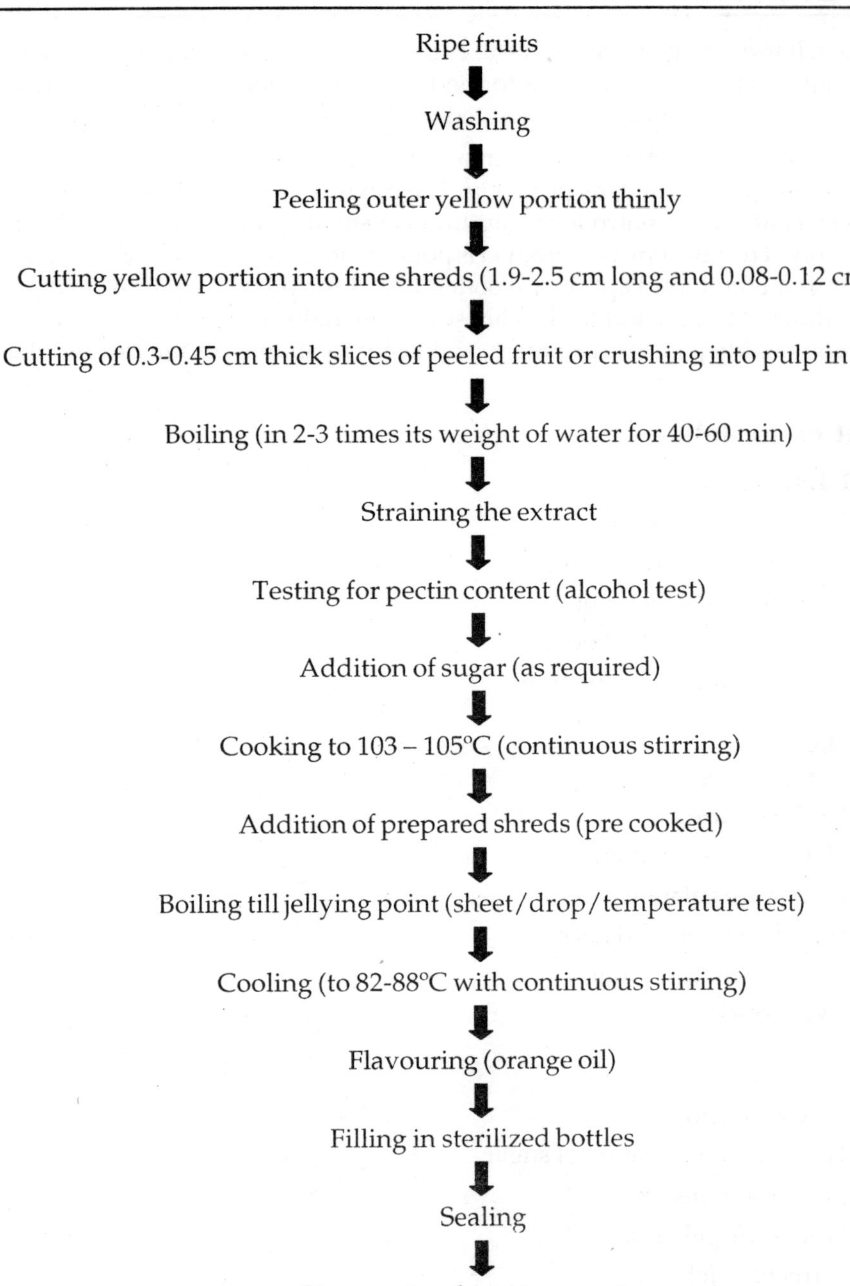

Figure 12.3: Process for the preparation of marmalade from fruits

natural flavours. After cooling, the marmalade is poured into containers, hermetically sealed and pasteurized at 80°C for 30 min. In jam marmalade, the pectin extract is not clarified before cooking (Figure 12.3).

Marmalades are made usually from citrus fruits and consist of jellies of the concerned fruit containing shreds of peels suspended in them. The mature fruits are selected, washed in cold water and peeled to remove only the outer thin portion of the skin leaving as much as possible the white pith (albedo). The extract of the mixture of peeled fruits is prepared without adding any acid. Peel is cut into fine shreds 2-3cm long. Before that, off-white portion of the peel is scrapped off and the shreds are boiled in water until soft. The fruit extract is boiled by adding the appropriate quantity of sugar according to the pectin content in it. After 15 to 30 min of boiling, the prepared shreds are added to the boiling mixture and boiling is continued till the end point (68.5°Brix), is attained. At this stage, flavour and edible colours are added and mixed them very well. The cooked mass is poured in sterilized jars, cooled to room temperature and sealed with sterilized lid and stored in cool, dry place.

13

Pickle, Chutney and Ketchup

Pickle

The process of preservation of food in common salt or vinegar is called pickling and the product is called pickle. Spices and edible oils are added for taste. Pickled foods are produced by adding edible acids such as lactic or acetic acid in the form of vinegar. A wide variety of fruits and vegetables are used for pickling. Although the preservation of fruits and vegetables in pickled form began as household art, at present most of the world supply of pickles is produced in commercial plants. Mango and lime pickles are important form of preserved products prepared from fruits. The cucumber is one of the most important raw materials used for pickles in USA and other countries. Green tomatoes, peppers, cauliflower and onions are common ingredients of mixed pickles "Chowchow".

Mango Pickle

Healthy, under-ripe, fully developed mangoes preferably of tart variety 'are taken and washed in cold water. The fruits are cut into longitudinal slices with a stainless steel knife and immersed in 2-3 per cent brine to prevent them becoming black. The slices are then removed from brine solution and surface moisture is removed by drying under fan or air-blow method. The typical recipe of mango pickles include, mango slices 1 kg, common salt 250 g, metha (coarsely ground) 30 g, and aniseed or fennel 30 g, and turmeric powder 28 g.. The slices are smeared with all the ingredients in recipe which are mixed well, cooked in oil and cooled. Then slices are transferred to dry glass or glazed jars and edible oil (boiled and cooled) is poured on the pieces so that no air pocket is left inside. This leaves a small layer of oil at the top. The product is stored for 2-3 weeks before use. During this period, the product is stirred every 2-3 days and oil is added if necessary (Figure 13.1).

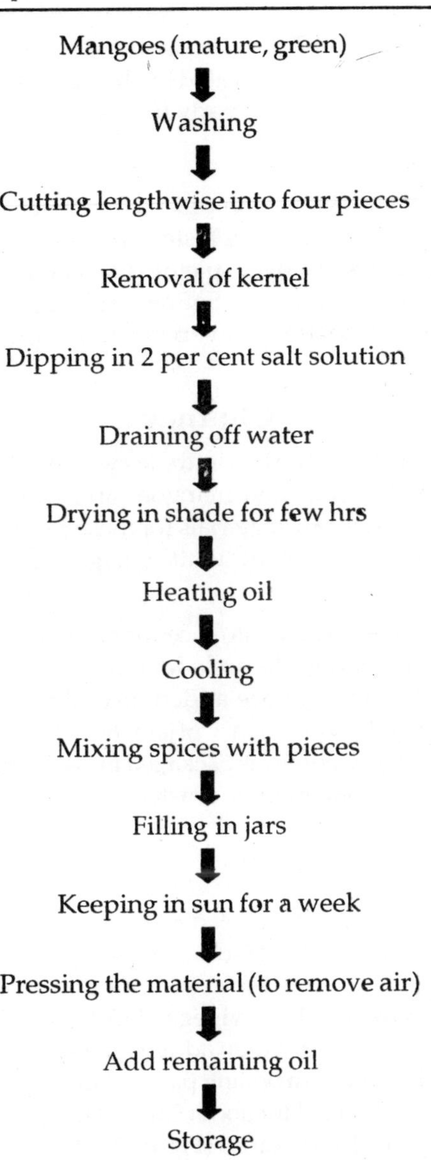

Figure 13.1: Processing of mangoes for pickles preparation

Lime Pickle

Sound, fully matured juicy limes having deep yellow skin are selected for making pickles. The fruits are washed thoroughly in running cold water and cut into halves or quarters depending upon their size. For every kg limes, 200-250 g of powdered salt is used and pieces are transferred to a clean sterilized wide mouth glass or glazed jar. The limes are partially squeezed and transferred to salt. The mass is well stirred to effect thorough mixing. If juice of lime is insufficient to cover the mass, cool boiled water acidified with citric acid (4-5 per cent) is added so that limes are covered by the liquid. The softening of lime skin is indicated by the fact that the skin of the lime turns light brown. The pickle at this stage is ready for use.

Spoilage of Pickles

Mango and cucumber pickles exhibit shriveling when placed in strong solution of salt and even in strong vinegar. Bitter taste in pickle is caused due to cooking the spices for a long time and also by over spicing. Blackening is caused by reaction of iron or spoilage due to microorganisms. Softness and slipperiness is developed due to action of bacteria. Scum formation is invariably found owing to the growth of wild yeast.

Chutney

Chutney and sauces of various kinds are very common in India. Fruits such as apples, peaches, plums, apricots, and mangoes and vegetables like cauliflowers, carrots, tomatoes, etc. are basic raw materials for these products. Onion, garlic, spices and herbs are used as flavouring agents while vinegar, salt and sugar are added for taste.

The fruits and vegetables are cut into slices or pieces of suitable size and softened by boiling in water. Onion and garlic are added at the start to mellow their strong flavours. Spices as well as vinegar are added just a little before the final stage of boiling. Chutneys are cooked to consistency of jam (68 to 69°Srix) to avoid fermentation of the product during storage. Chutney is packaged in clean dry bottles or jars, sterilized earlier in boiling water. The containers are pasteurized at 82°C for 30 min. The bottles are kept in cool dry place.

Mango Chutney

Slightly under-ripe mangoes are peeled and cut into thin slices. These are heated in small amount of water to make them soft. The typical recipe contains mango slices 0.9 kg, sugar 0.9 kg, salt 56.6 g, mixed spices (cardamon, cinnamon. cumin etc.) 28.3 g, garlic 6.0 g, red chilli powder 14.0 g, vinegar 113.4 g, onion (chopped) 28.3 g, and ginger 13.4 g (12). Sugar and salt are then added. Spices are loosely tied in cloth bag and the bag is placed in the slices in a boiling pan and mass is cooked to the consistency of jam. Vinegar is added and boiled for about 5 min. The spice bag is removed and hot product is filled into dry sterilized bottles (Figure 13.2).

Apple Chutney

Apples are peeled, cored and cut into slices. The slices are cooked along with other ingredients (onions, sugar, chilli powder, garlic, salt and vinegar) until they become tender. The product is bottled when it is hot. The final acidity of Chutney should not be less than 2.4 per cent.

Sauce

Sauces are mainly two kinds–thin and thick sauces. The thin sauce contains vinegar, extract of spices and herbs. Thick sauce is more viscous and does not flow easily. It has at least 3 per cent acetic acid for improving the keeping quality. The

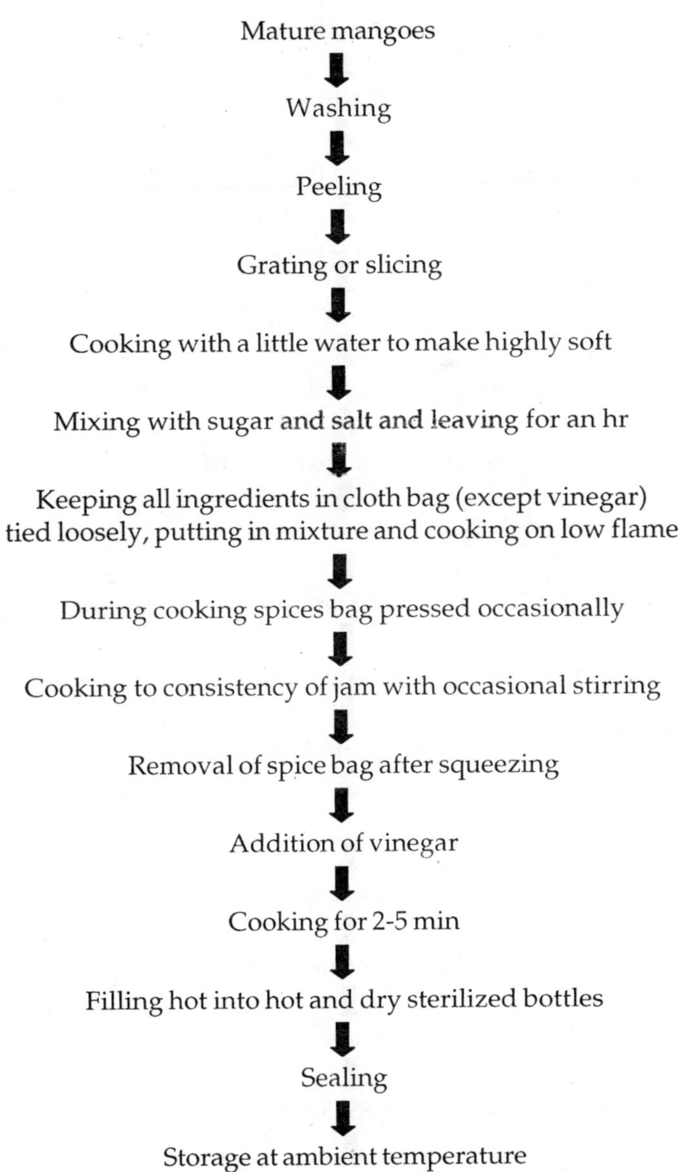

Figure 13.2: Process for preparation of mango chutney

procedure for making thick sauce is similar to that employed for chutney. Thickening agent such as maize or cassava starch is utilized for this purpose.

Ketchup

Commercial Process

Tomato ketchup can be made directly from fresh cyclone juice or from concentrated pulp or bulk stored tomato paste. The other constituents of ketchup are

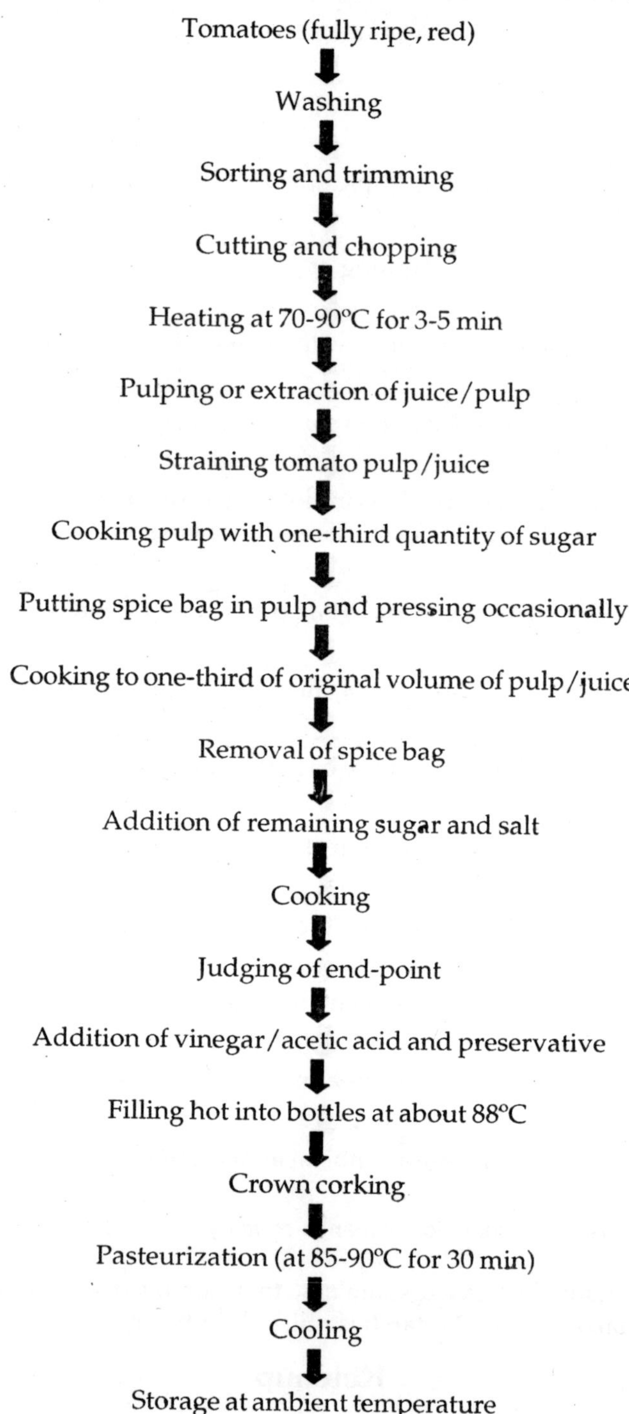

Figure 13.3: Processing of tomato for preparation of ketchup

sugar, salt, vinegar, onions, and spices. The ingredients are cooked for 30-45 min in kettles. The thickness of the bottled ketchup is an important aspect of its quality. Part of the thickness of the ketchup is due to the pectin from the tomatoes. Some manufacturers prefer a hot-break method before the tomatoes are cycloned in order to retain the largest possible amount of pectin. The hot-break method dissolves some of the mucilaginous material from the seeds; again contributing to the final consistency. The ketchup is bottled, deaerated and sealed at a temperature of 82-88°C and cooled to prevent loss of flavor (Figure 13.3).

Home-scale Process

Fully ripe and healthy fruits having well developed colour are selected and washed thoroughly in fresh water. These are cut into pieces. These are transferred to stainless steel open vessel and crushed thoroughly with wooden ladle. The crushed mass is cooked for about 5-10 min and mashed it well while cooking. When sufficiently soft, it is strained through fine stainless steel sieve. The seeds and skin are discarded and only pure pulp is taken for further processing. A typical recipe includes tomato pulp 3.0 kg, onion (chopped) 37.5g, garlic (chopped) 2.5g, cloves 1.0g, spices (coarsely powdered, cardamon and. blackpepper in equal quantities) 1.2g, mace 0.25g, cinnamon 1.75 g, red chilli (powdered) 1.25g, salt 31.2g, sugar (according to sweetness) 100- 150g, vinegar 150 ml, sodium benzoate 900mg and pectin 3g.

The pulp is taken in open vessel and 1/3 of the sugar given in recipe is added to the pulp. Spices (onion, garlic, cloves, cardamon, blackpepper, geera mace, cinnamon and red chillies) are placed in muslin bag and bag is immersed into the pulp. The pulp is heated until it thickens and is reduced to about 1/3 of its original volume. The muslin bag is then squeezed to extract aroma and flavour of the spices and it is removed from- the pulp. Vinegar, salt and remaining sugar is added and the mass is heated for few min so that the volume of the finished product is about 1/3 of the original pulp. To a small quantity of the finished product, sodium benzoate 295 mg/kg of finished product is added and mixed thoroughly. The dissolved sodium benzoate is added to the rest of the product and mixed well.. The finished product is poured into medium size sterilized bottles and sealed air-tight with a crown seal and bottles are pasteurized in boiling water for 30 min. The bottles are cooled in air and stored in dry place.

Black neck is a common problem in ketchup. A black ring is formed on the surface of the ketchup in the neck of the bottle. It is due to oxidation of iron compounds which get into ketchup from boiling equipment and from the metal of the cap through action of acetic acid. When iron comes in contact with the spice tannins, it forms ferrous-tannate which on oxidation forms the black ferric-tannate. Addition of ascorbic acid helps in prevention of black neck formation.

Enzymes in the Food Industry

- ☆ Enzymes are organic biocatalysts which govern, initiate and control biological reactions important for life processes.
- ☆ Amylase found in saliva promotes digestion or breakdown of starch in the mouth.

- ☆ Pepsin found in gastric juice promotes digestion of protein.
- ☆ Lipase found in liver promotes breakdown of fats.
- ☆ There are thousands of different enzymes found in bacteria, yeasts, moulds, plants and animals.
- ☆ Even after a plant is harvested or an animal is killed, most of the enzymes continue to promote specific chemical reactions, and most foods contain a great number of active enzymes.
- ☆ Enzymes are large protein molecules which, like other catalysts, need to be present in only minute amounts to be effective.
- ☆ All enzymes are proteins but all proteins are not enzymes.
- ☆ Enzymes are exceptional as catalysts in the following respects:
- ☆ They are exceedingly efficient, under optional conditions most enzymatic reactions proceeds 10^8 to 10^{11} times faster than the corresponding non-enzymatic reactions.
- ☆ Most enzymatic reactions are specific in terms of the nature of the reaction and the structure of the substrate.
- ☆ The spectrum of reactions catalysed by enzymes is very broad e.g., hydrolytic, polymerization, oxidation, reduction, dehydration etc.
- ☆ Enzymes themselves are subject to a variety of cellular controls. Even their biosynthesis is enzyme-catalyzed.

Important Properties of Enzymes in Fruit and Vegetable Technology

1. Enzymes control the reactions associated with ripening of fruits and vegetables.
2. After harvest, unless destroyed by heat, chemicals or some other means, enzymes continue the ripening process, in many cases to the point of spoilage such as soft melons or over ripe bananas.
3. Enter in to vast number of biological reactions in fruits and vegetables; they may be responsible for changes in flavour, colour, texture and nutritional properties.
4. The heating-process (pasteurization) not only to destroy micro-organisms but also to deactivate enzymes and so improve the fruit and vegetables storage stability.
5. During fermentation of food wine-yeast produces enzymes *i.e.* micro-organisms produce enzymes which are responsible for fermentation and alcohol production.
6. Enzymes can be produced from various biological materials and that can be added to foods to break down starch, tenderize meat, clarify wines, coagulate milk protein and produce many other desirable changes.
7. In fruit and vegetable storage and processing the most important roles are played by the enzymes classes of hydrolases (lipases, invertases, tannase, chlorophylase, amylase, cellulose), and oxidoreductases (peroxidae, tyrosinase, catalase, ascorbinase, polyphenol oxidase).

Enzymes Used in the Food Industry

Several types of enzymes are used for the preservation and processing of a wide variety of foods and beverages.

Enzyme	*Class*	*Source*	*Application*
Amylase	Hydrolases	Malt, fungi	Bread baking
		Malt, bacteria,	Mashing
		malt, fungi	Precooked baby food, Breakfast food
		Bactria, fungi	Syrups
		Fungi	Liquefying purees and soups
Cellulase	Hydrolases	Fungi, bacteria	Foods
Glucose isomerase	Isomerases	Fungi, actinomycetes	Sugar and starch
Glucose oxidase	Oxido-reductases	Fungi	Glucose removal, oxygen removal
Invertase	Hydrolases	Yeast	Soft-centres candies, High-test molasses
Lactase	Hydrolases	Bacteria	Milk concentrate, ice-cream, frozen desserts
Pectinase	Hydrolases	Fungi	Pressing, clarification, filtration, concentration, coffee bean fermentation, coffee concentrates.
Protease	Hydrolases	Fungi, bacteria, papain, bromelin	Bread baking
Rennin	Lyases	Fungi, animal	Cheese production
Diastase	Hydrolases	Fungi	Production of syrups

Immobilized Enzymes

1. Although enzymes are useful as catalysts in food processing, they may not always be suitable for practical application.
2. Conventionally, an enzymatic reaction is carried out in a batch process, by including the substrate with soluble enzymes.
3. But it is very difficult to recover the enzyme after the reaction, for recycling. Enzymes remain in the processed food may cause allergy. This can overcome by following two ways.
4. Use of a synthetic polymer having enzyme-like activity (synzyme).
5. The second approach is the modification of the natural enzyme by immobilizing it.
6. The best way to inactivate the enzyme is to immobilize the enzyme by attaching it to the surface of a membrane or another inert object in contact with the food being processed.

7. In this way reaction time can be regulated without the enzyme becoming part of the food.
8. Immobilized enzymes are defined as enzymes physically confined or localized in a certain defined region of space with retention of their activity and which can be used repeatedly.

Advantages of immobilized enzymes are as follows:

1. Recovery and reuse of enzyme is possible,
2. Stability of enzyme is increased,
3. Kinetic property of enzyme is enhanced,
4. Product is enzyme free,
5. Permit continuous operation,
6. Cost is lower and
7. Greater control of catalytic power is possible.

Immobilized enzymes are presently being used to hydrolyze the lactose of milk into glucose and galactose, to isomerize the glucose from corn starch in to fructose, and in many other industrial food processes.

Browning Reactions

Browning reactions occur vary widely in food materials. The colours produced range from pale yellow to dark brown or black, depending on the type of product and the extent of the reaction. In some foods browning is considered desirable, e.g., honey, chocolate, brown crust of backed product etc., while in other foods it is detrimental, as in darkening of dehydrated fruits, and vegetables etc. It may be enzymatic or non-enzymatic. Many of the enzymatic reactions are seen in fruits and vegetables, and involve the oxidation of polyphenolic compounds by oxidative enzymes in plant cells. The non-enzymatic browning reactions frequently involve sugars or sugar-related compounds.

Enzymatic Browning

1. Many fruits and vegetables turn to brown after damage or cut due to exposed to air.
2. The formation of brown colour is due to the action of the enzyme phenolase (also known as polyphenol oxidase, tyrosinase or catecholase) on phenolic substances.
3. Normally the phenolic substances are separated from phenolase in intact tissues and browning does not occur.
4. When foods containing such substances are cut and exposed to air rapid browning of cut surface takes place.

   Polyphenols + Oxygen ——→ Brown colour (orthoquinones/melanin)

5. Tyrosine is the major phenolic substrate for phenolase action in foods. Other phenolic substances are caffeic acid, protocatechuic acid, and chlorogenic acid.
6. Enzymes involved in browning reactions are phenolase, peroxidases and others.
7. Orange, lemon, grape fruit, straw berry, tomato do not contain these enzymes so no browning occurs.
8. Melanin formation is undesirable during processing of fruits and vegetables.
9. Melanin/brown pigment formation can be eliminated by inhibiting enzyme action.
10. The exclusion of oxygen is also used for preventing browning reaction.
11. By immersing in brine or syrup solution or processing under vacuum.

Non-Enzymatic Browning

1. Non-browning reactions are responsible for the colour and flavor of foods.
2. Sometimes this reaction produce desirable flavours e.g., chocolate flavor of cocoa beans.
3. The presence of reactive reducing sugars is responsible for browning in foods.
4. On heating the sugars undergo ring opening, enolization, dehydration and fragmentation.
5. The unsaturated carbonyl compounds that are formed react to produce brown polymers and flavor compounds.
6. Heat-induced browning reactions can be divided into two groups; Maillard reaction and carmelization.

Maillard Reaction (French Man Maillard)

1. Maillard reaction also known as maillard browning is a colour, flavor, odour and sometimes texture change which results from a chemical reaction between proteins and carbohydrates.
2. The maillard reaction is a complex reaction between nitrogenous compounds and sugars, nitrogenous compounds and organic acids, sugars and organic acids, among organic acids themselves.
3. The carbonyl group of acyclic sugars readily combines with the basic amino groups of proteins, peptides and amino acids, resulting in sugar amines.
4. The set of various reactions that sugar-amines undergo resulting in browning is known as the Maillard reaction.
5. The sugar-amines have a brown colour at a low temperature.
6. It has been found that histidine, threonine, phenylalanine, tryptophan and lysine are the most reactive amino acids.

7. The initial reaction is thought to be between the aldehyde group of the sugar and the amino group of the amino acid.
8. Maillard browning is responsible for the desirable browning of most heated foods such as crusts, roasted meats and roasted coffee beans.
9. This browning is accelerated by heat and very quickly in ovens, slowly at room temperature and very slowly at refrigerator temperatures.
10. Very small amounts of the protein or carbohydrate substances are needed for maillard browning to occur.
11. Maillard browning is also accelerated by low moisture content.
12. Maillard browning becomes more pronounced with increased storage time.
13. Light does not accelerate maillard browning but changes other colours.
14. Vitamin B_1 (thiamine) contains amino acid so reacts with sugar.
15. Hence, maillard reaction products predominate in browned foods.
16. The condensation product of sugar and amine undergoes enolization and rearrangement and then condensation and polymerization to form re-brown and dark-brown compounds.
17. The brown to black, amorphous, unsaturated heterogeneous polymers are called "melanoids".

Control of the Maillard Browning Reaction

1. This can be accomplished by keeping the pH below the isoelectric pH of the amino acids, peptides and proteins and keeping the temperature as low as possible during processing and storage.
2. Use of non-reducing sugars, sulphur dioxide and sulphites used in extending the storage life of dehydrated foods, fruit juices and wines also inhibit the maillard reaction.

Caramelization

1. Sugar in dry condition (or their syrups) when heated beyond their melting point decompose and form a brown mass known as caramel, which has a bitter, astringent taste.
2. The process of caramelization occurs at a high temperature, while maillard reactions develop brown colour at low temperature.
3. With the use of suitable catalysts it is possible to carry out caramelization to provide either flavouring or colouring caramel for food use.
4. For flavouring purposes, sucrose as concentrated syrup is caramelized.
5. For the manufacture of caramel colours for use in beverages, glucose syrup is treated with dilute sulphuric acid and then partially neutralized with ammonia.
6. Besides reaction of sugars and amino acids, break down of ascorbic acid during storage of the products may be another possible reason for the development of browning.

7. Ascorbic acid → dehydroascorbic acid → 2, 3-diketogulonic acid ?furfural ? polymerization → Brown pigment (browning).
8. Browning may also be due to metallic contamination, mainly by iron and copper salts.
9. The tannin in fruits and vegetables reacts with the iron of the tinplate to form ferric tennate, which is black in colour and spoils the appéarance.
10. Sometimes hydrogen sulphide gas is liberated (due to reaction between fruit acids and tinplate) from the canned product which, in turn, reacts with the iron of the can and forms black iron sulphide.
11. Browning is also caused by unsuitable vessels, *e.g.*, traces of copper (1 ppm) coming in contact with hydrogen sulphide form black copper sulphide.
12. Metallic contamination can be avoided by using glass containers or avoiding use of iron and copper vessels.

Fermentation

1. Decomposition of carbohydrates by microorganisms or enzymes is called fermentation.
2. Fermentation → carbohydrates, microorganisms or enzymes → organic acids, alcohol.
3. The derivation of the word fermentation signifies a gentle bubbling condition.
4. The term was first applied to the production of wine more than a thousand years ago.
5. The bubbling action was due to the conversion of sugar to carbon dioxide gas.
6. Fermentation came to mean the breakdown of sugar into alcohol and carbon dioxide.
7. Pasture later demonstrated the relationship of yeast to this reaction and the word fermentation became associated with microorganisms and still later with enzymes.
8. The term fermentation refers to break down of carbohydrate and carbohydrate like materials under either anaerobic or aerobic conditions.
9. Conversion of lactose to lactic acid by *Streptococcus lactis* bacteria is favoured by anaerobic conditions and is true fermentation; conversion of ethyl alcohol to acetic acid by *Acetobacter aceti* bacteria is favoured by aerobic conditions and is more correctly termed an oxidation rather than fermentation.
10. Only selected organisms are encountered and their metabolic activities and end products are highly desirable.
11. The increasing application of biotechnology and genetic engineering techniques to food production is brining added importance to food fermentations.

12. Acetic, lactic and alcoholic are the three important kinds of fermentation involved in fruit and vegetable preservation. The keeping quality of vinegar, fermented pickles and alcoholic beverages depends upon the presence of acetic acid, lactic acid, and alcohol, respectively.
13. Care should be taken to exclude air from the fermented products to avoid further unwanted or secondary fermentation.
14. Wines, cider, vinegar, fermented pickles and other fermented beverages etc., are prepared by these processes.

Some industrial fermentation in fruits and vegetable industries:

Acetic Acid Fermentation (Acetic Acid Bacteria)

Wine, cider, malt, honey or any alcoholic and sugary or starchy products may be converted to vinegar.

Lactic Acid Fermentation (Lactic Acid Bacteria)

1. Cucumber → dill Pickles, sour pickles, salt stock
2. Tomato → pickles
3. Lemon → pickles
4. Mango → pickles
5. Cauliflower → pickles
6. Olives → green olives, ripe olives
7. Cabbage → Sauer kraut
8. Turnips → Sauerruben
9. Lettuce → Lettuce kraut
10. Mixed vegetables, turnips, radish, cabbage → paw Tsay
11. Mixed vegetables in Chinese cabbage → Kimchi
12. Vegetables and milk → Tarhana
13. Vegetables and rice → Sajurasin
14. Coffee cherries → Coffee beans
15. Vanilla beans → Vanilla

    Lactic acid bacteria with other microorganisms:

    With yeasts → Nukamiso pickles

    With mould → tempeh, soy sauce

Alcoholic Fermentation (Yeasts)

1. Fruit → wine, vermouth
2. Malt → beer, ale, porter, stout, bock, pilsner
3. Wines → brandy
4. Grain mash → whiskey

Acetic Acid Fermentation

1. The production of vinegar (acetic acid) from fruit juices is perhaps one of the oldest organic acid fermentations known.
2. Acetic acid is produced by the oxidation of ethyl alcohol by bacteria such as *Acetobacter aceti, Acetobacter orleansis* and *Acetobacter schutzenbachi.*

$$2CH_3CH_2OH + O_2 \longrightarrow 2CH_3CHO + 2H_2O$$

$$2CH_3CHO + O_2 \longrightarrow 2CH_3COOH$$

Some acetobacter species do not stop at the stage of acid production but continue the oxidation to carbon dioxide.

$$CH_3COOH + 2O_2 \longrightarrow 2CO_2 + 2H_2O$$

3. Theoretically, 100 parts of sugar (sucrose or maltose) should yield about 51 parts of ethyl alcohol or 67 parts of acetic acid.
4. In actual practice, however, even under the most favourable conditions 43 to 48 parts of alcohol and 49 to 56 parts of acetic acid only are produced.
5. Reasons for losses in yield may be due to: the consumption of sugar in the solution by the yeast, loss due to alcohol and acetic acid due to evaporation and oxidation, loss due to utilization by acetic acid bacteria for their growth and small quantities of alcohol may also remain unconverted.
6. Ten per cent sugar must in juice for vinegar preparation and 5 per cent acetic acid strength.
7. After conversion of alcohol into acetic acid bacteria attack the acid itself. This can be prevented by filing the containers upto the brim and sealing them air tight.

Lactic Acid Fermentation

1. Lactic acid fermentation as a good method of preservation is another ancient art of unknown origin.
2. Lactic acid fermentation is an anaerobic intra-molecular oxidation-reduction process. Both homofermentative and heterofermentative lactic acid bacteria participate in food fermentation. In some cases, yeasts and moulds also participate along with lactic acid bacteria.
3. Lactic acid bacteria grow in low or no acid media and it also grow in 8 to 10 per cent salt.
4. The growth of undesirable organisms is inhibited by adding salt, while allowing the lactic acid fermentation to proceed.
5. At 20 per cent salt all fermentation stops.
6. Lactic acid bacteria are most active at about 30°C.
7. In practice, 2-3 kg of salt is mixed with every 100kg of material and the mixture allowed standing for 12 to 14 hrs, when sufficient juice comes out from the material to form the brine.

8. When sufficient lactic acid is formed the lactic acid bacteria cease to function and any further change in the composition of the material is prevented.
9. Lactic acid fermentation mostly occurs in the preparation of pickles.

Alcoholic Fermentation

1. Ethyl alcohol can be produced by fermentation of any carbohydrate containing a fermentable sugar or a polysaccharide that can be hydrolyzed to a fermentable sugar.
2. The equation that describes the net result of alcoholic fermentation by yeast.

$$C_6H_{12}O_6 + \text{Yeast} \longrightarrow 2C_2H_5OH + 2CO_2$$

3. Here a sugar is the substrate and the process is anaerobic. *Saccharomyces cerevisiae* is commonly employed for fermentation. It is imperative that the yeast must have a high tolerance for alcohol and must grow vigorously and produce a large quantity of alcohol.
4. Hundred grams of hexose sugar should yield 51.1g of ethyl alcohol and 48.9g of carbon dioxide. Besides alcohol a number of other substances are also formed in small quantity. The alcohol content of wine is usually expressed as volume per cent *i.e.* cc of alcohol per 100 cc of wine.
5. The percentage of alcohol will be approximately equal to the °Brix (total soluble solids) of the crushed material multiplied by a factor of 0.57.
6. Example: a crushed material containing 22 per cent total soluble solids should give theoretically a dry wine of about 22 x 0.57 = 12.5 volume per cent of alcohol (V per cent).

14

Preserves, Candies, Toffees and Cheese

Preserve

A preserve is prepared by cooking mature fruits in heavy syrup until they become tender and transparent. Fruit preserves however lose their appearance during long term storage. Preserves are made by cooking the entire fruit or its pieces in sugar syrup of high concentration.

Ripe and firm fruits are selected and washed in fresh water. A wash of dilute hydrochloric acid may be given to remove the spray residues in case of fruits that are used without peeling. The fruits are peeled, cored and cut into pieces, whenever necessary. When whole or destoned fruits are used, pricking is done by using stainless steel fork or bamboo prickers. Fresh pieces of fruits are blanched for few minutes in boiling water to make them soft which assists in absorption of sugar. Highly juicy fruits can be steeped in sugar syrup without blanching. Fruit pieces are boiled in sugar syrup and kept for 24 hours in that syrup. On the next day, syrup strength is raised by adding the required quantity of sugar so that the syrup gives 68°Brix. The mass is boiled again for 5 minutes and fruit/fruit pieces are left in syrup for 3-4 days. The fruits are boiled along with syrup for few minutes and filled hot into dry jars and sealed air tight.

Candy

The process of making fruit candy or candied fruit is practically the same as that employed in the case of preserves, with only difference that the fruit is impregnated with a higher percentage of sugar or glucose. A certain amount of invert sugar or glucose is substituted in place of cane sugar. The total sugar content of the impregnated fruit is kept at about 75 per cent to prevent fermentation.

Fruits and sugar are the main raw materials required for candying. Slightly unripe fruits are taken because fully ripe and over-ripe fruits develop jam-like consistency in the syruping process. Canned fruits of good quality can also be used. A process for preparation of ber candy standardized at Mahatma Phule Krishi Vidyapeeth, Rahuri has been given in Figure 14.1. Thus, fruits can be preserved by using sugar syruping process.

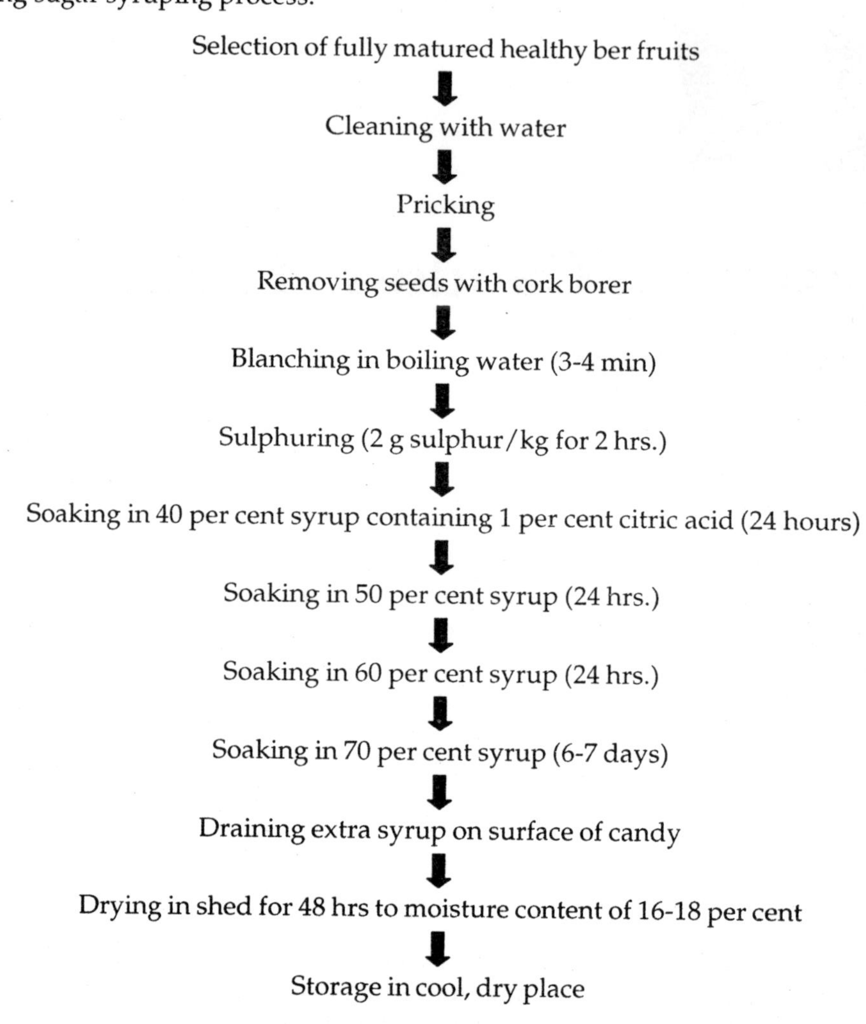

Figure 14.1: Process for preparation of ber candy

Fruit Toffee

Pulpy fruits like banana, mango, papaya, guava etc. can be utilized for the preparation of fruit toffees. Pulpy and ripped fruits are washed thoroughly in ample quantity of fresh water and then peeled. The peeled fruits are dipped in 0.1 per cent solution of potassium metabisulphite for 5 min to check enzymatic browning. Then peeled fruits are crushed and passed through a stainless steel sieve. The typical recipe of fruit toffee contains fruit pulp 1.0 kg, sugar 1.0 kg, glucose 50 g, skim milk powder 50 g, fat (vanaspati) 120 g, essence and colour if required. Except milk powder, all other ingredients given in recipe are mixed very well and the mass is cooked up to 80°Brix. Milk powder is dissolved in a small quantity of water and made a thick paste and mixed in above mass and cooked again up to 82-84°Brix. The cooked mass is transferred in stainless steel plate which is already smeared with fat and the product is spread into a thin sheet of 1 to 2 cm thickness (Figure 14.2). This is allowed to cool and set for two to three hours. Then the solid sheet is cut into cubes of different shapes with a stainless steel knife and wrapped in coloured butter paper.

Pulp + juice/Juice extracted pulp

↓

Addition of sugar fat, skim milk powder and corn flour

↓

Heating up to 80 °Brix

↓

Addition of citric acid and salt by dissolving in small quantity of water

↓

Heating again up to 82 °Brix.

↓

Spreading in the aluminium tray up to 1/2 to 1 inch thickness
in trays already smeared with butter/fat

↓

Cooling to room temperature

↓

Cutting into attractive pieces

↓

Packaging in butter paper

↓

Storage in glass jars

Figure 14.2: Preparation of toffee from fruit pulp

Fruit Butter

Fruit butters are very delicious product. The butters can be prepared from any fruit. The most popular butters are prepared from apple, pear, plum, peaches, apricot and grapes. Sometimes combinations of fruits are used to make a fruit butter. The homogenized pulp is used for making fruit butter. About 1/2 kg of sugar, 1 g of cinnamon, 1 g of cloves, 1 g of ginger in the form of powder are taken for one kg of pulp. The sugar and pulp are mixed thoroughly. The mass is boiled and cooked until product is thickened enough when cooled. Fruit butter is cooked sufficiently so that no liquid is separated from the butter when allowed to stand in a plate. Finally, the ground spices are added a little earlier than the end point. The boiling mass is stirred and mixed thoroughly. The butter is packaged in clean sterilized hot dry jars and stored in cool place (Figure 14.3).

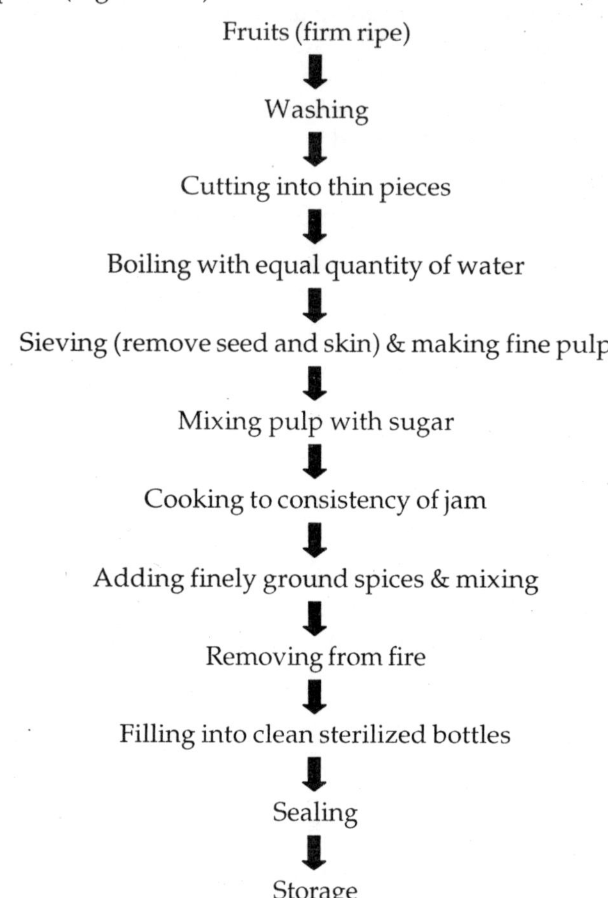

Figure 14.3: Processing of fruits for preparation of fruit butter

Fruit Cheese

Fruit cheese can be prepared from fruits like guava, apple and pear. Fully ripened, sound, firm and disease-free fruits are selected and washed thoroughly to remove dirt and they are cut into small pieces. The small pieces of fruits are taken in stainless steel container and equal quantity of water is added to the cut fruits. These are then boiled until the fruits become soft. The boiled mass is passed through a stainless steel sieve. The typical recipe includes pulp 1 kg, sugar 1.5 kg, butter/fat 120 g, citric acid 2 g, salt 1 tea-spoonful, and other ingredients as per requirement. Pulp, sugar and butter are mixed and heated until it becomes sufficiently thick. The remaining ingredients are dissolved in a small quantity of water a little before the end point. The mass is cooked until it reaches 82°Brix and then poured in aluminum tray or stainless steel plate which is already smeared with butter. Then it is allowed to cool and set. It is subsequently cut into pieces of the required and desired size. The individual pieces are wrapped in butter paper and stored in sterilized glass container (Figure 14.4).

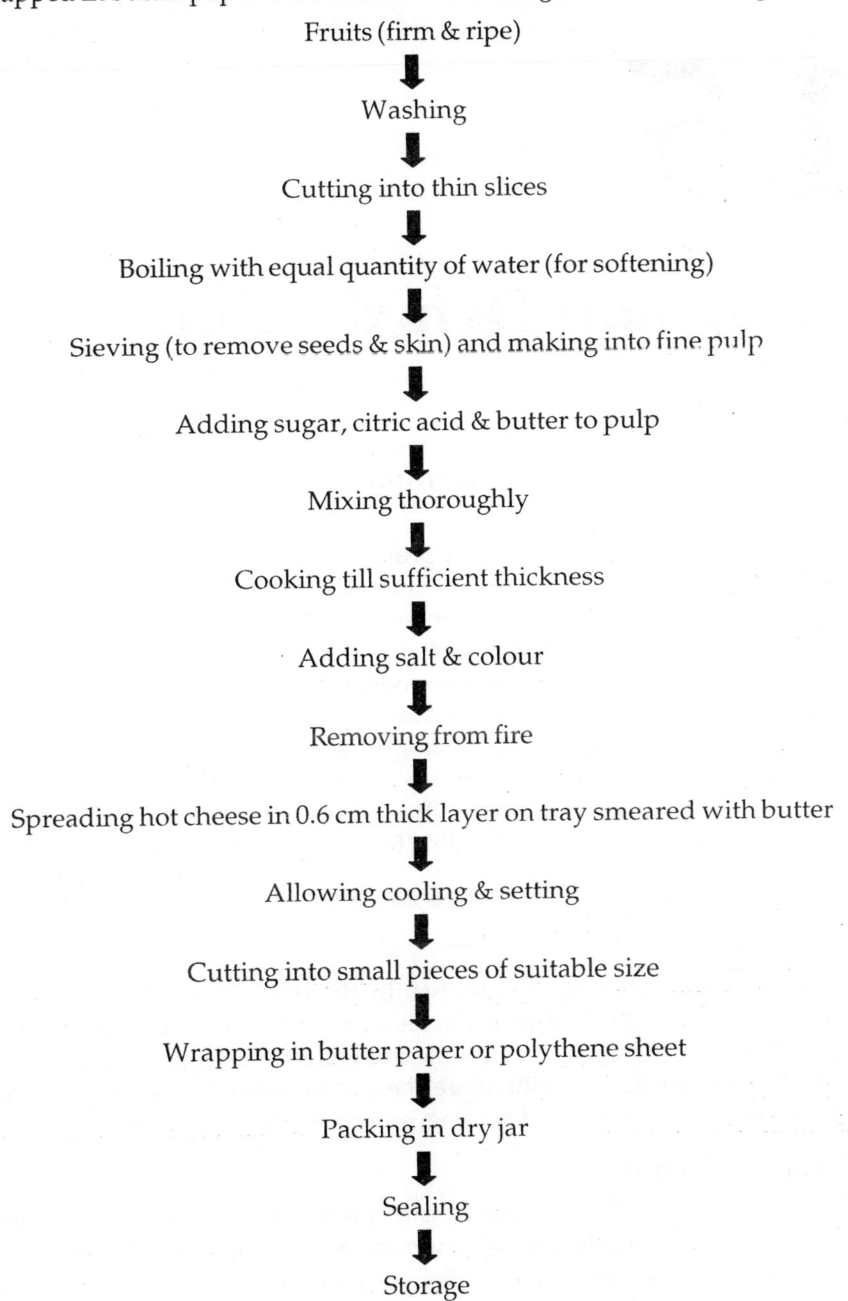

Figure 14.4: Processing of fruits for preparation of fruit cheese

15

Drying and Dehydration

Introduction

Several fruits and vegetables are dried in the sun in different parts of the world. This is probably the oldest method of preservation of food. Drying of fruits and vegetables under controlled conditions has assumed importance in recent years. Although, the first patent for a dehydration process was applied by Grayler in 1780. Dehydrated products had generally low consumer acceptance. The production and consumption of air dried vegetables increased rapidly since World War I. However the dehydrated products during World Wars were not very well received by fighting men. Since that time, there have been several technological advancements in dehydration technology, and with the result, now almost any food can be dried successfully by one or the other method. It is assumed that the dehydration industry will one day be the most important among food preservation industries.

Methods of Drying

Sun Drying

Fruits and vegetables can be successfully dried to a desired moisture level by exposing them to direct or indirect sun. However, this method is very slow and is influenced by the prevailing environmental conditions (light intensity, relative humidity, and temperature). Furthermore, the quality of sun dried product is generally poor because of the uncontrolled heating, scorchinq, or sun scalding.

Mechanical Drying

Modern methods of dehydration consist of removal of moisture by various fruits and vegetables products, improvements in the methods of heat transfer, standardization of optimum time for dehydration, selection of suitable variety or kind of raw materials for dehydration, and improved dehydration techniques have greatly improved the quality and consumer acceptance of the dehydrated products. Major methods of dehydration of fruits and vegetables are outlined in this section.

Hot-Air Drying

Hot-air drying is normally used for solids, in this method, warm air is passed over the product on perforated trays, and the exact conditions vary from product to product. The initial temperature is reduced to around 55°C in later stages to prevent damage to surface. The solids move to the surface during the drying process, causing slight denaturation of proteins. The drying imparts a bright shining to the surface. Dried products are slow to reconstitute upon rehydration because they are extremely hard and shrunken irregularly.

Puff-Drying

This method is used for vegetables. In this method the product is dried by conventional method to about 40-50 per cent moisture and then partially dried material is sealed within a chamber with a pressure raised to about 30 to 60 psi. The chamber is opened instantly to the atmosphere to release the pressure and dried vegetable is discharged in the puffed form ready for the completion of drying process. This process imparts a porous structure to the dried product.

Foam-Mat Drying

This is used for liquid or liquidized products, which when converted to stabilize foams dry rapidly in air due to moisture movement by capillary action in the liquid films separating the foam bubbles. It imparts a honey comb structure to the dried products, so it is very easy to reconstitute the product to its original texture. Foam stabilizers suitable for this purpose include soya proteins, albumin, sucrose, fatty acid esters, and glycerol mono-stearate. This method is used for potato, baby food, and fruit purees.

Drum Drying

This method is also suitable for liquid products in which the liquid product is run as a thin film over a heated roller. The products get dried instantaneously and are removed by scraping. This method has been replaced by other newer improved methods such as spray drying.

Spray Drying

This is a preferred method for liquids, paste, and slurries. In this process, the product is dispersed as small droplets and suspended in drying air. The major advantage of this method is short drying period required to obtain a better quality end product. Short drying time and milder heat treatment minimizes the loss in flavour, colour, and nutritive value of the product. A new technique involving passing of droplets of tomato puree down a large tower against the stream of warm air has been developed in Italy. Another modification (foam spray-drying) of this technique, which involves vacuum concentration of products and mixing with air before forming the droplets, has also been developed to overcome certain shortcomings of this method.

Belt-Trough Drying

In this process, warm air is blown over the products placed on a trough of wire mesh and kept moving on a conveyor belt. Fresh material is fed to the trough at one end and uniform drying is achieved by mixing due to tumbling action of the trough. This prevents overheating of only one portion.

Vacuum Drying

This method utilizes lower drying temperatures and hence, results in better quality end products. Vacuum drying can be adopted for large pieces of products. In this process, the product is sandwiched in a vertical bank of horizontal hollow steel plates. Hot water at controlled temperature is circulated through these plates to provide heat for evaporation and plates can be moved together during drying to maintain good thermal contact on both top and bottom surface of product. This process is used for obtaining friable material from concentrated fruit juices (60 per cent soluble solids).

Microwave Drying

Microwaves are electromagnetic waves used for drying of produce. As microwaves pass through food material, the molecules within food attempt to align themselves with the electric field direction. As they oscillate around the axis, heat is produced by the intermolecular friction with the amount of heat varying depending on structure, shape, composition and mass of the product. Microwave driers that operate under vacuum permits short drying times at relatively low temperatures. Fruit powder prepared by using this type of driers retains flavour, aroma and colour of the original product.

Freeze Drying

Freeze drying involves the removal of water from the frozen products by sublimation of ice crystals under vacuum with careful application of heat to keep the temperature of the tissue sufficiently low to avoid thawing. Since the water in the product cannot migrate from the inner tissues to the surface, solids, and salts are not transported to the surface. This prevents damage to storage proteins. As the ice crystals at the surface sublime, the drying front recedes to the inner tissues and a highly porous structure is achieved. The products dried this way can be reconstituted easily. The only disadvantage of this method is its high cost, which limits its application in fruit and vegetables processing industry.

In recent years, a new technique of accelerated freeze-drying with several modifications, has been developed to allow shorter drying time and lower drying temperatures. These techniques produce better quality end products and allow larger vegetable and fruit products such as brusseles sprouts, berries, and apricot halves to be handled without any difficulty. Freeze drying is probably the best method for obtaining top quality product.

Osmotic Drying

Osmotic drying is based on removal of moisture from a fruit pieces by placing them in contact with granular sugar or a concentrated sugar solution. The product is reduced to about 50 per cent of its original weight by osmotic dehydration after which it may be frozen or dried further in an air or vacuum drier. The osmotic agent used is dry sugars, sugar syrups, and sugar-starch mixture. Sucrose is considered to be the best for all round drying of substance. Many times, the fruits are partially dried by osmosis in sugar or syrup followed by vacuum drying. The product retains a large percentage of flavour volatiles of fresh fruit and has a crisp texture. Banana slices are dried by using these techniques. A sulfurdioxide treatment is sometimes used to help in preservation of colour. The economics of this process probably depends upon the availability of cheap sugar and on the possibility of using spent sugar solutions in canning, bottling and soft drink plants.

Drying of Fruits and Vegetables

Raisins

Raisins are the second most important product prepared from grape berries, wine being the first. The quality of raisins depends on the size of berries, the uniformity and brilliance of the berry colour, the condition of the berry surface, the texture of the skin, pulp in the berry, moisture content, chemical composition and presence of decay, mould, yeast and foreign matter. Based on the method of preparation and variety of grape used for raisin making, they are called natural, Sultana, golden bleached, sulfur bleached, soda oil dipped, Black Corianth and Valencians (Figures 15.1–15.3).

Predrying Treatments

Grapes are immersed in an alkaline solution prior to drying. Dipping the berries facilitates drying by forming cracks in skin. A sodium hydroxide (0.5 per cent NaOH) is used at a temperature usually ranging from 93 to 100°C. In Australia and India, cold dip solutions such as potassium carbonate (2.4 per cent) or lye with olive oil or ethyl oleate or commercial dipping solutions are used. These dips accelerate moisture loss by causing wax platelets on the grape skin to dissociate thus facilitating water diffusion. Raisins produced by cold dip process are light in colour. Other researchers have used acid dip (ascorbic acid + malic acid) instead of sulphuring as a method of obtaining light coloured raisins. However, this product would undoubtedly have to be held at reduced temperature to prevent darkening during storage.

Sulfuring

The use of sulfurdioxide (SO_2) is common in food industry. Grape berries are exposed to SO_2 before drying. These can be sulfured by placing them in compartment containing burning sulfur. Recently, bottled gas of SO_2 is injected into chamber containing the fruits. The bottled gas system offers numerous advantage such as better ability to control the quantity of SO_2 absorbed by the fruit, less air pollution, dependability and it is cost effective. The absorption and retention of SO_2 are influenced by the temperature, duration of sulfuring and concentration of SO_2 as well as the size, condition, maturity and cultivar of produce being sulfured. The levels of residual SO_2 in dried fruits depend upon duration of storage and storage temperature. The permissible level of SO_2 in raisins in India is 750 ppm.

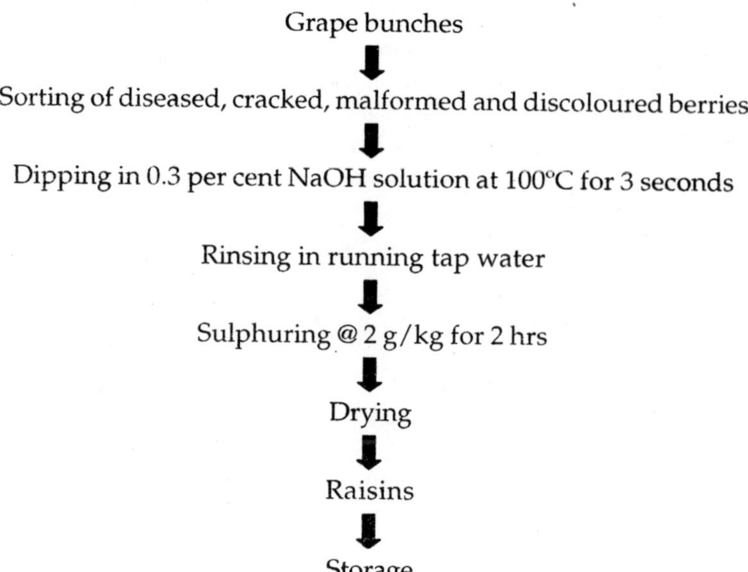

Figure 15.1: Process for preparation of grape raisins by golden bleach method

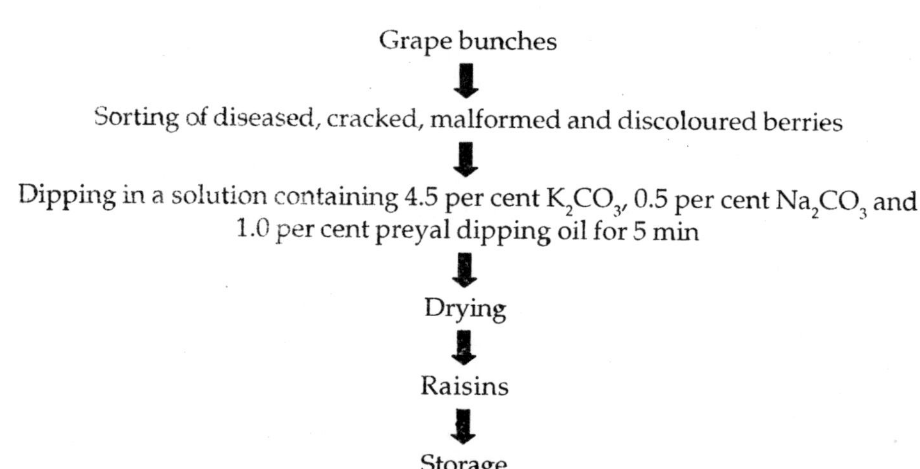

Figure 15.2: Process for preparation of grape raisins by soda oil dip method

Drying

The bulk of the world's raisins are dried in yards, outdoors by spreading clusters on damped earth floors that are exposed to sunshine. After 7-8 days, the clusters are turned by hand to complete drying. They are called natural raisins. In Maharashtra, grapes are dried under shed with intermittent spray of ethyl oleate. In Iran and Afghanistan the grape clusters are threaded on a vertical string suspended from

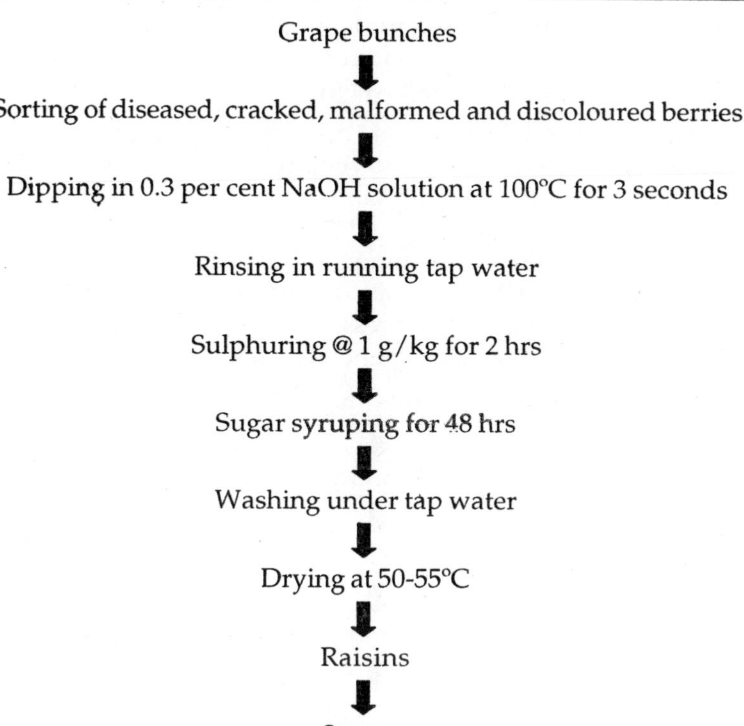

Figure 15.3: Process for preparation of grape raisins by sugar syruping method

ceiling in home, drying very slowly in the shed and it producing light coloured raisins of excellent quality. This has lead to construction of special drying houses in vineyards slitted walls for air circulation and yet protected from rain. After drying the fruits about 15 per cent moisture, it is bulked in large containers to equalize content and riddled to remove most of the dried cluster stems. The raisins are then passed through sizing screens for size grading.

Dried Figs

In recent years, the demand has been increasing for dried fruits with higher than usual moisture level (15-20 per cent). Such fruits are softer in texture and more readily prepared for eating. Figs are dried as whole fruits. In United States Symrna, which is a large white fig as well as Adriatic fig which has pink flesh are used for drying. The fruits are allowed to ripen on the tree and gathered when they drop. They are then spread thinly on drying yard for 3-4 days for drying. After drying, they are sorted and packed.

In California, Calimyrna figs are treated with lime and salt (1 kg each per 1000 liter water) to remove the hairs from skin and also to soften the flesh. They are then dried without sulfuring until there is exudation of juice upon pressing the dried fig between the fingers. In another method, fruits are dipped in boiling salt water for half a min then dried for few hours in sun and for 8 days under shade. At the end, the weight is reduced little over 1/3 of fresh weight.

In India, Daulatabag and Poona figs are dried by subjecting fruits to Ethrel (5000 ppm) treatment for 3 min followed by sun drying. In another method, fruits are dipped in 5000 ppm Ethrel, and then kept at room temperature for 4 hours, blanched in boiling water for 4 min and sulfur fumigated (4 g/kg) for 2 hour. Such fruits are then steeped in sugar syrup (40°Brix) containing 1 per cent citric acid to raise the level of fruit TSS to 26 °Brix. These are then partially dried and converted into flat and round shaped product by employing kitchen press. These are further dried to 16 per cent moisture in cabinet drier and packaged in polyethylene bags (Figure 15.4).

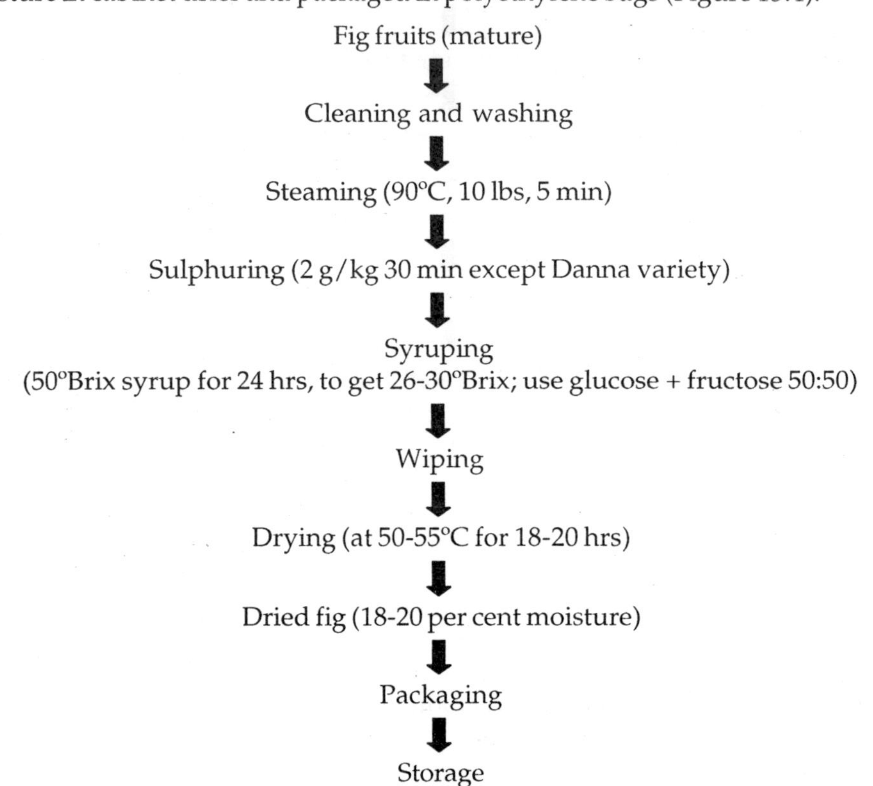

Figure 15.4: Process for preparation of dried fig

Potato Chips

Potatoes are washed, peeled and cut into slices of 0.4 to 0.6 mm thick in slicing machine and slices are placed in cold water or may be kept in 0.05 per cent KMS or 2 per cent NaCl solution. The slices are blanched for 1 min in boiling water and spread on trays. The trays are then loaded on the trolleys which are pushed into dehydrator tunnel. The drying temperature ranges from 60 to 66°C. The drying time is 7-8 hours and drying ratio is about 7: 1 in case of unprepared potatoes. These are subsequently fried and used as snack (Figure 15.5).

Potatoes (fully mature)
↓
Washing
↓
Sorting (removal of defected and small tubers)
↓
Peeling (by abrasion peelers)
↓
Trimming (removal of residual skin, discoloured area, green portion)
↓
Slicing (1/15 – 1/25 inch thick slices by rotary slicer)
↓
Drying (optional process may be done by drum driers)
↓
Frying (in continuous or batch fryers using refined oil)
↓
Salting and addition of flavouring materials
↓
Cooling and inspection
↓
Packaging
↓
Storage

Figure 15.5: Process for preparation of potato chips

Dried Peas

Only tender green peas are dehydrated after blanching in boiling water for 1-2 min and drying at a temperature not exceeding 63°C. Many times, peas are dehydrated by puncturing skin mechanically to facilitate uniform and quicker drying of peas. The dehydrated peas packed in aluminum and polyethylene pouches have recently become highly popular.

Dehydrated Onion

Onions are dried after removing outer leaves and cutting into 0.25 cm thick slices. The slices are immersed in 5 per cent solution of common salt for about 10 min and then drained. They are dried at 60-65°C for 11-13 hours. It is preferable to keep drying temperature below 57°C. The drying ratio is about 10: 1. The dried product can be powdered and packed for use in several ways. Recently, freeze drying of onion shreds is commonly followed in many parts of this country. For drying, onions of white variety with high pungency are preferred.

16

Freezing

Introduction

Freezing is one of the important techniques of preservation of fruits and vegetables. The use of freezing as a means of food preservation had to wait for the development of mechanical refrigeration. The vegetables were mechanically frozen first time in 1937. The industry has continued to grow over the years to have nearly 4 million tons of frozen fruits and vegetables in the year 1984 in U.S.A. Freezing technique is commonly used in the developed countries. However, this is also becoming increasingly popular in the developing countries.

Developments in Freezing Techniques

The use of low temperature for storing food materials began with the invention of refrigerator. Although. the application of refrigerator in food storage was invented very early, it was not until the discovery of ammonia compressor by Carl Linde that practical refrigerator equipped with air compression device was first used commercially to carry cargo load of frozen foods. Soon after the invention of mechanical refrigerator in the mid-nineteenth century, steamships and cargos carrying frozen meat were fitted with refrigeration plants. This really revolutionized the eating habit of Western people and since 1940's frozen foods became the significant competitors to other preserved foods.

One of the major problems with slow freezing (as in conventional refrigerator) is that the moisture in the product tends to form large crystals. These ice crystals lacerate delicate tissues so that when the produce is thawed again, intercellular nutrients are drained away with moisture. Quick freezing does not allow the formation of ice crystals so that when the product is thawed, it resumes its original condition without significant loss of nutrients. The product is made to pass through the zone of crystallization, the most critical stage, as quickly as possible. For most perishable plant produce, this zone lies in the range of 0 to 4°C. Today the products can be cooled to that temperature in a very short time (in a matter of few minutes as against two or three hours in the conventional method). Tremendous success in quick freezing and high quality of the frozen foods is largely due to the innovative research efforts conducted by chemists and physicists all over the world to understand the basic principles of freezing of perishable products.

Process

There are mainly three freezing processes that are being commercially used for preserving perishables: individual quick freezing, in-container freezing, and immersing the produce in a frozen solution. In the individual quick-freezing technique (IQF), the product is frozen by a fluidized-bed process and then packed in suitable containers. This method is quick and allows for large quantity of materials to be handled at a time.

The second process involves placing the unfrozen products in a container before freezing them. The produce is commercially frozen with cold static or moving (blast freezing) air. In blast freezing process, the air movement is accelerated by fans and the accelerated air is passed over the produce. In contact plate system, the freezing coils are placed in a plate and the food is placed between these plates. In all these systems, the air temperature is maintained around -34 to–40°C. In the third process, the produce is immersed in a liquid such as salt solution or any liquid (liquid nitrogen, liquid carbon dioxide, and Refrigerant 12) having very low intrinsic temperature. The produce in contact with the liquid freezes very rapidly.

Preparation of produce for freeze drying is very important. Recent developments in this area have emphasized the use of improved blanching and cooling systems. Evaporative cooling using cold water sprays has replaced the immersion cooling in flood washers. These improvements have resulted in faster freezing of produce and better quality end products. Another area which received lot of attention is the packaging technology for frozen foods. Most striking improvement is the introduction of boil-in-the-bag technique, in which the product is frozen in hermetically sealed, flexible plastic containers resistant to high temperatures. This requires a longer (as much as doubled) blanching operation. Temperature resistant plastic laminates (e.g., saran and propylene) have been developed.

Freezing Systems

Sharp air-blast, fluidized-bed immersion, and plate freezing are some of the more common freezing techniques. The choice of freezing technique depends on the kind of fruits and vegetables, cost, desired quality, volume of produce to be frozen and stored, and space available. For example, for diced carrot, shelled corn, green peas, and lima beans, either a belt or fluidized-bed freezing is used. Large vegetables such as asparagus, cauliflower and broccoli are frozen in cartons with plate freezers. Contact freezing technique has been used for almost all vegetables. Other methods like liquid nitrogen or fluorocarbon freezing, because of high cost, are used only for high cost items.

Blast-Freezing

Blast freezing, a commonly used freezing method, involves packing the produce in cartons which are loaded on trays and placed on portable racks which are moved to an insulated room. Cold air at -29°C or lower is blown at a high velocity through these cartons. Fruit and vegetable products usually freeze in about 3-12 hours depending on the produce, air temperature, and air velocity. This method is being replaced by other more rapid freezing methods.

Plate-Freezing

In this method, individual cartons are sandwiched between two refrigerated plates. This method takes about 30 to 90 min for freezing the produce. This is a very common method of freezing fruits and vegetables that are packed in cartons.

Belt-Tunnel Freezing

In this method, clean dry produce is placed on a wire-mesh belt or a stainless steel conveyor in an insulated enclosure and cold air at -32°C or lower is blown up through the belts. The produce is spread uniformly on the belt by shaker-spreader and kept moving on the conveyor so that the unfrozen material enters at one end, the frozen product leaves at the other end of the freezer. The produce is frozen within a period of 3 to 5 min.

Fluidized-Bed Freezing

This method is similar to the belt-tunnel freezing except that the cold air is blown at much higher velocity to partially fluidize product so that it is completely surrounded by a thin film of moving air." The product temperature can be reduced from room temperature to–20°C in about 3-10 min depending on the product and air temperature. It is adapted to products like diced carrot, peas, cut beans, limabeans, and corn. The produce that is frozen with this method is known as individual quick frozen products.

Cryogenic-Freezing

In this method, a cold refrigerant (liquid nitrogen, Refrigerant 12) is brought in direct contact with the produce to be frozen. This technique is very efficient and can freeze the produce within few seconds.

Dehydrofreezing

This method combines both dehydration and freezing process, in which the produce is first dehydrated and then frozen. Dehydrofreezing reduces weight of a fruit or vegetable to about 1/2 its original weight by warm air drying before freezing. A wet surface exposed to a hot air current will lose moisture rapidly, resulting in evaporative cooling effect which will maintain the surface temperature considerably lower than the temperature of the surrounding atmosphere. Product temperature is thereby maintained below the level that favours deteriorative changes in food. Drying may be carried out in cabinet, tunnel, continuous conveyor or continuous mixing dryer.

Technique

Fruits and vegetables to be frozen are harvested either mechanically or by hand and quickly transported to processing plants. The products are then cleaned, graded before being conveyed to the peeler, pitter and slicers. Vegetables such as carrots and potatoes are peeled while others such as peas, green beans etc. go directly to the sorting or inspection. Most fruits and some vegetables such as onions and green peppers at this stage are ready for packaging. Most vegetables next go through blanching operation to inactivate enzymes that could cause off-flavour in the product during storage. The products are then either packaged or frozen first and then packaged. Granular sugar or syrup is usually added to the packaged fruit to retard oxidative reaction and improvement in product quality. Freezing is accomplished by conveying the packaged product through a low temperature freezer to freeze the product quickly after which it is conveyed into storage freezer. Such products are frozen before packaging by individual quick freezing (IQF) procedure. This procedure provides a very rapid freezing rate and fruits and vegetables are frozen as individual units so that they are free-flowing in the frozen state. The frozen products are stored at -18 to -20°C.

Storage of Frozen Products

Frozen products are best stored at temperatures of -18 °C or lower. The stability of flavour and texture is adversely affected when stored at temperatures above -18°C. Thawing of frozen products and then refreezing them can degrade the quality of such products. Therefore, it is essential to store or transport frozen materials in the frozen state.

17

Processing of Minor Fruits

Aonla

Fresh aonla fruits are highly acidic and unsuitable for direct consumption. Hence, the fruits are processed into products such as preserve, pickles, and powder. Aonla fruits are widely used in the *Unani and Ayurvedic* systems of medicines. They are used in the form of fresh fruit or powder in various preparations such as *Chavanprash, Triphala and Arishtha.* A number of products including preserve, sauce, candy, dried chips, powder, chutney, diuretic, and laxative are prepared from aonla fruits (Figure 17.1).

Preserve can be prepared by following the method described in Chapter 8. Preserve prepared by using optimum-maturity fruit keeps longer and has better organoleptic qualities. The optimum stage of maturity appears to be from mid-December to the third week of January, when the fruits have 15.6-15.9 per cent TSS, 2.55-2.46 per cent acidity, 1.09-1.05 per cent tannins, and vitamin C content of 500-564 mg/100g.

Aonla pickles are becoming increasingly popular in India. Pickling results in changes in chemical composition of fruits. A substantial quantity of the ascorbic acid of aonla may be lost in the ordinary method of pickling, but much of it can be retained by boiling the fruit in water for a few min and then putting them in a heavy salt solution. Powder is prepared by mincing the fresh fruits and powdering slices after natural drying in the sun. The product contains about 10- 16 mg ascorbic *acid/100g* powder, which may be increased by refinement of the process. Steeping fruits in 2 per cent brine for few days removes most of the astringency and almost 1/3 of the ascorbic acid. The extract can be preserved with potassium metabisulfite or by adding sugar to the extract.

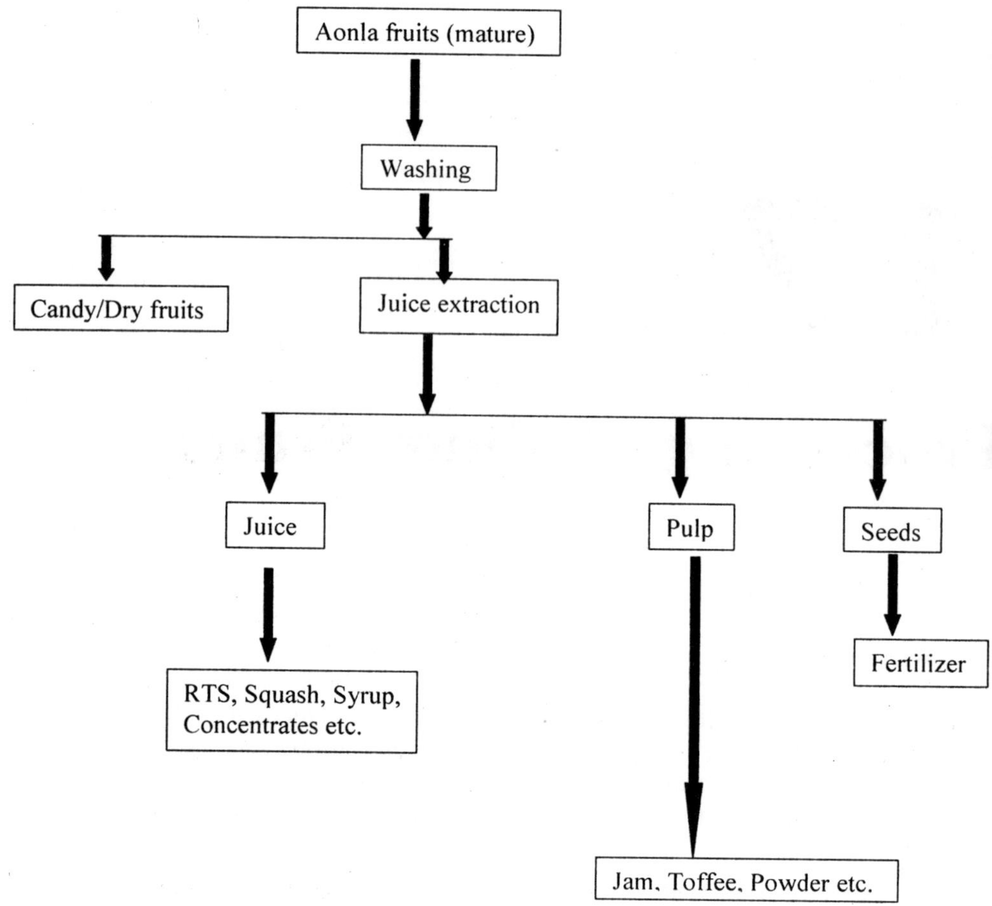

Figure 17.1: Processing of aonla fruits in to various products

Jamun

Jamun has never been exploited commercially for preparation of different products. However, it is reported that the fruits are used for making products such as jam, jelly, beverages, wine, vinegar and pickle (Figure 17.2).

Maximum yield of jamun juice with a high level of anthocyanins and other soluble constituents can be obtained by grating the fruit, heating it up to 70°C and passing the heated mass through the basket press. The jamun juice thus obtained is again heated to 85°C and then cooled to room temperature. Sodium benzoate is added to the juice before it is stored. Pure jamun juice can also be preserved by heat pasteurization. Concentration of the jamun juice is done either in open pan evaporator or vacuum concentrator. However, it has been found that in the open pan evaporator, concentrates only up to 30°Brix are acceptable; on the other hand, jamun juice can be concentrated to 60°Brix in the vacuum concentrator.

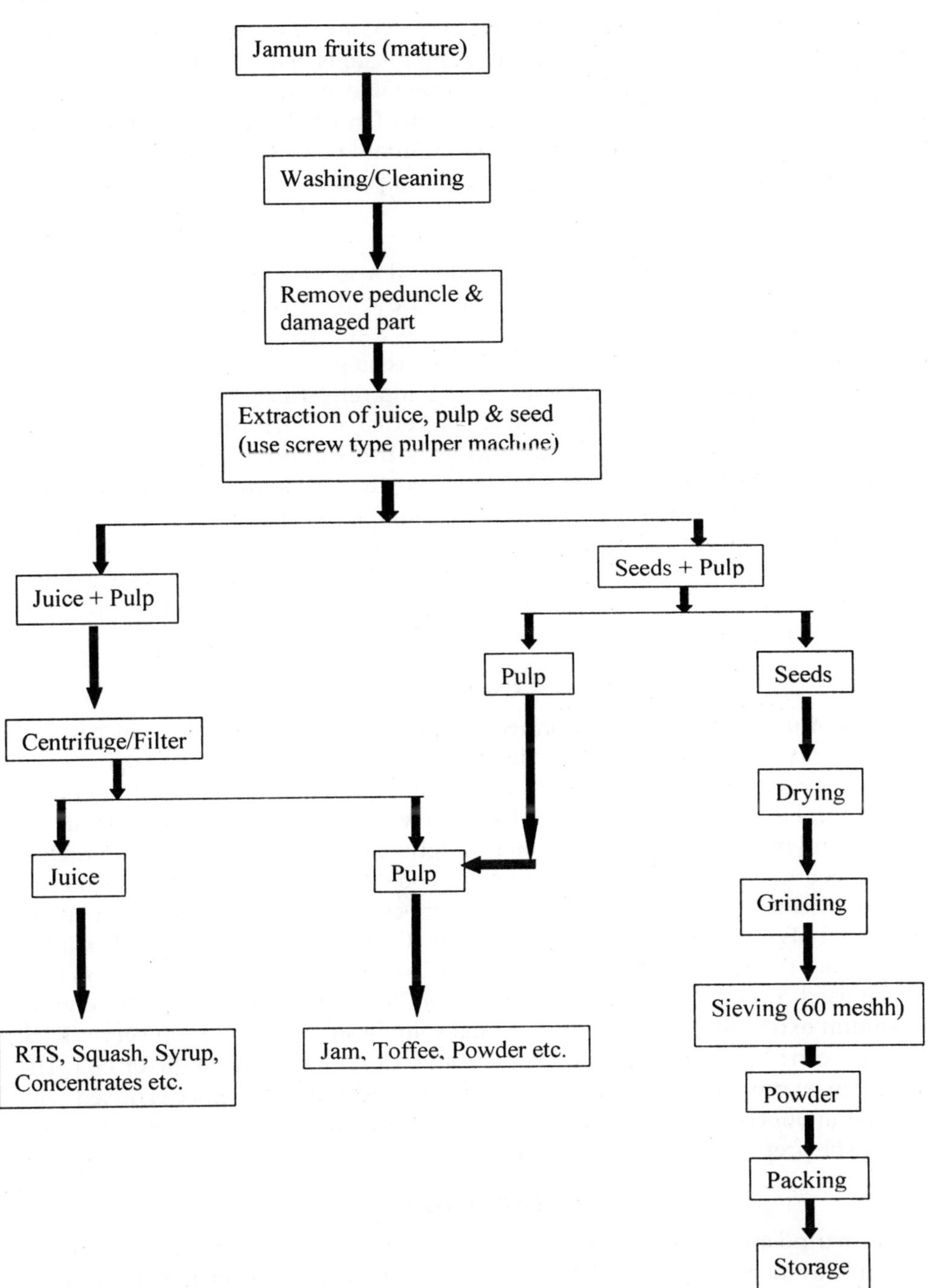

Figure 17.2: Processing of Jamun fruits in to various products

A ready to serve beverage (nectar) can be prepared from jamun juice with 25 per cent juice, 18°Brix and 0.6 per cent acidity. This nectar is delicately flavoured, has an appealing colour and is found to be highly acceptable. Acceptable dry table wine can be prepared from jamun juice. In the preparation of must, dilution of whole fruit pulp in the ratio of 1: 1 with water has been found suitable. The use of pectinase enzyme was found to be more beneficial in getting a clear product.

Kokum

The ripe fruit has an agreeable flavour and acid taste. It contains substantial amount of malic acid and little tartaric and citric acids. The fresh ripe rind is made into acid drink called 'Solkade' which serves as a substitute for butter milk. From fruits, kokum syrup locally called 'Amrit Kokum' is prepared which makes an excellent "Sherbet". It is also useful in fever as cooling and refreshing drink. The dried rind "Aamsol" is used as condiment in curry. The kernel, which forms about 60 per cent weight of seed, yields 33 to 44 per cent oil. On the basis of seed weight, the oil content is about 23 to 26 per cent. The oil is popularly known as 'Kokum butter'. Kokum fat is now much in demand as a substitute and extender of cocoa butter in chocolates and other confectionery products. On account of high stearic acid content, it could be used as source of stearic acid. It is suitable for ointment and other pharmaceutical preparations. The cake left over is used as a cattle feed. Most of the kokum fat produced in India is exported to Netherland, Italy, Japan and Singapore.

Phalsa

Phalsa juice has a deep, crimson-red colour and a pleasing flavour, which makes the juice very popular. In addition, the juice is extremely refreshing in summer. A method of extraction of juice from phalsa fruit has been developed at Indian Agricultural Research Institute, New Delhi, in which fully ripe and sound fruits are crushed and mixed with two-thirds their weight of water in a Waring blender. Then the crushed mass is heated quickly to 80°C, cooled, and finally pressed through a piece of muslin cloth to obtain a clear, crimson-coloured juice. In another method, crushed phalsa fruits are heated to 60-70°C for 10 min. This method gives clear juice. The pomace (about 30 per cent) left after the extraction of the juice from phalsa fruit can be used to extract pigments. Addition of 75 per cent water to pomace gives maximum extraction of pigments and total soluble solids. A ready-to-serve (RTS) beverage can be prepared by using 25 per cent juice and °Brix-acid ratio of 25: 1. The syrup can be prepared from phalsa fruit juice in which the clear juice is mixed with an equal amount of sugar and preserved with sodium benzoate. A carbonated beverage can also be prepared from phalsa juice.

Tamarind

Tamarind is one of the important fruit crops of India. It is mostly used as souring agent in various recipes in South India. The separation of pulp from tamarind seed is rather difficult due to low water content. In pulping process, shell is removed manually followed by agitation of seeds in water in order to disperse the pulp and separate it from seeds. The pulp can also be separated from seeds by covering with water placed in a steam bath for several hours, allowing to stand overnight, filtering through a filter paper and washing with hot water. The industrial process for the manufacture of tamarind juice concentrate in India is also based on extraction of pulp with boiling water.

A tamarind beverages containing 9-12 per cent pulp adjusted to 21.5°Srix with sugar and 0.35 to 0.44 per cent titratable tartaric acid can be prepared. Tamarind pulp is the fruit base in preparation of jams, jellies and ice-cream. In South American countries, it is often enjoyed in the form of refreshing drink.

Tamarind juice can be clarified by treatment with 0.12 to 0.15 per cent gelatins. After agitation of fruit pulp' with gelatin, the mash is left standing for 10-15 days at 6-10°C. During this period, colloidal particles are precipitated and removed by decantation and filtration in vacuum, using asbestos. The mash is adjusted to 0.75 to 0.80 per cent tartaric acid content and 18°Brix and pasteurized at 80-85°C for 5 min. It maintains its distinctive colour and flavour while being absolutely transparent. Syrup can be prepared from tamarind pulp and sugar. The recommended concentration of tamarind pulp in syrup is 20-24 per cent with 56.7 per cent total soluble solids and 1.11 per cent acidity which makes a beverage after dilution with distinct flavour and desirable acidity. A tamarind juice concentrate can be prepared by clarification of fruit extract and concentration of clear extract under vacuum to 65-68 per cent total soluble solids. The products can be filled into sterile glass containers. The products set to a jam like consistency upon cooling.

Woodapple

The pulp of the ripe fruit is eaten as such or with sugar. It can be used for making "sherbet". The pulp is used in making chutney. The fruit is very rich in acid and pectin. The fruit has a thick hard shell inside which is a darkish-brown and acid-sweet pulp, in which a large number of small seeds are embedded. The acid content of the pulp varies from 7.6 per cent in unripe fruits to 2.3 per cent in fully ripe ones. It contains 3.5 per cent pectin and forms an excellent material for jelly. The shell is broken and the pulp is boiled with water, in the proportion of 1:3 (pulp:water) for about 30 min. On cooling, it is strained, and to the liquid is added about its own volume of clean crystalline sugar. On adding sugar, the liquid is boiled to about 105°C, and then it is poured hot into sterilized container for setting. Woodapple jelly resembles to that of black current or apple jelly. It is clear and bright purple in colour with firm consistency. In addition to jelly, wood apple syrup can also be prepared and used as a drink. The chutney of this fruit if standardized may also be turned into a commercial product.

18

Waste Management and Value Added Products

Introduction

The present day society has become increasingly concerned about the environmental safety and social costs involved in treating the wastes generated in the processing industry. These public concerns have led several countries to pass legislations that severely curtail past practice of disposing the wastes in flowing streams. The new legislations require strict compliance to proper waste management procedures in every processing operation. Production of wastes can greatly be reduced by improving the efficiency of the operation. Characterization of wastes in processing operation is probably the step in any waste management programme. These methods identify and take inventory of waste materials generated in a particular processing industry, produce primarily the solid wastes having higher BOD than the dairy or the meat processing industries. Characterization of waste materials can be used as tool for developing more efficient processing technologies. The pest example of this is the introduction of lye peeling to increase the recovery of salable product and thereby significantly reducing the production of wastes during juice extraction. These newer technologies ensure safety, reduce energy use, maintain and/or enhance nutritional quality, reduce food losses and wastes, and increase the turnover of the operation.

Peels, rags, rind, core, trimmings, and seeds left after juice extraction, canning, or preparation of squash, jam, jelly etc., can be used to produce certain products of commercial importance. For example, apple pomace can be dried and utilized for the production of pectin. Apricot kernels can be added to apricot jam after removing the outer coat to improve its appearance. Similarly, stems and pomace left after the preparation of grape wine can be utilized for producing tartar and oil, respectively. Other byproducts of fruit and vegetable processing industry are guava cheese (made from core, seeds, and peelings), jackfruit jelly (from thick rind) vinegar (from fermented mango peels), pectin (from rind of passion fruit and citrus), oil (kernels of peach, apricot, tomato), citric acid and citrus oil.

Fruit Processing Waste

Some fruit waste material is used to produce methane gas. Other fruits, such as dried culls or surplus grapes, apples and oranges have been used for the production of brandy. Juice from fruits can also be used to produce citric acid.

Apple

Peels and cores from apple sauce canneries and apple driers can be utilized for producing vinegar and for jelly stock. In jelly manufacturing, the dried material is cooked with water to produce a jelly juice that can be combined with other juices, sugar, and citric acid and boiled to form a gel. The product can be artificially flavoured to produce a broad spectrum of flavouring products. Because of the ready availability of powdered pectin, less dried apple waste is used in this manner. Apple peel and the waste left after the juice extraction can also be used to produce natural flavourings.

Citrus

Citrus processing produces a large amount of waste materials which can be broadly divided into three categories; animal feed, a raw material used for further extraction of marketable components, and food products. Dried citrus meal that is used for animal feed is probably the main waste recovery product. Citrus seeds can be used for oil extraction and also the production of a citrus seed meal for animal feed. The raw material that is further extracted produces citrus peels oils, citrus pectin, flavonoids, and citrus seed oils. Food items produced are brined and candied peels, marmalades, syrups, and peel products used in food seasonings.

Pineapple

Pineapple peels and cores from canning operations are used for making syrups for addition to canned pineapple, producing juice and for making vinegar. The juice has also been fermented to produce industrial alcohol and can be used to produce citric acid by treating with lime in order to produce citrate from which citric acid can be recovered.

Mango

In the mango canning industry, 25 to 30 per cent of the fruit is lost in the form of peels. These peels can be extracted with water and fermented into fruit vinegar. The kernel of the stone can be dried, powdered and utilized for edible purposes. The possibility of utilizing mango stones which form a fairly large proportion of the fruit, as a source of food at the time of scarcity of cereal foods is of considerable importance.

Banana

In processing of banana for canning and dehydration, the banana peel is a waste product. The pulpy portion scraped from the thick peel of the banana can be utilized for the preparation of banana cheese similar to guava cheese. The pseudostem of the banana plant which is cut down after harvesting the bunch can be utilized as raw material for the preparation of paper pulp.

Vegetable Processing Waste

Tomato

Tomatoes represent a major tonnage product that is processed in many parts of the world. It is estimated that about 1/3 of the tomatoes delivered to a processing plant end up as waste. Tomato seeds contain oil that can be recovered and even the seed has potential for food use since its amino acid composition is similar to soyabean protein. Most of the tomato processing waste, attention has been centered on the utilization of the tomato pulp and pomace as animal feed.

Peas

The vines and pea hulls can be dehydrated and used in the preparation of stock feeds by suitable blending with other materials.

In the future, environmental and economic concerns will necessitate the adoption of integrated waste management technology involving a significant reduction in processing waste, better use of raw materials, and processing of food residues to new high value products.

19

Food Quality Control and Regulations

A family spends a large share of its income on food. However, the diet of many families in India is not well balanced in nutrients such as proteins, fats, carbohydrates, vitamins and minerals. All these nutrients are essential for health and need to form a regular part of the diet. Due to rising prices, many people are not in a position to buy nutritious food and that too in adequate quantity. Many times the food purchased at high cost is often found adulterated by substances which are injurious to health.

What is Food Adulteration?

A food is said to be adulterated if its quality is lowered or affected either by addition of substances which are injurious to health by addition of cheaper commodities resembling the food, or by removal of substances which are nutritious. Foods are adulterated by unscrupulous elements and greedy traders to become rich quickly. They take advantage of the shortage of food articles and mix fake material and sometimes material which is injurious to health. These anti-social elements have even set up factories to manufacture adulterants (materials used to adulterate) and sub-standard articles (Table 19.1). Adulteration deprives the food of its nutritional value. That is bad enough, but worse is the fact that intake of adulterated food affects health sooner or later in many ways. It may cause to name a few, blindness, paralysis or tumor. If the adulterant is contaminated or infected, it may give rise to diseases like cholera, typhoid etc. For instance, milk is adulterated with water, this lowers its quality.

Table 19.1: Adulterants used for adulteration of food products

Product	*Adulterant Used*
Besan and dhal from arhar and channa	Khesari dhal
Chilli powder and other spices	Saw dust
Edible oils like coconut oil, mustard oil	Mineral oil, Castor oil, Argemone oil
Sweets and aerated water	Inedible colours
Turmeric powder	Lead chromate (yellow) or Metanil yellow
Alcoholic beverages	Poisonous adulterants
Ghee or butter	Vanaspati
Milk	Water, urea, starch
Groundnut	Rock salt
Ice-cream and serbet	Metanil yellow dye and non-permitted colour
Maida	Maize or tamarind seed powder
Corn flour	Tamarind seed powder
Vanilla essence	Acetic acid
Baking powder	Cereal flours
Yeast	Grain flour (white colour)
Khoa	Maida
Wheat semolina	Maize semolina
Tea	Black gram hulls, and sawdust

But if dirty and contaminated water from a roadside pond is used for adulterating milk, it is not only diluted but becomes a serious danger to health, Butter milk or lassi can be similarly contaminated by dirty water. Many of our common food items like milk, atta, khoa, ghee, butter, edible oils, tea, coffee, dhals, turmeric powder, chilli powder, spices etc. may be adulterated with substances which cause ill effects on health.

Food quality is generally defined as the regulation by law of food manufacture, distribution and sale in order to prevent health hazards and fraud to the consumer. There are three main aspects to the application of food quality control:

1. Moral
2. Commercial, and
3. Legal.

The commercial aspect requires that standard quality products are economically produced.

The legal viewpoint demands that the quality of products conforms to national and international standards.

The control of food quality by law leads to:

1. Improved quality of product.

2. Achievement of greater consumer satisfaction.
3. The promotion of quality consciousness.
4. Increased consumption and sales.
5. Employment opportunities for scientific and technical personnel.
6. Avoidance of controversy and litigation in marketing at the national and international levels.
7. Promotion of national and international trade.
8. Provision of the means for the intelligent comparison of prices in relation to quality and grade.
9. Greater confidence in the minds of consumers.

The specific responsibility of quality control is to ensure that the system used produces a standard product with acceptable quality in respect to nutritional, purity, wholesomeness and palatability.

The specific responsibilities of quality control assigned to a department or to an individual include:

1. Standardizing procedure for sampling and examining raw materials.
2. Development of test procedures.
3. Establishment and implementation of quality standards for fresh and processed products.
4. Setting up preventive quality control methods for in-plant liaison between manufacturing section and test laboratories.
5. Examination of finished products.
6. Storage controls.
7. Recording and reporting.
8. Special problems, including attendance to consumer complaints by locating their cause and eliminating them.
9. Research and development into new products and their packaging.

The sequence of operations in quality control is as follows:

1. Raw material control.
2. Process control, or the control of the manufacturing process.
3. Production inspection, including the inspection of the finished product, packaging and storage.
4. Sensory evaluation or evaluation of the acceptability of the final product.

Raw Material Control

1. The quality of food material is judged in terms of its nutritional value, its purity, its wholesomeness and its palatability.
2. The definition given to a "raw material" in the food industry is anything purchased by the manufacture for direct or indirect use in food processing?

3. Raw materials include: food ingredients, water and packaging materials.
4. Raw materials examinations generally include tests for genuineness and composition, freedom from contaminants and conforming to official or factory standards.
5. Examinations carried out vary with the nature and type of ingredients and its expected use.
6. Where fats and oils are included in the ingredients used, they are examined to determine their identity, purity, freshness and keeping qualities because chemical changes or rancidity taste and odour if the fat is used below the required quality.
7. The shelf-life of the end-product, signs of corrosion of equipment must be checked.
8. Approval for processing is only given after all quality specifications on the sample run have been met.

Process Control

1. All treatment given during the processing are standardized, ingredients are used in the correct amounts, accurate methods of preparation and mixing are employed, checks are made on the containers used to make sure that they are sound, and processing times and temperatures are standardized to make sure that the desired results are obtained.
2. There may be change in shape or structure: Cracks, cause hardening, a browning reaction, and the oxidation of unstable components due to physical, chemical or biological processes.
3. Satisfactory hygienic conditions need to be maintained.
4. For the routine bacteriological control of the plant or factory, counts on utensils, equipment, working surfaces, walls and floors are regularly carried out.
5. Periodic inspection of the plant is made by a trained inspector to make sure that adequate hygiene standards are maintained.

Inspection of Finished Product

1. To determine to what extent the desired quality specifications have been achieved.
2. At the end of the production line, packets are weighed individually. Manufactures are expected to supply customers with finished products of the correct weight and volume as specified.
3. Chemical analysis are carried out of the essential ingredients (to check if there are variations from the limits set) and
4. To check that the composition conforms to the set legal requirements for contaminants, such as heavy metals.

5. Physical properties: Such as crispness, colour, viscosity and texture must be checked because they can affect the palatability and acceptability of the product.
6. Microbiological examinations are carried out to check whether the finished product is safe to keep and to eat.

Sensory Evaluation

1. Palatability or sensory quality is of great importance to both processors and consumers.
2. Palatability attracts consumers; to the consumer, palatability satisfies his aesthetic an gustatory senses.
3. Sensory quality is a combination of different senses of perception which come into play in choosing and eating a food.

Sensory Properties

1. Colour and appearance
2. Texture
3. Flavour
4. Taste

Expert Sensory Judgement

These tests determine the significance of variation of average scores. This small group of people works in the rigorously controlled environment of the quality control laboratory.

Market Testing

By selecting people and testing is very time consuming and costly so more laboratory tests are useful for this purpose.

Three major survey techniques:

1. Summarizing market data on what consumers buy.
2 Surveying consumer opinions about products of different quality.
3. Setting up experiments designed to test preferences on the spot.

Packaging

1. Protect the food product.
2. Preserve the flavour, aroma until it reaches to the consumer.
3. Paper and paper board: Paper and paper board are used in a variety of package types and forms. These include paper wrappers, sacks and labels. Others are fibreboard cases, boxes, folding cartons, paper and carrier bags etc.
4. *Metal:* Tin-plated steel cans and boxes. The cylindrical and open-top variety, general line cans aluminium boxes. Other forms in which metal is used are aluminium foil, aerosols, collapsible tubes, steel drums, boxes and crates.

5. *Plastics:* Polyethylene, polystyrene, polyvinyl chloride and polypropylene as packaging materials.
6. Glass: Glass is a high inert material of great cleanliness. They are, ideal for the storage of solid and liquid foods.
7. *Wood:* Boxes, crates, casks, kegs, pallets and a few other types of containers made from wood are used for food product packaging.
8. *Textiles:* Cotton bags, sacks and bales are also used in the shipping of food products.
9. *Barrier packages:* To maintain a favourable climatic environment around food products which are normally subject to deterioration from moisture, oxygen, light or heat, any of several barrier materials are employed. These include waxed paper, metal foil, plastic film or any of the flexible materials.
10. *Labelling and storage:* Labelling can reflect on the quality of a product. Clear and informative, name and address of the manufacture, list of ingredients and additives, net weight or volume, expire date.

Prevention of Food Adulteration Act

Food adulteration is an offence punishable under the Prevention of Food Adulteration Act, 1954. This act was amended in 1976. According to the amended act, a person found guilty of adulterating a food article with a non-injurious substance is liable to a minimum punishment of six months imprisonment. This may extend up to three years with a fine of not less than Rs. 1000. In cases where the adulterant is of an injurious nature, the punishment can rise to an imprisonment for six years with a fine of not less than Rs. 2000. If an article of food when consumed by any person is likely to cause death or harm to his/her body as would amount to grevious hurt, the imprisonment can be extended up to life with a minimum fine of Rs. 5000.

Food adulteration has become a major public health problem. This is also a social evil and can be effectively checked only if people are alert and co-operate with the Government in curbing the menace. Whenever a consumer suspects any defect in food article, either in taste or appearance, he should report the matter immediately to the local health authority or to the health officer. Besides food inspectors, even a purchaser can have an article of food analyzed by public analysts on payment of the prescribed fee. If the food is found to be adulterated, the purchaser is entitled to the refund of the fees paid.

Every state has food analysts and local health authorities. There addresses can be had either from health officer of the area or the Director of Health Services of the State or the Food and Drugs Department where such organization has been established.

Quality Control

Statutory provisions for quality control in India; Reasons for statutory provisions:

1. To maintain the quality of food produced in the country.
2. To prevent exploitation of the consumer by the sellers.

3. To safeguard the health of the consumers.
4. To establish criteria of quality of food products, since more and more foods were eaten in processed, rather than in natural forms. This has resulted in the inability of the consumer to identify the quality of the contents that could be identified easily.

Prevention of Food Adulteration Act 1954 and Rules 1955 (PFA Act)

Primarily intended to check adulteration of food staffs available in the country. The rules and regulations that are formulated to make available pure food materials devoid of adulterations and contaminants to the Indian population at large. Any food not conforming to these standards is said to be adulterated. The various provisions of PEA Act formulated at Central level have to be implemented at State and Local bodies' levels.

Fruit Products Order 1955 (FPO)

Regulates the manufacture, storage and sale of fruit and vegetable products. The fruit products order 1955 was issued by the Department of Food, Ministry of Food Processing Industries under the powers vested in the Government under the Essential Commodities Act to ensure the quality of fruit and vegetable products. This order controls production, distribution and quality of the fruit and vegetable products manufactured in the country as well as Registration, Licensing and Operation of manufacturing units. Under the FPO Central Government has constituted an expert committee called Central Fruit Products Advisory Committee to advise the Government on standards and policy matters pertaining to Fruit and Vegetable Preservation Industry. The various provisions of the FPO are being implemented by the Director, Fruit and Vegetables Preservation. Zonal Deputy Directors stationed at Chennai, Mumbai, New Delhi and Kolkata with technical staff. FPO mark is given to each professor after the grant of licence for manufacturing fruit or vegetable products, after the inspection of the factory for hygiene and sanitation. FPO mark and licence number is required by law to be exhibited on labels of each processed item along with other information as laid down in the FPO Rules. Continuous inspection schedule and pre shipment inspection schedule are being exercised under the FPO especially regarding export oriented Fruit and Vegetable Products. The FPO specifications cover methods of preservation, permissible colours in the preparations and also the minimum quality requirements of the final products. Fruit and vegetable products which do not conform to the FPO specifications are considered adulterated.

Agricultural Produce (Grading and Marketing) Act 1937 (AGMARK)

This is one of the oldest Food Laws promulgated in the country to provide for quality control of agricultural produce through grading and marketing. This Act provides compulsory standards for export and voluntary standards for vegetable oils, ghee, butter, cream, essential oil, guar, egg, groundnuts, potatoes, fruits, pulses, rice, condiments, spices etc. These standards are formulated on the physical and chemical characteristics of food, both the natural as well as those acquired during processing. This Act is administered by the Directorate of marketing and inspection with Agricultural marketing advisor to the Government of India, Faridabad, as its Chairman. This act includes all Agricultural produce and food drinks. The AGMARK seal issued by the Directorate of Marketing and inspection of the Government of India is thus a stamp of good quality. Any organization or packer who wants to use "AGMARK" symbol on the label of the sample containers can apply to the authorities. The authorities after inspection and ensuring that the necessary facilities like equipment, laboratory etc., are available allows them to use AGMARK symbol which ensures that the quality standards laid down for that product under AGMARK have been complied with. This gives a sort quality assurance to the consumers and at the same time a fair return to the farmers.

Sugar (Control) Order 1956

Indigenous Production of raw sugar and its quality are controlled primarily under this order., The Directorate of Sugar, Ministry of Agriculture and Rural Development is the operating authority of this order. The quality of sugar is specified in terms of its colour and size. Detailed quality standards have been specified under the provisions of PFA Act 1954 and rules thereof.

Vegetable Oil Products (Control) Order 1947; The Solvent Extraction Oil, Deoiled Meal and Edible Flour (Control) Order 1967; Vanaspati Control Order 1975

These orders have been promulgated under the Essential Commodities Act 1955 by the Central Government and covers commodities like vanaspati, margarine, bakery shortening, edible vegetable oils, edible flour, meal etc. The Directorate of Vanaspati, Ministry of Food and Civil Supplies is the operating authority, registration, licensing, manufacture and distribution are all controlled by this Directorate.

Meat Food Products (Control) Order 1975

This order covers both raw and processed meat and meat products and has been promulgated under essential commodities act 1955. This order is operated by the meat food advisory committee with agricultural marketing advisory to the Government of India as its chairman.

Rice Milling Industry (Regulation) Act 1958 and Regulation and Licencing Rules 1976

Modernisation of rice mills and safe preservation of stocks.

Export (Quality Control and Inspection) Act 1963 and Rules 1964

Certain notified exportable commodities are subjected to compulsory quality control and inspection before shipment. This Act is intended to provide for sound and development of export trade from India. Under this Act, the Export Inspection Council of India is exercising quality control and preshipment inspection on export oriented commodities including marine products.

Insecticide Act 1968

Provides for compulsory registration of all insecticides. Regulates import, manufacture, transport, storage, sale and use of all pesticides under this act.

Standards of Weights and Measures Act 1976

It is an act enacted by the parliament (No. 60 of 1976) to established standards of weights and measures to regulate inter-state trade or commerce in weights, measures and other goods which are sold or distributed by weight, measures or number and to provide for matters connected therewith or incidental there to. Director of legal metrology appointed under section 28 of the Act, additional, joint, Deputy or Assistant Directors and other staff appointed by the Director implements the provisions of the Act. The Central Government in consultation with the State Governments may controller of legal metrology for speedy and strict enforcement of the provisions of the Act and Rules thereof including packaged commodities.

State Licensing Order Governing Grain Dealers

State Governments Control grain dealers so as to protect stock against deterioration through moisture, insects, rodents etc.

The Consumer Protection Act 1986

This Act operated by the Ministry of Food and Civil Supplies is enacted recently where consumers through recognized consumer organizations can play the role of food samples. Rules are yet to be formulated for implementation at state levels.

Food Standardization and Regulatory Agencies in India

Central Committee for Food Standards

It is concerned with the prevention of Food Adulteration Act. It specifies standards for food items to check food adulteration and fraudulent practices.

Since 1941, CCFS has been functioning to advise the Central and State Government on matters arising out of the administration of the PFA Act. The PFA Act, 1954 and the PFA Rules, 1955 amended in 1976, 1980, 1981 and 1985, provide guidelines for the minimum basic requirements of food quality such as handling, storage, preparation and serving of food under sanitary conditions; freedom from extraneous matter, use of approved food additives such as preservatives, flavours, colours etc., proper packaging, branding and declaration of net weight as well as the date of manufacture and packing on the package. The guidelines are primarily intended to protect consumers from the health hazards of poisonous food. They also serve to prevent consumer exploitation by malpractices such as misbranding, adulteration, incorrect labelling, false claims, excessive and indiscriminate use of food additives etc.

There are at present four Central Food Laboratories (CFL):

1. Central Food Laboratory, Kolkata.
2. Food Research and Standardization Laboratory, Ghaziabad.

3. Public Health Laboratory, Pune.
4. Central Food Technology Research Institute (CFTRI), Mysore.

Central and State Food Departments

This Department come under the Ministry of Food and Civil Supplies, formulates specifications for cereal food grains/pulses and selected food stuff for purchase and procurement operations.

State Food Laboratories/Food and Drug Administration

Generally under the department of health–there is law enforcing authorities to check food adulteration.

Bureau of Indian Standards

Constituted under the Act of Parliament which deals with standardization of various articles including food. Agriculture Food Division Council (AFDC) deals with quality standards of various articles of food. The certification scheme under ISI is a voluntary one and any manufacturer who wants to use ISI mark on his food product may apply for the same to the authority. The authorities after inspection and other formalities permit the manufacture to use ISI Mark on their product. ISI symbol on any product ensures quality to the consumers. Recently, ISI has been renamed as Bureau of Indian Standards (BIS).

Food Corporation of India

Food Corporation of India has laid down specifications for several food commodities for internal purchase and procurement. To meet the essential requirements of the people in respect of rice, wheat, other cereals, pulses and gram products, the FCI procures these items in season, stores them and then distributes according to need the stored food stuffs to the public through fair price shops as well as other outlets. Different types of godowns, silos, cover and plinth are built by the FCI. Modern rice mills, maize mills, soybean processing units and solvent extraction (of oil seeds) plants are maintained by the FCI to help in food processing. The FCI has provision to make nutritious foods as fortified atta (wheat flour) and Balahar, on a very limited scale. Portable grain driers and purifies developed by the corporation serve to furnish clean and dry grains to the consumers in addition to ensuring better prices to the farmer.

Army Supply Corps

Army supply corps has stipulated A.S.C. specifications for supply of food commodities to Army purchase organizations.

Central Insecticide Board

Central insecticide board under the Department of Agriculture controls the use of insecticides as applied to food crops. Under the Insecticide Act, all insecticides are required to be compulsorily registered with the registration committee established under section 5 of the Act. Registration certificates are issued after the committee is satisfied that the insecticide is effective in pest control and at the same time safe to human beings, domestic animals and wild life.

F.P.O. Specifications for Fruit and Vegetable Products

Specifications for fruit syrup, crush, squash, nectar, cordials, juice, aerated water containing fruit juice or pulp, ready-to-serve beverage, fruit juice concentrate, jam, jelly, marmalade, cheese, preserve, chutney etc.

Product (Prepared from any suitable fruit and variety)	*Specifications*	
	Minimum per cent of Total Soluble Solids in Final Product (w/w)	*Minimum per cent of Fruit Juice or Fruit in Final Product (w/w)*
Fruit syrup	65	25
Crush	55	25
Squash	40	25
Fruit nectar (excluding orange & pineapple nectars)	15	20
Orange and pineapple nectars	15	40
Mango nectar	15	20
Cordial	30	25
Unsweetened juice	Natural	100
Sweetened juice	10	85
Ready-to-serve fruit beverages including aerated water containing fruit juice	10	10
Barley water	30	25 (minimum of barley starch 0.25)
Fruit juice concentrate	32	100
Jam and fruit cheese	68	45
Fruit jelly and marmalade	65	45
Fruit preserve	68	55
Fruit chutney	50	40
Synthetic syrup/sharbat (prepared from herbs, flowers or essences)	65	–
Candied and crystallized or glazed fruit and peel	T.S. 70 per cent and above	R.S. 25 and above
Tomato juice	5	–
Tomato soup	7	–
Tomato puree	9	–
Tomato paste	25	–
Tomato sauce/ketchup	25	–
Sauces	15	–

Storage Life of Vegetables

Vegetable	Temperature (°C)	Relative Humidity (per cent)	Storage Life (Weeks)
Asparagus	0.0	95	3-4
Beet, bunched	0.0	90	1.5
Beet, topped	0.0-1.7	90-95	8-14
Bitter gourd	0.6-1.7	85-90	4
Brinjal	10.0-11.1	92	2-3
Cabbage Early	0.0-1.7	92-95	4-6
Cabbage Late	0.0-1.7	92-95	12
Carrot topped	0.0	95	20-24
Cauliflower (snow ball)	0.0-1.7	85-95	7
Celery	0.0-0.6	92-95	8
Colocasia	11.1-12.8	85-90	21
Coriander leaves	0.0-1.7	90	5
Cucumber	10.0-11.7	92	2
Hyacinth bean pod	0.6-1.7	90	3
Garlic	0.0	65	28-36
Lettuce leaves	0.0	95	1
Ginger	7.2-10.0	75	16-24
Lima bean pod	4.4-7.2	90-95	1.5-2.0
Musk melon, Cantaloupe	1.7-3.3	85-90	1.5
Musk melon, Honeydew	7.2	85	4.5
Okra	8.9	90	2
Onion bulbs	0.0	70-75	20-24
Pea green	0.0	88-92	2-3
Pepper green	7.2	85-90	3-5
Pepper ripe	5.6-7.2	90-95	2
Potato	3.0-4.4	85	34
Pumpkin	1.7-11.6	70-75	24-36
Radish topped	0.0	88-92	3-5
Squash winter	12.8-15.6	70-75	24-36
Sweet corn	0.6-1.7	90-95	1
Sweet potato	10.0-12.8	80-90	13-20
Tapioca root	0.0-1.7	85	23
Tomato unripe	8.9-10.0	85-90	4.5
Tomato ripe	7.2	90	1
Turnip	0.0	90-95	8-16
Water melon	7.2-15.6	80-90	2
Yam	26.7	66-70	3-5

Storage Life of Fruits

Sl.No.	Fruit	Temperature (°C)	Relative Humidity (per cent)	Storage Life (Weeks)
I.	**Tropical and sub-tropical**			
1.	Banana			
	(*a*) Green (for ripening)	12-13	80-5	1-2
	(*b*) Ripe	12-13	85-90	3
2.	Chiku/Sapota	3-4	85-90	6-8
3.	Citrus fruits			
	(*a*) Mandarin			
	i. Coorg orange	6-8	85-90	10-12
	ii. Nagpur orange	5-6	85-90	10-14
	(*b*) Sweet orange			
	i. Malta blood red	2-5	85-90	16
	ii. Malta common	5-6	85-90	16
	iii. Valencia late	5-6	85-90	20
	iv. Mosambi	7-8	85-90	16
	v. Sathgudi	7-8	85-90	16
	(*c*) Pummelo lime	8-10	85-90	3-6
	(*d*) Lemon	8-9	85-90	8-12
	(*e*) Grape fruit (Marsh seedless)	8-9	85-90	12
4.	Custard apple	5-7	85-90	5-6
5.	Dates ripe	7-8	85-90	2
6.	Figs			
	Fresh	0-2	85-90	4
	Dry	0-2	65-70	52
7.	Grapes	0-2	80-85	6-8
8.	Guava	9-10	85-90	3
9.	Jackfruit	11-13	85-90	6
10.	Litchi	0-2	85-90	10
11.	Mango	8-9	85-90	4-7
12.	Mangosteen	5-7	85-90	6-7
13.	Papaya	9-10	80-85	1-2
14.	Passion fruit	7-8	80-85	4-5
15.	Pineapple			
	Mature, green	11-13	85-90	3-4
	Ripe	8-9	85-90	4-6
16.	Pomegranate	0-2	80-85	4-6

Contd...

Sl.No.	Fruit	Temperature (°C)	Relative Humidity (per cent)	Storage Life (Weeks)
II.	**Temperate fruits**			
17.	Apple	0-2	85-90	16-32
18.	Apricot	0-2	80-85	2
19.	Cherry	0-2	85-90	2
20.	Peach	0-2	80-85	2-4
21.	Pear	0-2	85-90	12-26
22.	Persimmon	0-2	85-90	7
23.	Plum	0-2	85-90	4-8
24.	Straw berry	0-2	80-85	2-3

Permissible Limits of Preservatives in Food Products

Sl.No.	Food Product	Preservative	Parts per million (ppm)
1.	Fruit, fruit pulp or juice for conversion into jam, crystallized, glazed, cured fruit or other products		
	(*a*) Cherries	Sulphur dioxide	3000
	(*b*) Strawberries, raspberries	Sulphur dioxide	2000
	(*c*) Other fruits	Sulphur dioxide	1000
2.	Fruit juice concentrates	Sulphur dioxide	1500
3.	Dried fruits		
	(*a*) Apricots, peaches, apples, pears and other fruits	Sulphur dioxide	2000
	(*b*) Raisins, sultanas	Sulphur dioxide	750
4.	Squashes, crushes, fruit syrups, cordials, fruit juices and barley water	Sulphur dioxide or Benzoic acid	350 or 600
5.	Jam, marmalade, preserves, canned cherry and fruit jelly	Sulphur dioxide or Benzoic acid	40 or 200
6.	Crystallized, glazed or cured fruit	Sulphur dioxide	150
7.	Fruit and fruit pulp not other-wise specified in the schedule	Sulphur dioxide	350
8.	Sweetened ready-to-serve beverage	Sulphur dioxide or Benzoic acid	70 or 120
9.	Pickles and chutneys made from fruit or vegetables	Sulphur dioxide or Benzoic acid	100 or 250
10.	Tomato and other sauces	Benzoic acid	750
11.	Dehydrated vegetables and dried ginger	Sulphur dioxide	2000
12.	Tomato puree and paste	Benzoic acid	250

Contd...

Sl.No.	*Food Product*	*Preservative*	*Parts per million (ppm)*
13.	Syrups and sharbats	Sulphur dioxide or Benzoic acid	350 or 600
14.	Cider	Sulphur dioxide	200
15.	Wines	Sulphur dioxide	450
16.	Beer	Sulphur dioxide	70
17.	Brewed ginger beer	Benzoic acid	120
18.	Sugar, glucose, gul and Khandsari	Sulphur dioxide	70
19.	Corn flour and such like starches	Sulphur dioxide	100
20.	Corn syrup	Sulphur dioxide	450
21.	Gelatine	Sulphur dioxide	1000
22.	Coffee extract	Sulphur dioxide	450
23.	Cheese or processed cheese	Sorbic acid or its Na, K, salts or Nisin	1000
24.	Flour confectionary	Sorbic acid or its salts	1500
25.	Hard boiled sugar confectionary	Sulphur dioxide	350
26.	Sausage, cereals, condiments	Sulphur dioxide	450
27.	Cooked pickled meat including ham and bacon	Sodium or potassium nitrate	200
28.	Smoked fish (in wrappers)	Sorbic acid	Only wrappers may be impregnated with sorbic acid

Additives to be Used with Caution

Additive	*Use*	*Possible Adverse Effects*
Artificial flavourings	In soft drinks, breakfast cereals, baked goods, vegetable and fruit products, ice-cream, custards, desserts, alcoholic beverages.	Hyperactivity in some children, not adequately tested for safety.
Butylated hydroxy-anisole (BHA)	Antioxidant in cereals, chewing gum, potato chips, edible oils etc.	Appears to be safer than BHT but needs more testing.
Butylated hydroxy-toluene (BHT)	Antioxidant in cereals, chewing gum, potato chips, edible oils etc.	Cancer, allergic reactions, stored in body fat.
Caffeine	Stimulant in soft drinks	Insomnia and other adverse effects at high levels of intake. Not recommended for children and pregnant mothers.
Coal tar dyes	Colourant in vegetable and fruit products, soft drinks, candy, desserts, pastry, sausage, baked food, ice-cream, hot dogs, hamburgers, sweetmeats, snacks, confectionary, alcoholic and other beverages.	Allergic reactions, cancer and pathological lesions in vital organs.

Contd...

Additive	Use	Possible Adverse Effects
Monosodium glutamate	Flavour enhancer for soup, poultry meat preservations, sauces, stews and cheese.	Damages brain cells in infants' mice, so not recommended for children, headache, tightness of head, neck and arms in sensitive adults.
Phosphoric acid and phosphates	Acidifier, chelating agent, buffer, emulsifier, nutrient, discolouration inhibitor used in baked goods, cheese, cured meat, soft drinks, dried potatoes.	Dietary imbalance that may cause bone thinking (osteoporosis) on prolonged use.
Saccharin	Non-calorie sweetener in food products also as adulterant	Bladder cancer reported in animals. Not recommended for normal people.
Sodium nitrite and nitrate	Preservative to prevent growth of bacillus. *Clostridium botulinum* and colourant for beacon, ham, meat, smoked fish, corned beef.	Formation of small amounts of cancer-producing nitrosamines.
Sulphur dioxide and bisulphites	Preservative and bleach for sliced fruit, wine, grape juice, dried potatoes, dried fruits, vegetables and fruit products.	Destroys vitamin B_1, but otherwise safe at prescribed levels.
Talc and kaolin	Making dry powdery foods free flowing and as dusting agent for rice, confectionary chewing gum.	Absorbed and stored in vital organs, cancer if asbestos is present.

Food Adulteration, Food Standards and Labelling

Food is consumed to provide energy and nutrition and as such it should be wholesome and not have deleterious substances. Adulteration is a term used to designate that a product is not what it should be from biochemical, cleanliness and hygienic point of view. Consumers are rightly concerned about the safety, wholesomeness, nutritional adequacy, palatability and cost of foods, that they buy for their use. Prevention of food adulteration Act 1954 and the rules under this Act were made in 1955. The standards laid down under the PFA Act and rules are minimum standards of purity and are based on the agricultural practices followed, climatic conditions prevailing, economic conditions and nutritional status of the people in the country.

List of Adulterants Generally Used in Different Food Products

Food	Adulterants
Asafoetida	Resins, gum, starch
Cardamom and cloves	Extracted fruits
Chilli powder	Oil soluble dyes, added colours, brick powder, sawdust
Coffee powder	Date seed, tamarind seed powder
Coriander	Dung powder

Contd...

Food	Adulterants
Ghee or butter	Hydrogenated oil
Honey	Sucrose, jiggery
Milk	Water, starch, urea
Pulses	Colouring matter
Spices and condiments	Dirt, filth, sand etc.
Tea	Iron filling
Turmeric powder	Lead chromate, metanil yellow
Vegetable oil	Argemone oil, castor oil, mineral oil
Wheat	Chalk powder

Monosodium-glutamate should not be added to any food for use by the infant below 12 months. Broadly the quality enforcement under the PFA Act can be categorized under three heads: 1. Enforcement, 2. Analysis, and 3. Prosecution. Packing material should not cause any health hazard to the consumer or does not contaminate the food.

Prevention of Food Adulteration Tips to Consumer

While shopping following care should be taken:

1. Read label before purchasing
2. Purchase food articles from licenced vendors and insist on bill or cash memo.
3. Prefer foods sold in packed containers even at higher cost.
4. Prefer foods certified by Government agencies like AGMARK, ISI, certification mark and F.P.O.
5. Avoid coloured foods especially sweetmeats or sharbats or ice candy.
6. Buy foods from reputed firms.
7. Do not buy cut or exposed fruits or vegetables.
8. Do not use containers or packages used for insecticide chemicals or non-edible items.

While preparing or serving food following care should be taken:

1. Wash hands with soap and water before the start of preparing food and after evening interruption.
2. Cover cuts in hand by bandage.
3. Cut mails short and keep them clean.
4. Cover head with hair net or band.
5. Wear cleans over cloths.
6. Keep all kitchen surfaces meticulously clean.

7. Wash food grains or vegetables, fruits, eggs, fish, meat thoroughly before cooking or eating or storing in refrigerator.
8. Avoid contact between raw foods and cooked foods.
9. Cook food thoroughly at boiling temperature.
10. Serve cooked hot food immediately.
11. Store cooked food carefully, preferably below 10 °C or above 60 °C.
12. Protect foods from flies, insects, rodents and other animals.
13. Keep the refrigerator door closed, deforest or clean refrigerator every week.
14. Do not consume stored prepared food if having off (rancid) flavour or smell or food in which froth has set in.
15. Use pure and clean water while preparing food.
16. Baby care is a difficult job and following care should be taken.
17. Mothers milk is best for baby.
18. After 4 months of age start have made weaning food.
19. Do not give left-over food to the baby.

Food Laws

The more important is that of safety; the health hazards that may arise on account of adulteration, contamination, microbiological deterioration, decays and other factors. Food is an item which could account for 50 per cent or 60 per cent of the budget of a family, and thus it is all the more important that the Government should give some protection to the hard-pressed consumer. The quality of food in India lies with organizations related to agriculture and food products, through different orders and acts like the Agriculture marketing products and grading Act 1973, prevention of food Adulteration Act 1954, vegetable oils control order 1947, Fruit products order 1955, solvent extracted de-oiled meal edible flour 1967, and so on. The meat products statutory order 1976, functioning under the directorate of marketing and inspection (Ministry of Agriculture), enables control of quality in production of raw and processed meat. There is also the export products control order.

Food Standards

A wide range of processed foods are now being manufactured in the country which include canned jams, curries, meat, fish, processed breakfast, cereals, papads, soups, noodles, potato wafers, of different varieties, protein textured foods, frozen pizzas etc.

- ✰ BIS: Bureau of Indian Standards (earlier ISI).
- ✰ ISO: International Organization for Standardization 1947.
- ✰ ICUMSA: International Commission for Uniform Methods of Sugar Analysis.
- ✰ CODEX: Codex Alimentarius Commission was set up in 1962.

Prevailing Food Laws, Acts and Implementing Agencies for Food Standards

Act Order	*Implementing Agency*	*Year of Introduction*	*Kind of Legislation*
Agricultural produce Act (grading x marketing)	Directorate of marketing inspection	1937	Voluntary: Compulsory for export only
ISI (Certification Indian market Act)	Indian standards institution (BIS)	1952	Voluntary
Prevention food Act adulteration	Ministry of health	1954	Compulsory for internal trade
Fruit products order	Department of food, Ministry of Agriculture and Rural Reconstruction	1955	Compulsory for export/ internal trade
Export (quality Act control & inspection)	Export inspection council	1963	Compulsory for export only
Solvent extracted oil, de-oiled meal edible flour control order	Directorate of vanaspathi	1967	Compulsory for export/ internal trade
Meat products order and inspection	Directorate of marketing	1973	Compulsory for export/ internal trade

- ☆ *Market standards:* Economic status and quality consciousness of the consumer.
- ☆ *End use standards:* Depend on the product preparation.
- ☆ *Health Ministry Standards:* Mandatory in nature quality denotes the degree of excellence of a product.
- ☆ *Quality and its down gradation:* Quality of an agricultural commodity can be divided into two categories, viz; intrinsic quality and acquired quality.

A. The *intrinsic quality* is a genetic factor grouped in to four categories.

1. Biochemical: Protein, fat, fiber, vitamins, minerals, carbohydrates and enzymes as well as anti-nutritional factors.
2. Physical: Colour, texture, size, shape, aroma, and weight tec.
3. Processing: Milling yield, cooking quality, yield of juices, pulses and other items.
4. Storability: Storability of agricultural produce.

B. *Acquired quality*: Those characteristics are acquired during pre-and postharvest conditions and practices such as discolouration, infection due to field fungi *e.g.*, ergot, smut, bunt, sprouting, admixture, contamination etc.

Quality Centers

At the international level, the WHO and FAO of United Nations are the quality centers. For ensuring safety and health of consumers, the Government of India has enacted a few Acts and detailed Rules have been framed. They are i. agricultural produce (Grading and marketing) Act 1937, ii. Prevention of Food Adulteration Act 1954 (PFA), iii. Fruit Products Order 1955, iv. Bureau of Indian Standard Act 1986 and v. Export (Quality Control and Inspection) Act 1963.

Prevention of Food Adulteration Act

Broadly the control of the quality can be categorized under three heads viz: 1. Enforcement, 2. Analysis and 3. Prosecution. The Act deals with preservatives, poisonous metals, naturally occurring toxic substances, anti-oxidants, emulsifying and stabilizing agents, flavouring agents, colouring matter and other food additives, insecticides and pesticides, solvent extracted oils and edible flours, non-alcoholic beverages, starchy foods, spices and condiments and their mixtures, honey, jiggery, saccharin, coffee, tea, milk and milk products, fruit products, edible oils, cereals, backed products, sweets and confectionaries and a range of similar products.

AGMARK

The agricultural produce (Grading and Marketing) Act was enacted by the Government of India in 1937. Central Agmark Laboratory is at Nagpur in Maharashtra (21 Labs in the Country). It checks 140 commodities from agriculture produce.

The FPO covers all types of fruit products such as juices, pulp concentrates, squashes, nectar, aerated water containing fruit juices, bottled and canned fruits and vegetables, jams, jellies, fruit cheese, preserves, chutneys, soups, ketchup, paste, dehydrated vegetables and dehydrated onion. In addition to specifications for various products, limits for poisonous metals, list of permitted and harmless food colours, limits for permitted preservatives and other additives are given in the schedules.

FAQ (Fair Average Quality) Specifications for Food Grains

About 20-25 million tones of both wheat and rice procure per year for network distribution for fair price shops in the country. The procurement is done through Food Corporation of India, State Government and co-operative organizations.

Bureau of Indian Standards

Bureau of Indian Standard also prescribes quality standards for agricultural products and processed foods such as milled wheat, maize, barley, pulse products, corn flakes, macaroni, spaghetti, biscuits, bread etc. The BIS standards are quality standards and more rigid than PFA standards.

Consumer Protection Act

The consumer protection Act 1986 and the central consumer protection rules 1987 were enacted by the Government of India. Consumer councils at various levels have also been established to advise the consumers for their rights.

Export Inspection Council

Export inspection council has been set up under the export inspection act to have a check on quality of food articles exported.

Other Quality Control Legislations

Essential commodities Act of 1954 in addition to the fruit products order 1955, vegetable oils product council order 1947 and meat product control order 1973 were formulated to regulate manufacture, commerce and distribution of vegetable oils and their products and meat products respectively.

Rule: A-18.06 Food Grains

1. Foreign matter: Inorganic matter up to 1 per cent (in paddy 3 per cent); organic matter up to 3 per cent.
2. Damaged grains up to 5 per cent by weight.
3. Insect damaged grains not more than 10 per cent.
4. Uric acid content should not be more than 10 mg/100g.
5. Rodent hair and excreta should not more than 5 pieces/kg sample.

Rule: A-07.03 Honey

1 Moisture 25 per cent or less.
2. Ash 0.5 per cent
3. Sucrose 5 per cent
4. Reducing sugars 65 per cent
5. Fructose: glucose not less than 0.9 per cent

Food Labelling

Part-I: Covers general guidance on labelling.

Part-II: Covers claims and lays down

Part-III: Covers guidelines on labelling with respect to nutritional information. To increase the sales, attractive design, name and colours etc. Mention FPO, ISI, AGMARK makes sure of standard quality.

Agro-Chemical Residues in Indian Items

To meet the needs of growing population, we require massive food production programme and this in turn is impossible without the use of pesticides. Even small quantities of these residues consumed over long periods can build up to high levels in the body fat.

Safe Levels

In order to regulate contamination of food with pesticide residues to safe levels, government has laid down principles for arriving at maximum residue limits in food commodities.

Tolerance Limit of Common Pesticides

Pesticide	*Food*	*Tolerance Limit (mg/Kg)*
Aldrin, dieldrin	Fruits and vegetables	0.4
Carbaryl	Okra, leafy vegetables, potato	10.4
Chlordance	Vegetables, fruits	0.2, 0.1
Diazinon	Vegetables	0.5
Dichlorvos	Vegetables, Fruits	0.15, 0.1
Dicofol	Fruits and vegetables	5.0
Dimethoate	Fruits and vegetables	2.0
Endosulfan	Fruits and vegetables	2.0
Feniltrothion	Fruits and vegetables	0.5, 0.3
Heptachlor	Vegetables	0.05
Inorganic bromide	Fruits	30.0
Lindane	Fruits and vegetables	3.0
Parathion	Fruits and vegetables	0.5
Parathion methyl	Fruits and vegetables	0.2
Phosphamidon	Fruits and vegetables	0.2
Pyrethrine	Fruits and vegetables	1.0

Safety Periods

Every pesticide has some safety period or waiting period. Safety period is the number of days to lapse before the pesticide reaches the tolerance limit.

Effects of Processing

The extent of removal depends on the nature of the chemical, type of food, length of contact with food, and environmental conditions. No residues are found in cauliflower when harvested 9 days after spray. Washing residues 50-75 per cent residues in cauliflower and cooking reduces almost completely. Washing, soaking in 2 per cent salt solution, cooking in slightly acid medium like lime juice or tamarind extract removes more than 75 per cent residue. In green leafy vegetables it is difficult to remove the residues with one washing.

Removal of Insecticides by Common Processing Produce in Vegetables

Vegetable/ Insecticides	*Washing with 2 per cent Salt Water and Cooking for 15 min.*	*Washing and Steam Cooking*
Beans		
Malathion	60	69
Monocrotophos	42	47
Carbaryl	58	69

Contd...

Vegetable/ Insecticides	Washing with 2 per cent Salt Water and Cooking for 15 min.	Washing and Steam Cooking
Chillies		
Monocrotophos	28	30
Quinalphos	22	29
Tomatoes		
Carbaryl	67	75
Monocrotophos	31	32
Quinalphos	29	30

Biological Effects

Pesticide residues causes acute and long term toxic effects in humans, animals, fish and birds. DDT accumulates in the fat and causes more toxicity in malnourished population. Due to sensitization of body, allergic reaction can occur and in such causes, each additional exposure to matter how small will cause serious health problems. Immediate effects may be dizziness, burning eyes and skin rash. Continuous exposure for longer periods can cause liver or kidney damage or damage to nervous system; it can also cause mutations resulting in birth defects. WHO has indicated that maximum safe limit for DDT to be 0.01 mg/kg body weight/day, new born consuming breast milk from mothers exposed to pesticide sprays are consuming 100 mg DDT/day.

Points for Action

1. Random checking of market samples.
2. Educate farmers, labourers and public etc.
3. Enforcement of regulation on pesticides utilization.
4. Create facilities for analysis of food samples.
5. To see that food produced is safe for human consumption.

Fruit Products Order (FPO) 1955

Every manufacturer shall manufacture fruit products in conformity with the sanitary requirements and the appropriate standard of quality and composition specified in the second schedule to this order or according to the standard of quality and composition laid down by the licensing officer.

S.R.O. 1952. The powers conferred by section 3 of the (Essential Commodities Act, 1955), the Central Government makes the order.

1. a) This order is called the fruit products order, 1955,
   b) It extends to the whole of India except the state of Jammu and Kashmir
2. In this order unless the context otherwise requires
   a) The act means the Essential Commodities Act, 1955 (10 of 1955)

b) "Committee" means the central fruit products Advisory Committee constituted by the Central Government under clause-3
c) 'Form' means a form setforth in the first schedule
d) "Fruit product" means any article prepared from fruit and vegetable and named according to their specification.

The First Schedule

FORM 'A'

Application for license under the Fruit Products Order, 1955

1. Name and address of the applicant

2. Address of the Factory

3. Description of the fruit products which the applicant wishes to manufacture

4. Period for which the licence is required

5. Plan of the factory and list of equipments

6. Whether any power is used in the manufacture of fruit products. If so, state the exact horse power used

7. Licence fee paid during the previous year

8. Total value of fruit products manufactured during the previous year

9. I/we hereby undertake to comply with all the provisions of the fruit products order, 1955

10. I/we have forwarded a sum of rupees in respect of the licence fee due, according to the provisions of fruit products order, 1955

Signature (s) of the applicant (s)

FORM 'B'

Government of India,
Ministry of Food Processing,

Government of India
Emblem

Licence under Fruit Products Order, 1955
Licence No. FPO ————————————

1. Name and address of licensee

2. Address of authorized premises for manufacture

3. Change of premises if any

This licence is granted under and is subject to the provisions of F.P.O., 1955 all of which must be complied with by the licensee.

Place:

Date:

Licensing Officer
to the Government of India

Validation and Renewal

Period of validity	Categories fee of fruit products authorized to manufacture	Rate of Licence fee per category	Licence paid

Signature of licensing officer

FORM 'C'

1. Name and address of licensee

2. Address of the authorized premises for the manufacture of fruit products

3. F.P.O. Licence No.

4. Statement showing quantities of fruit and vegetable products manufactured in kg. with their sale value during the term

Sl.No.	*Name of the Fruit Product*	*Size of Can or Bottle*	*Quantity in Kg*	*Sale Price per kg or per Unit of Packing*	*Value*

Quantity Exported (kg)	*Name of the Country or Part of Export*	*Rate per kg or per Unit of Packing (C.I.F./F.O.B.)*	*Value*	*Remark*

Signature of the licensee

A register in the form of the above table shall be maintained by each licensee for inspection.

Preservatives

Preservatives means a substance which when added to food is capable of inhibiting, retarding or arresting the process of fermentation, acidification or other decomposition of food.

Class-I Preservative

1. Common salt
2. Sugar
3. Dextrose
4. Glucose (syrup)
5. Spices
6. Vinegar or acetic acid
7. Honey
8. Edible vegetable oils

Addition of class-l preservatives in any food is not restricted.

Class-II Preservative

1. Benzoic acid and their salts
2. Sulphurous acid and their salts
3. Nitrites of sodium or potassium
4. Sorbic acid including its sodium, potassium and calcium salts
5. Nisin
6. Sodium and calcium propionate
7. Methyl or propyl parahydroxy-benzoate
8. Propionic acid and its esters or salts
9. Sodium diacetate
10. Sodium, potassium and calcium salts of lactic acid.

The use of more than one class-II preservative is prohibited by Prevention of Food Adulteration Act, 1954. The use of class-II preservatives shall be restricted to the following group of foods in concentration not exceeding the proportions given in Table 19.2.

Food Colours

These are the food additives added to the product to improve the sensory properties of the processed products. Food colours used are mostly the permitted coal-tar-dyes which are classified as synthetic colour. The natural colours are extracted from plant products and are used in foods. In India, the Fruit Product Order (F.P.O.) and Prevention of Food Adulteration (PFA) Act have laid down stringent regulation regarding the use of artificial colours in foods. The Indian Standard Institution has examined the possibilities of certifying food colours used in India. The list of important colours used in fruit and vegetable products is given in Table 19.3.

Table 19.2: Permissible level of SO_2 in processed products of fruits and vegetables

Product	*Preservative*	*Permissible Level (ppm)*
Fruit, fruit pulp or juice, jam, cherries, strawberries and raspberries	Sulphurdioxide	3000
Fruit juice concentrate	Sulphurdioxide	1500
Dried fruits	Sulphurdioxide	2000
Raisins and sultanas	Sulphurdioxide	750
Non-alcoholic beverages, wines, squashes, crushes, fruit syrups, cordials, fruit juices	Sulphurdioxide	350
Jam, marmalade, preserve, canned cherry and fruit jelly	40 or 200	Sulphurdioxide or Benzoic acid
Cured fruit	Sulphurdioxide	150
White sugar, dextrose and jaggery	Sulphurdioxide	70
Cider	Sulphurdioxide	200
Alcoholic wines	Sulphurdioxide	450
Pickles and chutneys	Sulphurdioxide or Benzoic acid	100 or 250
Tomato and other sauces	Benzoic acid	750
Dehydrated vegetables	Sulphurdioxide	2000
Tomato puree and paste	Benzoic acid	25 0
Syrups and sherbets	Sulphurdioxide or Benzoic acid	350
Fruits and vegetable flakes, powder, figs	Sulphurdioxide	600

Table 19.3: Food colours and their uses in fruits and vegetables products

Colours	*Use in Fruit/Vegetable Product*
Tartrazine NSS	Lemon jelly, lime juice cordial
Sunset Yellow FCF	Apricot jam
Orange AG	Orange jelly, Orange squash
Ponceau RNS	Canned straw berries, Strawberry jam and jelly
Carmoisine WNS	Raspberry jelly
Caramel W	Vinegar, Sauces
Black Currant A	Black current jam and jelly
Erythrosin AS	Canned Cherries, Victoria plums
Pea Green B	Canned Fresh and processed peas
Red BR	Canned beet root

The PFA act defines the quantitative limits of these colours in the products. Recently, some colours have been banned for use in foods. Acid Magnets II and Blue VRS have been deleted from the list of permitted colours. The naturally occurring colours are freely allowed to be added to foods. According to FPO (1955), the natural colours allowed to be used in food products are, carmani, carotene and carotenoids, chlorophyll, lactoflavin, caramel, annatto, saffron (safflower) and curcumin (turmeric).

The PFA act defined [illegible] Recently some [illegible] VF [illegible] [illegible]

Bibliography

Anderson, R. A. (1962). A note on wet-milling of high amylose corn containing 75 per cent-amylose starch, *Cereal Chem.*, 39, 406-408.

Anderson, R. A. (1963). Wet-milling Properties of grains: Bench- Scale Study, *Cereal Sci. Today*, 8, No.6, 190-92, 195, 221.

Anon. (1964). Joint United States-Canadian tables of feed composition, *Natl. A cad. Sci.*, Nath. Res. Council Publication 1232.

Anonymous, (1982). Prevention of Food Adulteration Act, Government of India Publication, New Delhi.

Arumughan, C., Gopalkrishnan, N., Thomas, P.P., Narayanan, C. S. and Mathew, A.G. (1985). Refining and bleaching of indigenous palm oil at pilot plant scale. J. Food Sci. Technol. 22: 330-333.

Arya, P.R. and Rastogi, P.P. (1984). Fruit-Vegetable Preservation at Home (In Hindi). Directorate of Translation and Publication, G.B. Pant University of Agriculture and Technology, Pantnagar, Nainital.

Athalye (1992). Plastics in Packaging. Tata McGraw-Hill Publishing Company Limited, New Delhi.

Atkins, T. A., and Geddes, W. F. (1939). The relationship between protein content and strength of flours in glutenenriched flours, *Cereal Chem.*, 16, 223-31.

Au, P. M. and Fields, M. L. (1981). Nutritive quality of fermented sorghum. J. Food Sci. 46: 652-654.

Autrey, H. S. (1953). Effect of variables upon milling yields, *Rice J. Annual.*

Autrey, H. S., Grigorieff, W. W., Attschul, A. M., and Freeman, E. E. (1955). Effect of milling conditions on breakage of rice grains, *J. Agr. Food Chem.*, 3(7), 593.

Badi, S. M. and Hoseney, R. C. (1976). Use of sorghum and pearl millet flours in cookies. Cereal Chem. 53: 733-738.

Bailey, A. E. (1948). Cotton Seed and Cotton Products; Inter-science Publishers, Inc., New York.

Bailey, A. E. (1951). Oil and Fat Products. Inter-science Publishers, Inc., New York.

Baird, P. D., Mac Masters, M. M., and Rist, C. E. (1950). Studies on a rapid test for the viability of corn for industrial use, *Cereal Chem.;* 27, 508-13.

Banerjee, S. N. (1975). Report on the evaluation of the modern dal mill, Department of Food, Govt. of India.

Bankar, J. R., Chavan, J. K. and Kadam, S. S. (1986). Effect of grain conditioning on physico-chemical properties of sorghum bhakari. J. Maharashtra Agric. Univ., 11: 221-223.

Banwart, G.J. (1979). Basic Food Microbiology. The AVI Publishing Company, Inc., Westport, Connecticut.

Bau, H. M., Mohatadi-Nia, D. J., Mejean, L. and Debry, G. (1983). Preparation of colourless sunflower protein products: Effect of processing on physiochemical and nutritional properties. J. Am. Oil Chem. Soc. 60:1141-1148.

Bender, A. E. (1973). Composition of infant foods, In Nutritional and Dietetic Foods, International Textbook Co. Ltd., London, U. K., p. 64.

Bhat, C.M., Sharma, R.N. and Sehgal, S. (1982). A Manual on Food Preservation at Home. Directorate of Publication, Haryana Agricultural University, Hissar.

Bishop, W., and Dustin, J. (1951). Topnotch wet Corn mill house, *Food Ind.,* 23, No.3, 121-24, 211-12.

Blession, C. W., Inglett, G. E., Garcia, W. J. and Deatherage, W. L. (1974). An edible defatted germ flour from a commercial dry-milled corn fraction. Cereal Sci. Today. 19:224-255.

Board, P.W. (1988). Quality Control in Fruit and Vegetable Processing. FAO Food and Nutrition Paper 39. Food and Agriculture Organisation of the United Nations, Rome.

Borasio, L., and Gariboldi, P. (1957). Illustrated glossary of rice processing machines, Food. and Agricultural Organization of the United Nations, Rome.

Bose, T K. and M. K. Mitra, Eds. (1986). Fruits of India Tropical and Subtropical. Naya Prokash, Culcutta.

Bose, T K. and M.G. Som., Eds. (1986). Vegetable Crops in India, Naya Prokash, Culcutta.

Boundy, J. A., Woychik, J. H., Dimler, R. J., and Wall, J. S. (1967). Protein composition of dent, waxy. and high-amylose corns, *Cereal Chem.,* 44, 160-169.

Bradbury, D., Wolf, M. J., and Dimler, R. J. (1962). The hilar layer of white com, *Cereal Chem.,* 39, 72-78.

Brautlecht, C.A. (1953). *Starch, Its sources, Production and Uses,* Reinhold Publishing Corp., New York.

Brekke, O. L., and Weinecke, L. A. (1964). Corn dry-milling, A comparative evaluation of commercial degermer samples, *Cereal Chem.*, 41, 321-28.

Brekke, O. L., Weinecke, L. A., Boyd, J. N., and Griffin, E. L., Jr. (1963). Corn dry milling: Effects of first-temper moisture, Screen perforations, and rotor speed on Beall degerminator throughput and products, *Cereal Chem., 40,* 423-29.

Brekke, O.L, (1967). Corn dry milling: Pretempering low-moisture corn, *Cereal Chem.*, 44, 521-31.

Brekke, O.L. (1968). Corn dry-milling: Stress crack formation in tempering low-moisture corn, and effect on degerminator performance, *Cereal Chem.*, 45, 291-303.

Brekke, O.L. (1970). Dry milling artificially dried corn: Roller milling of degerminator stock at various moistures, *Cereal Science Today,* 15, 37-42.

Brekke, O.L. and Kwolek, W.F. (1969). Corn dry milling: Cold tempering and degermination of Corn of various initial moisture contents, *Cereal Chem.*, 46, 545-49.

Burton, W.G. (1989). Specific gravity as a guide to the content of dry matter and of starch in potato tubers. In 'The Potato'. 3rd Edn. Longman Scientific and Technical. Essex, England.

C.F.T.R.I. (1981). Home-Scale Processing and Preservation of Fruits and Vegetables. Central Food Technological Research Institute, Mysore.

Canella, M. and Bernardi, A. (1983). Changes in phenolics and oligosaccharides during sunflower seed germination. Riv. Ital. Sost. Grasse. 60: 761-762.

Chadha, K. L. and G. Kalloo, Eds. (1993). Advances in Horticulture. Vol. 6-7, Malhotra Pub. Co. New Delhi.

Chadha, K. L. and O. P. Parekh, Eds. (1993). Advances in Horticulture. Vol. 1-5, Malhotra Pub. Co. New Delhi.

Chakraverty, A. (2007). Postharvest technology of cereals, pulses and oilseeds. Oxford & IBH Publishing Co. Pvt. Ltd. New Delhi.

Chandra, P. K. (1972). Methodology and Procedures for evaluating performance of commercial rice mills, Unpublished M. Tech. Thesis, L I. T., Kharagpur.

Chavan, J. K. and Kadam S. S. (1989). Nutritional improvement of cereals by fermentation. CRC in Food Sci. and Nut. 28: 349-399.

Chavan, J. K. and Salunkhe, D. K. 1984. Structure of sorghum grain. In Nutritional and processing quality of sorghum. Pp. 21-31.

Chavan, J. K., Chavan, U. D. and Nagarkar, V. D. (1989). Effect of malting and fermentation on nutritional quality of sorghum. J. Maharashtra Agric. Univ., 14: 246-247.

Chavan, U. D. and Chavan, J. K. (1991). Utilization of malted sorghum, mung bean and black gram in preparation of bhakari. J. Maharashtra Agric. Univ., 16: 141-142.

Chavan, U.D. (1987). "Effect of Fermentation on the Nutritional Quality of Grain Sorghum (*Sorghum bicolour* L. Moench). "M. Sc. (Agric.) Thesis, Mahatma Phule Agriculture University Rahuri, Maharashtra, India. pp.i-95.

Chavan, U.D. and Chavan, J.K. (1992). Utilization of groundnut flour in preparation of sorghum bhakari. J. Maharashtra Agric. Univ., 17 (20): 346-347.

Chavan, U.D., Chavan, J.K. and Kadam, S.S. (1988). Effect of fermentation on soluble proteins and *in-vitro* protein digestibility of sorghum, green gram and sorghum + green gram blends. J. Food Sci. 53: 1574-1575.

Chevassua-Agnes, S., Favier, J. C. and Joseph, A. (1976). Traditional technology and nutritive value of sorghum beer from Cameroon (in French) Cah. Nutr. Diet. 11: 89-104.

CSIR, (1980). The Wealth of India: Raw Materials, Council of Scientific and Industrial Research, New Delhi.

Dandekar Machinery Works, Catalogue for rice and pulse milling machinery, *Mis.* G. G. Dandekar Machine Works Ltd., Bhiwandi, Thana, Maharashtra, India.

Dauthy, M.E. (1995). Fruit and Vegetable Processing. FAO Agricultural Services Bulletin 119. Food and Agriculture Organisation of the United Nations, Rome.

Desikachar, H. S. R. (1977). Processing of sorghum and millets for versatile food uses in India. In Sorghum and millets for human food. Pp. 41-45. Symp. Proc. IACC, Vienna, 1976. Tropical Products Institute, London.

Desrosier, N.W. and Desrosier, J.N. (1987). The Technology of Food Preservation. CBS Publishers and Distributors, New Delhi.

Desrosier, N.W. and Tressler, D.K. (1977). Fundamentals of Food Freezing. The AVI Publishing Company, Inc., Westport, Connecticut.

Earle, F.R., Curtis, J. J., and Hubbard, J. E. (1946). Composition of the component parts U1 the Corn kernel, *Cereal Chem.*, 23, 504-11.

Elias, D. G., and Scott, R. A. (1957). British flour milling technology, *Cereal Sci. Today*, 7, 180-84.

Ezaki, H. (1973). Paddy husker, Group training course-Fiscal, Institute of Agricultural Machinery, Japan.

Fennema, O.R. Ed. (1976). Principles of Food Science, Part-l. Food Chemistry, Marcel Dekker Inc, New York.

Finney, K. F., and Barmore, M. A. (1948). Loaf volume and protein content of hard winter and spring wheats, *Cereal Chem.*, 25, 291-311.

Frazier, W.E. and Westhoff, D.C. (1983). Food Microbiology. Tata McGraw-Hill Publishing Company Limited, New Delhi.

Garibaldi, F. (1974). Rice Milling Equipment Operation and Maintenance, Agricultural Services Bulletin No. 22, FAO, Rome.

Gauri Shanker (1984). Practical Manual in Horticulture. Kitabistan, Allahabad.

Girdhari Lal, Siddappa, G.S. and Tondon, G.L. (1986). Preservation of Fruits and Vegetables. Publications and Information Division, Indian Council of Agricultural Research, New Delhi.

Godin, V. J. and Spensley, P. C. (1971). Oils and oilseeds crop and product Digest No. 1. London: Tropical Products Institute.

Goel, A. K., Kumar, R., and Mann, S. S. (2007). Postharvest Management and Value Addition. Daya Publishing House, Delhi.

Gopalan, C, Rama Sastri, B.V. and Bala Subramanian, S.C. (1982). Nutritive Value of Indian Foods. National Institute of Nutrition, R.CM.R., Hyderabad.

Grist, D. H. (1959). Rice (3rd edn.), Longmans Green and Co. Ltd., London.

Haard, N.F. and D.K. Salunkhe, Eds. (1975). Postharvest Biology and Handling of Fruits and Vegetables, AVI, Westport, CT.

Hagenmaier, R. D. (1979). Experimental coconut protein products. J. Am. Oil Chem. Soc. 56:448-449.

Hagenmaier, R. D., Quinitio, P. H. and Clark, S. P. (1975). Coconut flour: technology and cost of manufacture. J. Am. Oil Chem. Soc. 52:439-443.

Hahn, R. R. (1969). Dry milling of grain sorghum. *Cereal Sci. Today*. 14: 234-237.

Hardenburg, R. E., A.E. Watada and C. Y. Wang, (1986). The Commercial Storage of Fruits, Vegetables and Florist and Nursery Stocks, Agricultural Handbook 66, US Dept. Agric. Washington D.C.

Harris, RS. and Karmas, E. (1975). Nutritional Evaluation of Food Processing. 2nd Edition. AVI Publishing Co. Westport, Connecticut.

Hartley, C. W. S. (1967). The oil palm, pp. 1-70; 608-692. London; Longman.

Haung, M. T, O. Toshihiko, C. T Ho and R.T. Rosen, (1994). Food Phytochemicals for Cancer Prevention, I. Fruits and Vegetables ACS Symposium Series 546, American Chemical Soc. Washington, D. C.

Hofman, V., *et al.* (1979). Sunflower Oil as a Fuel Alternative. North Dakota Ext. Bul. 13 AENG-735. 4 pp.

Hofman, V., *et al.* (1981). Sunflower for Power: North Dakota Ext. Circ. AE-735. 12 pp

Hogan, J. T. (1967). Rice bran oil and wax. FAO Agr. Eng. Informal Working Bull. 30:1.

Hogan, J.T. and Deobald, H. J. (1995). Measurement of the degree of milling of rice, *Rice* 1, 68 (10), 10.

Houston, D. F. (1973). Western Marketing and Nutrition Research Division, Agricultural Research Service, U. S. Department of Agriculture, Berkeley, California.

Hulme A.C. (1978). Biochemistry of Fruits and Their Products, Vol. I &. II. Academic Press, London.

ICAR (1970). Pulse Crops of India, I.C.A.R. Publication, New Delhi.

Ilyas, S. M. (2003). Importance of postharvest processing and value- addition for additional income and employment generation. Presented at National Symposium on Resource Management for Eco-friendly Crop Production, held at CSAUA&T, Kanpur, February, 26-28.

Ilyas, S. M., Wanjari, O. D. and Goyal, R. K. (2002). Prospects of income and employment generation through agro-processing and value-addition. Presented at Conference on Emerging Opportunities of Agro and Food Processing Industries in Jharkhand held at Indian Institute of Coal Management, Ranchi during October, 8-9.

Ilyas, S.M. and Coyal, R.K. (2002). Quality assurance in food processing-impact of WTO, In: Proc. National Seminar on Development of Food Processing Industries in U.P., held at Lucknow during January 16-17.

Jacob, T. (1987). Poisons in our Food. Publications Division, Ministry of Information and Broadcasting, Government of India, New Delhi.

Jasperson, H. and Pritchard, J. L. R. (1965). Factors influencing the refining and bleaching of palm oil. Paper presented at the TPI oil palm conference. P.96. London: Ministry of overseas Development.

Javaraman, J. (1981). Laboratory Manual in Biochemistry. Wiley Eastern Limited, New Delhi.

Jones, CR. (1940). The production of mechanically damaged starch in milling as a governing factor in the diastatic activity of flour, *Cereal Chem.*, 17, 133-69.

Kader A.A. Ed.(1992). Postharvest Technology of Horticulture Crops. University of California, Publication No. 3311.

Karel, M., O.R. Fennema and D.S. Lund, Eds. (1975). Principles of Food Science Part II. Physical Principles of Food Preservation. Marcel Dekker, Inc. New York.

Karim, A. and Rooney, L. W. (1972). Characterization of pentosans in sorghum grain. J. Food Sci. 37: 369-371.

Kazanas, N. and Fields, M. L. (1981). Nutritional improvement of sorghum by fermentation. J. Food Sci. 46: 819-821.

Kent-jones, D. W., and Amos, A. J. (1957). Modern Cereal Chemistry, 5th Edn., Northern Publishing Co. Ltd., Liverpool.

Kerr, R. W. (1950). Chemistry and Industry of starch, 2nd Edition, Academic Press, New York.

Kester, E. B., and Matz, S. A. (1970). Rice Processing, In: Cereal Technology, (Ed.) S.A. Matz. AVI Pub. Co., Westport, Conn.

Kewdar, A. A. (1986). Biochemical and physiological basis for effects of controlled and modified atmospheres on fruits and vegetables. Food Technol. 40: 99-104.

Klein, K. and Crauer, L. S. (1974). Further developments in crude oil processing. J. Am. Oil Chem. Soc. 51: 382A-385A.

Koga, Y. (1969). Drying, husking and milling in Japan IV and V., *Farming*, Japan.

Kordylas, J.M. (1991). Processing and Preservation of Tropical and Subtropical Foods. Macmillan Education Ltd., Houndmills, Basingstoke, Hampshire.

Krishnamurthy, K., Girish, G. K, Ramasivan, T., Bose, S.K., Singh, Karan and Tomar, R. P. S. (1972). A new process of removal of husk of red gram using sirka, Bulletin of Grain Technology, 10: 18l.

Kuprits, Y., ed. (1967). Technology of grain processing and provender milling, Tekhnologiya perer atbotki zernai Kombikormovoe, Izd. 'Kolos': Moscow (1 per cent 5). (Translation by Israel Program for Scientific Translation, Jerusalem, Israel)

Kurien, P. P. (1977). Grain legume milling technology. International Proceedings of FAO Expert Consultation on Grain Legume Proceeding. Rome, Italy, AGS/GLP/ 77/11.

Kurien, P. P. (1979). Pulses Milling in Food Industries. C.F.T.R.I., Mysore, pp. 3.1-3.20.

Kurien, P. P., and Parpia, H. A. B. (1968). Pulse milling in India, I, Processing and milling of tur and arhar *(Caianas cajan)*, *Food Sc. and Tech.*, 5(6): 203-207.

Kurien, P.P. and Parpia, H. A. B. (1968). Pulse milling in India 1- Processing and milling of tur, arhar *(Caianus caian linn.i*, CFTRI, Mysore, 11 June.

Lal, G., G.S. Siddappa and G. L. Tandon, (1986). Preservation of Fruits and Vegetables, Indian Council of Agricultural Research, New Delhi.

Lockwood, J. F. (1952). *Flour Milling*, 3rd edn, Northern Publishing Co. Ltd., Liverpool.

Lockwood, J. F. (1960). *Flour Milling*, The Northern Publishing Co., Liverpool.

Macrae, R., R. K. Robinson and M. J. Sadler, Eds, (1993). Encyclopaedia of Food Science, Food Technology and Nutrition, Academic Press, London.

Markley, K. S. (1950). Soybean and Soybean Products. Inter-science Publishers, Inc., New York

Matanhelia, K. G. (1980). Commercial *dal* milling in India. Proceedings of the International Workshop on Pigeon pea, Vol-I. ICRISAT, Patanchery 15-19 December.

Matz, S. A. (1959). The Chemistry and Technology of Cereals as Food and Feed, Avi Publishing Co., Westport, Conn.

Mayande, V. M. (1987). Pigeonpea *iCaianus cajan* L.) milling Technology: Biochemical and Engineering Studies, unpublished M. Tech. Thesis, G. B. Pant University of Agric. and Technology, Pantnagar.

Meena, RK and Yadav, J.S. (2001). Horticulture Marketing and Postharvest Management Pointer Publishers Jaipur (Raj.).

Miller, O. H. and Burns, E. E. (1970). Starch characteristics of selected grain sorghum as related to human food. J. Food Sci. 35: 666-668.

Misra, J.B. (1983). A simple arrangement for the accurate determination of the specific gravity of potato tubers. J. *Indian Potato Assoc.* 10: 121-128.

Modi B. S. (1972). Factors affecting the performance of paddy separator, Unpublished M. Tech. Thesis, 1. 1. T. Kharagpur.

Mudambi, S.R and Rao, S. (1986). Food Science. Wiley Eastern Limited, New Delhi.

Murty, D. S. and Subramanian, V. (1982). Sorghum roti. I. Traditional methods of consumption and standard procedures for evaluation. ICRISAT, 1982, Proc, Intl. Symp. On Sorghum grain quality. 28-31 October, 1981, Patancheru, A.P. India. Pp. 73-78.

Murty, D. S., Patil, H. D., Prasad Rao, K. E. and House, L. R. (1982). A note on screening the Indian sorghum collection for popping quality. J. Food Sci. Technol. 19: 79-80.

Nagy, S. and P.E. Shaw, Eds. (1980). Tropical and Subtropical Fruits AVI, Westport, CT.

Narsirnha, H. V, Ramakrishnan. N., Pratape, V M., Sashikala, V B. and Narsirnhan, K. S. (2000). Status of pulse milling industries in India. Presented in Workshop on Recent Development in Pulse Milling, CFTRI, Mysore on 14 February.

Negi, J.P., Singh, B. and Dagar, KS. (2000). Indian Horticulture Database-2000. National Horticulture Board, Ministry of Agriculture, Government of India, Gurgaon (Haryana).

Nissen, M. (1967). The weight of potatoes in water. Further studies on the relation between the dry matter and starch content. *Eur. Potato* J. 10: 85-99.

Novellie, L. (1977). Beverages from sorghum and millets in "Proc. of Symp. on Sorghum and Millets for Human Food". International Association for Cereal Chemistry. Symposium. Vienna. 1976. Tropical Products Institute, London. Pp. 73-77.

Odunfa, S.A., Nordsrom, J. and Adeniran, S.A. (1994). Development of starter cultures for nutrient enrichment of ogia West African fermented cereal gruel. Report submitted to HBVC research grants program. USAID, Washington, U.S.A.

Pandey, P.H. (1997). Postharvest Technology of fruits and Vegetables (Principles and Practices). Saroj Prakashan, 646-47, Katra, Allahabad.

Pantastico, Er. B. (1975). Postharvest Physiology, Handling and Utilization of Tropical and Subtropical Fruits and Vegetables. The AVI Publishing Company, Inc., Westport, Connecticut.

Patil, U. S, Dalvi, U. S. and Chavan, J. K.(2003). Studies on the production of starch from mould infected discolored sorghum. J. Food Sci. Technol. 40: 115-117.

Paulus, K. O. (1984). Modelling in Industrial Cooking. In: *Thermal Processing and Quality of Foods.* Zeuthen *et al.* (eds.), Elsevier Applied Science Publishers, London.

Peplinski, A. J, and Pfeifer, V. F. (1970). Gelatinization of corn and sorghum grits by steam-cooking, *Cereal Science Today*, 15, 144, 149-51.

Potter, N. N. (1984). Food Science. 3rd Edition. The AVI Publishing Company, Inc., Westport, Connecticut.

Potter, N.N. (1973). Food Science. The AVI Publishing Company, Inc., Westport Connecticut.

Potter, N.N. and Hotchkiss, J.H. (1996). Food Science. 5th Edition. CBS Publishers and

Potty, VH. (1988). Horticultural Industry in India. UNIDO Consultation Meeting, Beijing, Nov. 22-24.

Powar, C.B. and Daginawala, H.F. (1986). General Microbiology Vol II Himalaya Publishing Company, Bombay.

Primo, E., Barber, S., Tortosa, E., Camacho, J., Ulldemolins, J., Jimenez, A, and Vega, R, (1970). Chemical composition of the byproducts obtained in the different steps of the rice milling process (In Spanish), Rev. Agroquim Teenol, Alimentos 10': 244.

Pryde, E. H. (1981). Vegetable Oil vs. Diesel Fuel: Chemistry and Availability of Vegetable Oils. Alcohol and Vegetable Oil as Alternative Fuels. Proceedings of Regional Workshop: USDA Peoria, IL.

Quick, G. (1980). Developments in use of Vegetable Oils as fuel for Diesel Engines. ASAE Paper 80-1525r 15 pp.

Quick, Graeme R. (1980). An In-depth Look at Farm Fuel Alternatives. Power Farming Magazine, Feb. 10-17 pp.

Raghavandra Rao, S. N. and Desikachar, H. S. R. (1964). Pearling as a method of refining jower and wheat and its effect on their chemical composition. J. Food Sci. Technol. 1: 40-42.

Raghavandra Rao, S. N., Malleshi, N. G., Sreedharamurty, S., Viraktamath, C. S. and Desikachar, H. S. R. (1979). Characteristics of roti, dosa and vermicelli from maize, sorghum and bajra. J. Food Sci. Technol. 16: 21-24.

Rai, Mangla and Mauria S. (1999). Coarse cereal: Not for the poor alone. The Hindu Survey of Indian Agriculture. pp. 55-58.

Rai, Mangla. 2002. Amazing nutrient composition. The Hindu Survey of Indian Agriculture. pp. 59-62.

Ranganna, S. (1979). Manual of Analysis of Fruit and Vegetable Products. Tata McGraw-Hill Publishing Company Limited, New Delhi.

Raymond, W. D. (1961). The palm oil industry. Trop. Sci. 3:69-89.

Reddy, V. P. R (1981). Dehusking of arhar grain (*Cajanus cajan*). Unpublished M. Tech. thesis, G. B. Pant University of Agric. and Technology, Pantnagar.

Reichert, R. D. and Young, C. G. (1976). Dehulling cereal grains and grain legumes for developing countries. I. Quantities comparison between attrition and abrasive type mills. Cereal Chem. 53: 829-839.

Ryall A. L. and W. T. Pentzer, (1982). Handling Transportation and Storage of Fruits and vegetables, Vol. 1. Vegetables and Melons, AVI, Westport.

Ryall, A.L. and W.T. Pentzer, (1982). Handling, Transportation and Storage of Fruits and vegetables, Vo1. 2. Fruits and Tree Nuts AVI, Westport CT.

Saeed, M. and Cheryan, M. (1988). Sunflower protein concentrates and isolates low in polyphenols and phytates. J. Food Sci. 53:1127-1131.

Salunkhe D. K. and S.S. Kadam, Eds. (1995). Handbook of Vegetable Science and Technology: Production, Composition, Storage and Processing, Marcel Dekker lnc., New York.

Salunkhe, D. K. and B. B. Desai, (1984). Postharvest Biotechnology of Fruits (Vol. I & II), CRC Press, Boca Raton, Florida.

Salunkhe, D. K. and B. B. Desai, (1984). Postharvest Biotechnology of Vegetables (Vol. I & II), CRC Press Boca Raton, Florida.

Salunkhe, D. K. and S.S. Kadam, Eds, (1995). Handbook of Fruit Science and Technology: Production, Composition Storage and Processing, Marcel Dekker, Inc. New York.

Salunkhe, D. K., Chavan, J. K., Adsule, R. N. and Kadam, S. S. (1992). World Oilseeds. Chemistry, Technology and Utilization. AnaviBook, Van Nostrand Reinhold, New York.

Salunkhe, D. K., Kadam, S. S. and Chavan, J. K. (1977). Nutritional quality of proteins in grain sorghum. Qual. Plant. Pl. Foods Human Nutr. 27: 187-205.

Salunkhe, D.K., S.S. Kadam and S. J. Jadhav, (1991). Potato: Production, Processing and Products, CRC Press, Soca Raton.

Salunkhe, DK, H. R. Bolin and N. R. Reddy. (1991). Storage, Processing and Nutritional Quality of Fruits and Vegetables 2nd Edn. Vol. II. Processed Fruits and Vegetables, CRC Press, Boca Raton, Florida.

Sastry, M. C. S., Subramanian, N. and Parpia, H. A. B. (1974). Effects of dehulling and heat processing on nutritional value of sesame proteins. J. Am. Oil Chem. Soc. 51: 115-118.

Satake Engineering Co. Ltd. (1973). Rice Milling Machinery, Technical Note No. 601, Extension and Training Institute, Satake Engineering Co. Ltd., Tokyo, Japan.

Satake Owner Manual, Type-03, 91973). Satake Enzz, Co., Ltd., Japan.

Saxena, R P., Singh, B. P. N.; Singh, A. K. and Singh, J. K. (1981). Effect of chemical treatment on husk removal of arhar *(Cajanus Cajan)* grain. Paper presented at the Annual Convention of the Indian Society of Agricultural Engineering (ISAE), New Delhi, India.

Schoch, T. J. (1941), Physical aspects of starch behavior, *Cereal Chem.*, 18, 121-28.

Schrimshaw, N. S. (1974). Nutritional requirements for high protein foods, Proc. Int. Symp. Protein Foods and Concentrates, Central Food Technol. Res. Institute, Mysore, India, pp. 13-18.

Scott, J. H. (1951), *Flour Milling Processes,* Second ed., Chapman and Hall, London.

Seetharam A. (1997). Coarse cereals: Waiting for incentives. The Hindu Survey of Indian Agriculture. pp. 55-59.

Seetharam A., Kadalli, G G and Halaswamy B H. (2001). Results of front line demonstrations and Technology for increasing Prod uction of finger millet and small millets in India. All India Coordinated Small Millets Improvement Project, University of Agril. Sciences, GKVK, Bangalore 560 065.

Seetharama, N. and Rao, B.D. (2004) Sustaining nutritional security. The Hindu Survey of Indian Agriculture. pp. 37-38.

Shakuntala Manay, N. and Shadaksharaswamy, M. (1987). Foods: Facts and Principles. Wiley Eastern Limited, New Delhi.

Shepherd, A. D. (1974). How grain structure influences sorghum quality, presented to 5th E. African Cereals Res. Conf. Malawi.

Shibano, M. (1973). Construction and Function of Abrasive Roll Type Rice Whitening Machine, Technical note, JRMA, Japan.

Shreve, N. R. (1956). The Chemical Process Industries, McGraw-Hill Book Co. Inc., Tokyo.

Shrivastava, H. C. (1985). Oilseeds Production Constraints and Opportunities. Oxford & IBH Pub. Co., New Delhi.

Siew, W. L. and Mohammad, Y. (1989). Effects of refining on chemical and physical properties of palm oil products. J. Am. Oil Chem. Soc. 66:1116-1119.

Singh, A. (1986). Fruit Physiology and Production. Kalyani Publishers, New Delhi.

Singh, G. (1997). Data book on Mechanization and Agro-Processing Since Independence, CIAE, Bhopal-462038.

Smith, L. (1944). *Flour Milling Technology*, 3rd edn. Northern Publishing Co., Liverpool.

Sosulski, F. W., Sabir, M. A. and Fleming, S. E. (1973). Continuous diffusion of chlorogenic acid from sunflower kernels. J. Food Sci. 38: 468-470.

Sprague, G. F. and Dudley, J. W. (1988). Corn and corn improvement, 3rd Edn., Madison, WI: American Society of Agronomy.

Srivastasva, M (1985). Essentials of Food and Nutrition Vol.II (Applied Aspects). The Bangalore Printing and Publishing Company Limited, Bangalore.

Srivastava, R.K and Singh, S. (1987). Utilization of Mushrooms for Preparation of Different Types of Preserved Products. Government Fruit Preservation and Canning Institute, U.P., Lucknow.

Srivastava, R.P. (1992). Preservation of Fruit and Vegetable Products. Bishen Singh Mahendra Pal Singh, New Connaught Place, Dehra Dun.

Srivastava, V.; Mishra, D.P.; Laxmi Chand, Gupta, RK. and Singh, B.P.N. (1988). Influence of soaking on various biochemical changes and dehusking efficiency in pigeon pea *(Cajanus caian L.,)* seeds. Journal of Food Science and Technology 25,: 267-271.

Stermer, R A, (1968). Environmental conditions and stress cracks in milled rice, *Cereal Chem.*, 45, 365.

Stivers, T. E., Jr. (1955), American corn milling systems for degermed products. *Assoc. Operators Millers Bull.*, 2168-79.

Sugden, G. H. (1956). Various aspects of wheat conditioning, *Cereal Science Today*, I, 136-42.

Sukumaran, N.P. and Ramdass, C. (1980). A simple variable load potato hydrometer. *J. Indian Potato Assoc.* 7: 32-37.

Sullins, R. D. and Rooney, L. W. (1974). Microscopic evaluation of the digestibility of sorghum lines that differ in endosperm characteristics. Cereal Chem. 51: 134–142.

Sulllins, R. D., Rooney, L. W. and Rosenow, D. T. (1975). The endosperm structure of high lysine sorghum. Crop Sci. 15: 599-600.

Tauro, P., Kapoor, KK and Yadav, KS. (1986). An Introduction to Microbiology. Wiley Eastern Limited, New Delhi.

Thompson, R. A., and r-oster, G. H. (1963). Stress cracks and breakage in artificially dried corn, U. S. Dept. Agr., AMS, 631.

Tressler, D.K. and Joslyn, M.A. (1971). Fruit and Vegetable Juice Processing technology. The AVI Publishing Company, Inc., Westport, Connecticut.

Tuite, J., and Foster, G. H. (1993). Effect of artificial drying on the hygroscopic properties of corn, *Cereal Chem.*, *40*, 630-37.

Verma, P.; Saxena, R. P.; Sarkar, B. C. and More, P. K. (1993). Enzymatic pre-treatment of pigeon pea (Caajanus cajan) grain and its interaction with milling. Journal of Food Science and Technology. 30: 368-370.

Verma, S.c. (1991). Potato Processing in India-An Appraisal. Central Potato Research Institute, Shimla.

Wang, Y. D. and Fields, M. L. (1987). Germination of corn and sorghum in the home to improve nutritive value. J. Food Sci. 43: 1113-1115.

Watson, H. and Helmer, J. D. (1964). Cottonseed quality as affected by the ginning process-a progress report, ARS, USDA, Publ. 42, 107.

Watson, S. A. (1970). Wet milling process and products. In Sorghum production and utilization. J. S. Wall and Ross, W. M. Eds. The AVI Publishing Co. Inc, Westport, Connecticut USA. Pp. 602-626.

Watson, S. A. (1976a). Manufacture of corn and milo starches, In: *Starch: Chemistry and Technology*, Vol. II, R. L. Whistler, and E. F. Paschall (Editors), Academic Press, New York.

Watson, S. A., and Yahl, K. R. (1976b). Comparison of wet-milling properties of opaque-2 high-lysine corn and normal corn, *Cereal Chem.*, 44, 488-98.

Watson, S. A., Hirata, Y. (1962). Some wet-milling properties of artificially dried corn, *Cereal Chem.*, 39, 35-43.

Watson, S. A., Sanders, F. W., Wakely, R. D., and Williams, C. B. (1955). Peripheral cells of the endosperms of grain sorghum and corn and their influence on starch purification, *Cereal Chem.*, 32, 165-82.

Weichmann, J. (1987). Postharvest Physiology of Vegetables. Marcel Dekker Inc. New York.

Weinecke, L. A., Brekke, O. L., and Griffin, E. L., jr. (1963). Corn dry milling: Effect of Beall degerminator tailgate configuration on nroduct streams, *Cereal Chem., 40,* 575-81.

Weiss, E. A. (1983). Oilseed crops. New York: London.

Wills, R.HH, T.H. Lee, D. Graham, W.B. Megalasson and E. G. Hall, (1981). Postharvest: An Introduction to Physiology, Handling of Fruits and Vegetables, New South Wales University Press, Kensington, Australia.

Wolf, M. J., Buzan, C. L., Mac Masters, M. M., and Rist, C. E. (1952a). Structure of the mature Corn Kernel, Gross anatomy and structural relationship, *Cereal Chem.,* 29, 321-33.

Wolf, M. J., Buzan, C. L., Mac Masters, M. M., and Rist, C. E. (1952b). Structure of the mature Corn Kernel, II, Microscopic structure of pericarp, seed coat, and hilar layer of dent corn, *Cereal Chem., 29,* 334-48.

Woodroof, J. G. (1983). Peanut: Production, Processing, Products. Westport, CT: AVI.

Woodroof, J.G. and Luh, B.S. (1975). Commercial Fruit Processing. The AVI Publishing Company, Inc., Westport, Connecticut.

Woods, A.E. and Aurand, L.W. (1977). Laboratory Manual in Food Chemistry. The AVI Publishing Company, Inc., Westport, Connecticut.

Yamaguchi, M. (1983). World Vegetables: Principles, Production and Nutritive Values, AVI Westport CT.

Appendices

APPENDIX–I
Discoveries and Definitions of Food Science and Technology

The definitions given below are mainly drawn from the book Preservation of Fruits and Vegetables by G. Lal, G.S. Siddappa and G. L. Tondon, Indian Council of Agricultural Research, New Delhi, 1986 and other Food Science and Technology books.

Discoveries in Food Science

Scientist	*Year*	*Discovery*
Greek Physician Hippocrates	300 BC	Father of medicine had many ideas about nutrition and diets
Drummond	1400 BC	Noted night blindness
Carl Wilhelm & Scheele	1742-1786	Isolated & studied properties of lactic acid, Discovered chlorine, glycerol and oxygen 3 years before priestly but unpublished
Antoine Laurent & Lavoisier	1743-1794	Principles of Modern Chemistry
Marggraf	1747	Milk sugar
Needham	1749	First time he explained the cause of spoilage of stored food
James Lind	1753	Scurvy can be cured by fresh fruits and vegetables among sailors
Spallanzani	1765	Microorganisms are responsible for spoilage of fruits and vegetables during storage
Nicolas	1767-1845	Principles of Agricultural and Food Chemistry
Lavoisier	1770-1794	Nature of respiration, oxidation of organic compounds into CO_2, H_2O and heat (energy)
Josephlouis, Gay-Lussac & Louis-Jacques, Thenard	1778-1850	Percentage of carbon, hydrogen and nitrogen in dry vegetable substances
Jons Jacob Berzelius & Thomas Thomson	1779-1848 1773-1852	Beginnings of organic formulas
William Beaumont	1785-1853	Gastric digestion

Scientist	*Year*	*Discovery*
Michel Eugene & Chevreul	1786-1889	Listed O, Cl, I, N, S, P, C, Si, H, Al, Mg, Ca, Na, K, Mn, Fe, exist in organic substances
Fr.de. Fourcroy	1789	Identified proteins and fats in plants and animal tissues
Arthur Hill Hassall	1800	Differentiated pure and adulterated food stuffs
Jean Baptiste Dumas	1800-1884	Only protein, carbohydrate and fat are inadequate for support of life
Beaccornot	1802	First hydrolyzed protein with acid
Justus Von Liebig	1803-1873	Vinegar fermentation, First book on Food Chemistry (1847), Research on the chemistry of Food. He also discovered nitrogenous and non-nitrogenous compounds in food
M. Nicholas	1804	First to report the successful preservation of food in glass containers
Prout	1806	Glucose and malt sugar & protein analysis
Sir Humphry Davy	1807-1808	Isolated K, Na, Ba, Sr, Ca and Mg
M. Nicholas	1809	Awarded the prize; Book on "The art of Preserving Animal and Vegetable Substances for Many Years"
Wollaston	1810	Discovered cystine (amino acid)
Courtois	1811	Presence of iodine in burnt sponge (for the treatment of goiter)
Chevereul	1814	Fats are composed of fatty acids and glycerol
Fastier	1824	"Hole-and-cap" cans development; patent granted in 1839
Angilbert	1833	Full aperture containers
Prout	1834	Put forward the theory that there are three nutrients in food: CHO, Fats and Proteins
Mulder	1838	The word protein was coined
Leibig	1842	No organic compounds but carbohydrates, fats and proteins oxidized
Winslow and Raymond Chevalien Appert	1843	Canned foods could be processed by steam and water under pressure
-	1852	Development of pressure cookers

Scientist	Year	Discovery
Takaki	1857	Observed addition of meat, vegetables, condensed milk and barley in the diet of sailors prevent beriberi disease
	1857	Fruit and Vegetable processing first started in an organized manner
In Germany	1860	Established first Agricultural experiment station, Hanneberg, W. – Director & Stohmann, F. – Chemist
Papin	1861	First time cooking of foods by means of pressure
Isaac Newton	1862	Established United States Department of Agriculture (USDA), First Commissioner
Harvey Washington Wiley	1863	First Chief Chemist of the USDA
Louis Pasteur	1864	Role of microorganisms in food spoilage
Peltier and Paillard	1868	Used varnish for internal coating of cans
Shriver	1874	Autoclave provided with inlet for steam from external source
Howe	1876	Soldering of cans
Parry and Cobley	1882	Used sodium, potassium or calcium silicate and a serum made of proteinous materials
Baumann	1896	Presence of iodine in thyroid gland
Government Policy	1906	Pure Food and Drug Act (PFDA)
Hopkins, Osborne and Mendel	1906 1912	Showed zein a maize protein did not support for growth, deficient in lysine and tryptophan
Funk	1912	Coined the term Vitamin for the growth factors, Vitamin theory
McCollum & Davis	1913	Vitamin "A"
McCollum & Coworkers	1913	Water soluble Vitamin "B"
Cold berger	1915	Found that the addition of milk and eggs to poor maize diet to prevent occurrence of pellagra
-	1927	Canning of fruits and vegetables started for export

Scientist	Year	Discovery
Filby, F. A.	1934	A history of food adulteration and analysis
Cicely Williams	1935	Demonstrated Kwashiorkor disease in children due to protein deficiency and can be cured by feeding milk
Farrow and Green	1941	Classified lacquers into five groups
Government of India	1935	First fruit and vegetable processing factory was established at Bombay
Government of Uttar Pradesh	1949	Fruit Preservation and Canning Institute was established at Lucknow
Government of India	1950	Established Central Food Technological Research Institute, Mysore
Government of India	1955	Passed Fruit Products Order
Government of India	1973	For licensing, a Food and Nutrition Board was established
Darby	1977	Noted several deficiencies in human health in ancient Egypt
Van Leeuwenhock	-	Developed microscope
Harvey	-	Blood circulation in body
Rutherford	-	Discovered nitrogen
Priestly	-	Discovered oxygen
Black	-	Carbondioxide
Lavoisier	-	Showed that respiration is the life process
Von Liebig	-	Founder of the sciences of Agricultural chemistry and with Pasteur of Biochemistry
Lunin	-	Vitamins in milk
Eijkman	-	Vitamins in rice
Szent-Gyorgyi, King & Waugh	-	Isolation and identification of Vitamin "C"
Lavoisier	-	Established the law of conservation of mass. Father of modern chemistry and Nutrition
Scottish Physician	-	Lime and lemons are good source to prevent scurvy disease

Definitions/Terminologies

Terminology	*Definition/Explanation/Principles*
Abnormal behavior	Behavior that is characterized by the convergence of the four criteria of relative infrequency, social deviance, impaired social functioning and personal distress.
Absolute threshold	The minimum amount of stimulation necessary to produce a sensation.
Accommodation	The process by which schemata are revised completely or developed further and made more complex to fit new incoming information.
Acetyl value	The number of milligrams of potassium hydroxide required to neutralize the acetic acid liberated by the hydrolysis of 1 g of the acetylated fat or oil.
Achievement	Performance of children in mid-year and inspire examinations
Acid value	The number of milligrams of potassium hydroxide required to neutralize the free fatty acids in 1 g of fat or oil.
Acquisition	During conditioning, the process by which an organism learns a new response.
Active absorption	Absorption in which a carrier is used and ATP energy is expended.
Active listening	In humanistic psychotherapy, the process by which the therapist tries to grasp both the content of what the client is saying and the feeling behind it.
Active oxygen method (AOM)	It involves maintenance of the sample at 97.8 °C while air is continuously bubbled through it at a constant rate. The time required to obtain a specific peroxide value is determined.
Adaptation	The process of changing behavior to fit changing circumstances.
Additives	Chemical substance added to food during processing or packaging to improve their sensory properties, textural properties and extend their shelf life.
Adenosine triphosphate (ATP)	The main energy currency for cells. ATP energy is used to promote ion pumping, enzyme activity, and muscular contraction.
ADF (acid detergent fiber)	A fraction of crude fiber consisting mostly of lignocellulose.

Terminology	Definition/Explanation/Principles
Adjuvant	Substance which increase immunity response to an antigen.
Adolescence	The period that begins with the onset of puberty and ends somewhere around age 18 or 19 years.
Adsorption	The tendency of a molecule to attach itself to a finely.
Adult	An individual who is capable of assuming the responsibility of daily work and committed love.
Aerobic fermentation	The fermentation carried out in the presence of air/oxygen is known as aerobic fermentation.
Affinity chromatography	Separation of biological mixture based on highly specific biological interaction like receptor and ligand.
Affinity chromatography	Separation of biological mixture based on highly specific biological interaction like receptor and ligand.
Aflatoxin	Toxin produced in groundnut due to mold, *Aspergillus flvus*
Aftertaste	The experience that, under certain conditions, follows removal of the taste stimulus; it may be continuous with the primary experience or may follow as a different after a period during which swallowing, saliva, dilution, and other influences may have affected the stimulus substance. The result of the persistence of flavor note, particularly after swallowing.
Ageusia	Lack or impairment of sensitivity to taste stimuli.
Agnosia	Inability to recognize sensations; may be primarily in one sense, *e.g.* olfactory agnosia.
Agricultural marketing	Human activity directed at satisfying the needs and wants through exchange process. It can be also depended as comprising of all activities involved in supply of farm inputs to the farmers and movement of agricultural product from the farms to the consumers.
Alcohol yield	With corn containing 60 per cent starch, distillers traditionally obtain 19-19.7 L (5.0-5.2 proof gallons)/0.03 m (bushel). Theoretical yields as liters of absolute alcohol/100 kg of starch are, stoichiometric 72.0 L (100 per cent), maximum level 68.4 L (95 per cent), industrial standards 62-65 L (86-90 per cent). For the maximum level, 5 per cent or more of carbon substrate is consumed by yeast growth and by-product formation.
Alcoholism	Alcohol abuse characterized by psychological dependence, physical dependence and impaired social functioning.

Terminology	*Definition/Explanation/Principles*
Alkali refining or neutralization	A process in which the oil is heated with a calculated amount of sodium/potassium hydroxide to saponify and remove the free fatty acids from the crude oil.
Alpha (α) helix	A spiral shape constituting one form of the secondary structure of proteins, arising from a specific hydrogen-bonding structure.
Alveolvs	Smallest drogans involved in biosynthesis of milk, having lumen and provided with myoephithelial secretary cells.
Amino acid	An organic molecule possessing both carboxyl and amino groups. Amino acids serve as the monomers of proteins.
Amnesia	A psychological disorder in which memories for certain events are unavailable to recall.
Amylase	Starch digesting enzymes from the salivary glands or pancreas.
Amylopectin	A digestible branched-chain polysaccharide made of glucose units.
Amylose	A digestible straight-chain polysaccharide made of glucose units.
Analgesic	Any agent that produces insensitivity to pain without loss of consciousness.
Anemia	A condition in which the haemoglobin content of the blood is lower than normal
Anesthesia	Loss of sensation with or without loss of consciousness.
Angle of Nip	Angle formed by the tangents to the roll faces at the point of contact between particle and the rolls.
Angle of repose	When a grain mass is arranged in a heap, the angle between its horizontal base and the slope of cone formed is called angle of repose.
Animal fats	It mostly contains large amounts of C_{16} and C_{18} fatty acids (oleic and linoleic acids).
Anisidine value	A measure of oxidation of fats/oils beyond peroxide stage. With aldehydes, anisidine forms compounds (Schiff's base), which show a strong absorbance at 350 nm.
Anisidine value	In the presence of acetic acid, p-anisidine reacts with aldehydes producing a yellowish colour. The molar absorbance at 350 nm increases if the aldehyde contains a double bond conjugated to the carbonyl double bond. Thus the anisidine value is mainly a measure of 2-akenals. An expression termed the Totox or oxidation value (OV), which is equivalent to 2X peroxide value + anisidine value, has been suggested for the assessment of oxidation in oils.

Terminology	Definition/Explanation/Principles
Annealing	It is the process in which glass is heated up to 1000 F for 15 minutes and slowly cooled to impart stress and crack resistance is called annealing.
Annealing	Term annealing generally refers to removal of stress annealing temperature or point being defined as temperature at which stresses in glass are relived in few minutes.
Anorexia	It is the term used to describe the lack of appetite for food
Anorexia nervosa	An eating disorder involving a psychological loss of appetite and self-starvation, resulting in part from a distorted body image and various social pressures associated with puberty.
Anosmia	Inability to smell, either totally or a particular substance or group of substances.
Antagonist	It is a chemical substance that counteracts the effects of another chemical substance
Anterograde amnesia	A memory disorder in which humans lose the ability to store information.
Antetaste	A prior taste, or foretaste, usually of short duration, preceding the main taste or flavor characteristic.
Antherosclerosis	It is a disease in which the walls of the arteries become narrowed as a result of the deposition of lipid containing material
Anthropometry	Measurement of physical and gross composition
Antibiotic	A chemical agents produced by one organism that inhibits or harmful to other organism.
Anticaking agents	Chemical substance used to avoid formation of cakes during storage in powdered food.
Antifoaming agents	An agent that reduces foaming often caused by the presence of dissolved proteins and other stabilizers.
Antioxidants	Substances that retard the oxidative rancidity and thereby improve the shelf life of fats *e.g.* propyl gallate (PG), octyl gallate (OG), butylated hydroxyanisole (BHA), butylated hydroxy toluene (BHT), tertiary butyl hydroquinone (TBHQ) and tocopherol.
Antioxidants	The substances, which are responsible for the control of free radical-mediated lipid oxidation, are known as antioxidants.
Antioxidants	The substances which can prevent the occurrence of oxidation process in food and avoid rancidity as well as free radical formation are known as antioxidants.

Terminology	*Definition/Explanation/Principles*
Anxiety hierarchy	A rank ordering of anxiety-pro-voking situations used in systematic desensitization.
AOM (active oxygen method) time	The time required to obtain a predetermined peroxide value washed air is bubbled through a fat sample held at 97.8 °C.
Aphagia	It means the loss of the power of swallowing
Aphasia	Impairment in the ability to speak or to understand spoken language.
Aroma	The fragrance or odor of food, perceived by the nose by sniffing. In wines, the aroma refers to odors derived from the variety of grape, *e.g.*, Muscat aroma. It is the overall odor impression as perceived by the nasal cavity.
Arousal	An internal state of excitement or tension. Drive theorists believe that a state of arousal demands the reduction or elimination of tension.
Aseptic packaging	The technique in which food is sterilize the package and fill in the previously sterilized container and packed in it under sterile environment is known as aseptic packaging.
Aspartame	A sweetener made of two amino acids and methanol; it is 200 times sweeter than sucrose.
Aspiration	A process of cleaning grains by using a large volume of air through grain layer to separate lighter particles.
Aspiration	It is the process of separation of husk or hull by use of air currents.
Assignment problem	Scheduling jobs to person at the least cost so that one person gets one and only one job is called as assignment problem.
Astringency	Astringency is a taste-related phenomenon, perceived as a dry feeling in the mouth along with a coarse puckering of the oral tissue. Astringency usually involves the association of tannins or polyphenols with proteins in the saliva to form precipitates or aggregates.
Attention	The process that determines which sensations will be perceived; *i.e.*, which information will be transferred from sensory to short-term memory.
Auditory nerve	Axons from neurons within the cochlea that conduct neural impulses to the auditory areas of the brain.
Autosmia	Disorder of the sense of smell in which odors are perceived when none are present.

Terminology	*Definition/Explanation/Principles*
Autotrophic nutrition	A mode of obtaining organic food molecules without eating other organisms. Autotrophus use energy from the sun or from the oxidation of inorganic substance to make organic molecules from inorganic ones.
Autoxidation	The breakdown of fatty acids into peroxides and hydroperoxides upon addition of molecular oxygen across the double bonds during the storage of oil.
Auxotrophs	Bacterial mutant that require one or more growth factors that wild type or phototropic can synthesis.
Axon	The portion of a neuron that transmits neural impulses away from the soma toward the synapse.
Backset	Backset is the screened aqueous by-product from distillation. It is recycled and added to the cooked grain mash prior to fermentation.
Balling	Balling is a measure of the sugar concentration in a grain mash, expressed in degrees. It approximates percent by weight of the sugar in solution.
Basal metabolic rate (BMR)	The minimal number of kilocalories a resting animal requires to fuel itself for a given time.
Beer	Beer is the alcoholic product arising from the yeast fermentation of saccharified grain mash. It may or may not include stillage from a previous fermentation/distillation.
Beer Lambert's law	It states that the absorbance of a solution is directly proportional to the solutions concentration.
Behavior	Any activity that can be observed recorded and measured.
Beriberi	The thiamin deficiency disorder characterized by muscle weakness, loss of appetite, nerve degeneration and sometimes edema.
Beta (β) pleated sheet	A zigzag shape constituting one form of the secondary structure of proteins formed of hydrogen bonds between polypeptide segments running in opposite directions.
Beverage	It is a fruit juice considerably altered in composition before consumption. It may be diluted before it is served.
Biennial	These are defined as plant which requires two years or at least two growing seasons.
Bio functional membrane	Used in principle of bio sensors for molecules recognition in the process.
Bioavailability	The degree to which the amount of an ingested nutrient actually gets absorbed and so is available to the body.

Terminology	*Definition/Explanation/Principles*
Biochemical engineering	Biochemical engineering is concerned with conducting biological processes on an industrial scale, providing the link between biology and chemical engineering.
Biological value	Biological value of protein is the body's ability to retain protein absorbed from a food.
Bitter pit	Small brown dry area disfigures the flesh location below the skin.
Bitter taste	Some have speculated that bitter taste serves as a detergent from poisonous foods; however, we enjoy many bitter foods, *e.g.*, coffee. Many different types of molecules produce a bitter taste including divalent cations, alkaloids, and some amino acids. With >30 receptor systems for bitterness, it is the least discriminating of the taste modalities. Bitterness could arguably be broken down into several additional taste classifications. Due to the broad range of chemical structures, multiple analytical approaches may be necessary to analyzing bitter components, through HPLC is often applicable.
Blanching	Blanching is an important heating then holding and cooling process to inactivate enzymes.
Bland	Having no distinctive taste or odor property.
Bleaching	A process for the removal of colouring matter from the oil.
BV (biological value)	The percent of the absorbed nitrogen actually retained in the body.
Blind spot	A portion of the retina through which the optic nerve exits from the eye and travels to the brain. This area has no rods or cones and is not responsive to light.
Blood	Blood is the fluid which circulates through the body bringing nourishment too and removing waste products from the cells
Blood brain barrier	A specialized capillary arrangement in the brain that restricts the passage of most substances into the brain, thereby preventing dramatic fluctuations in the brain's environment.
Blood doping	A technique by which an athlete's red blood cell count is increased.
Blood pressure	The hydrostatic force that blood exerts against the wall of a vessel.
Body mass index	Weight (kg) divided by height squared (meter). A value of 30 or greater shows obesity related health risks.

Terminology	*Definition/Explanation/Principles*
Boiling point	A liquid boils when its vapour pressure is equal to the external pressure. The boiling point is thus constant for any external pressure. The normal boiling point refers to an external pressure, which is equal to the atmospheric pressure (760 mm Hg), which for water is 100 °C.
Bomb calorimeter	An instrument used to determine the Calories content of a food.
Bond	A sharing of electrons, charges or attractions. This links two atoms.
Bonded whiskey	Bonded whiskey is whiskey stored at least four years in wooden containers where the spirits have been in contact with the wood surface. It is unaltered from the original character by the addition or subtraction of any substance other than by filtration or chill proofing, is reduced in proof by the addition of water to 100 proof (5 vol. per cent) and bottled at 100 proof, and is produced at the same distillery in the same season (January through June or July through December).
Bound water	Water that is attached to organic substances. This is not available for microbial use.
Bound water	In certain foods some of the water may be bound very closely and very different to separate. Often it is extremely different to separate the water without decomposing other molecules present in the food sample.
Brainstem	The bottommost protein of the brain, an enlarged extension of the spinal cord consisting of the medulla, pons, midbrain and thalamus.
Bran	Corresponding fraction from the last break of grind in milling of wheat or any cereal grains or the brownish covering of cereal grains below the husk, that consists of pericarp, testa and aleurone layer is called bran..
Bread staling	Change in crust and crumb of bread.
Break-even point	The point at which the level of sales just equal to total expenses.

Terminology	*Definition/Explanation/Principles*
Browning	Certain enzymatic and non-enzymatic changes occurring in food mostly fruits and vegetable due to oxidation reaction cause change in colour of the food product is known as browning.

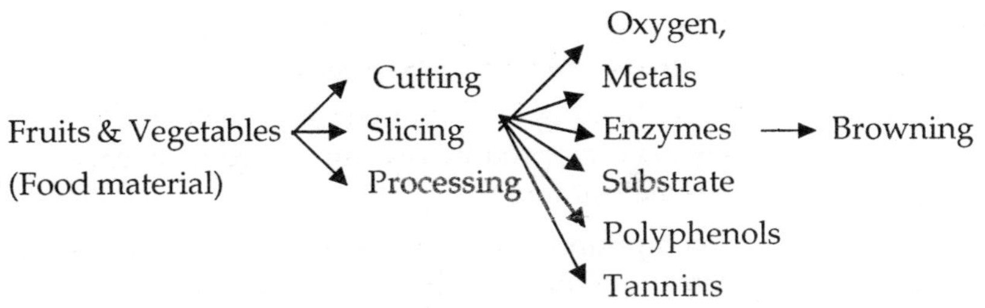

Fruits & Vegetables (Food material) → Cutting / Slicing / Processing → Oxygen, Metals, Enzymes, Substrate, Polyphenols, Tannins → Browning

Terminology	*Definition/Explanation/Principles*
Browning	Browning of foods is due to oxidative or non-oxidative reactions. Oxidative or enzymic browning is a reaction between oxygen and a phenolic substrate catalyzed by polyphenol oxidase. Nonoxidative or nonenzymic browning involves the phenomena of cartelization and or the interaction of proteins or amines with carbohydrates.
Browning	Browning reaction contributes to the aroma, flavour, and colour of many foods such as ready-to-serve cereals, toffees, roasted coffee, malted barley and baked goods.
Bubble gum	It is confection which can produce bubbles in mouth. It is prepared by using gum, resins and plasticizer with sugar and glucose syrup.
Buffer	Compounds that can take up or release hydrogen ions to maintain a certain pH value in a solution.
Bulk density	Mass of particles occupied by unit volume of bed.
Butter	It is a fat or cream separated from other milk constituents.
Cacogeusia	Persistent or intermittent unpleasant taste in the mouth.
Cacosmia	Perception of persistent or intermittent unpleasant odor.
Cake	A product obtained by baking a leavened and shortened batter containing flour, sugar, egg, milk, flavor and leavening agents.
Calories	A measurement of energy provided by foods or The amount of heat energy required to raise the temperature of 1 g of water 1 °C; the amount of heat energy that 1 g of water releases when it cools by 1 °C. The Calorie (with a capital C), usually used to indicate the energy content of food, is a kilocalorie.
Calorific value	The amount of energy (kcal) produced when one gram of substance is completely oxidized.

Terminology	*Definition/Explanation/Principles*
Cancer	A condition characterized by uncontrolled growth of abnormal body cells.
Canning of peas	Heat processing in hermetically sealed container.
Caramel	Sugar heated beyond their melting point decompose and form a brown mass known as caramel.
Caramelization	Direct heating of carbohydrates, particularly sugars and sugar syrups, produces a complex group of reactions termed Caramelization.
Carbohydrate	A compound containing carbon, hydrogen and oxygen atoms; most are known as sugars, starches and dietary fibers.
Carbon dioxide injury	In CA storages were sometimes CO_2 concentration goes higher then, the periphery of the internal tissue turns slightly brown initially then deep brown.
Cariogenic	A substance often carbohydrate-rich that promotes dental caries.
Carotenes	Pigment substances in plants that can often form vitamin A. Beta-carotene are the most active form.
Casein	Proteins in milk that form hard curds. These are difficult for infants to digest.
Casein	It is principle milk proteins, having heterogeneous mixture of various proteins.
Catabolism	Breaking down compounds.
Catalyst	A compound that speeds reaction rates but is not altered by the reaction.
Catalyst	A compound that speeds reaction rates but is not altered by the reaction.
Cavitations	If the pressure in the suction line is less than the vapour pressure, some of the liquid flashes into vapour or if the liquid contains gases, they may come out of the solution resulting into gas pockets. This phenomenon is known as cavitations.
Celiac disease	Also known as gluten-induced enteropathy. It is caused by an allergy to protein found in wheat, rye, oats and barley. If untreated, it causes a severe flattening of the villi in the intestine, leading to severe malabsorption of nutrients.
Cellulose	A straight-chain polysaccharide of glucose molecules that is undigestible because of the presence of beta carbohydrate bonds.
Cereals	Seeds of grass belonging to graminiae family

Terminology	*Definition/Explanation/Principles*
CGRE	Combine glass reference electrode used to measure the difference in pH in bioreactor.
Chalky grains	The grains which appear milky/white/opaque called chalky grains.
Cheilosis	Cheilosis is the condition characterized by lesions of the lips and at the angles of the mouth, which is caused by a deficiency of the vitamin riboflavin
Chemesthesis	Chemesthesis in the mouth is the chemical irritation (e.g., pain, heat, cooling) due to stimulation of the trigeminal nerve. Chemesthesis also occurs in other parts of the body including the eyes, nose, and throat. Some examples of chemesthesis are burning from jalapenos, cooling from mint, and pain from carbonation.
Chemical score	The percentage of the ratio of the most limiting amino acid in the test protein to that in the standard (whole egg) protein.
Chilling injury	Commodities (mainly tropical and subtropical) held at temperature above their freezing point and below 5 to 15 °C depending on the commodity.
Chilling injury	Exposure of tissues of commodities to temperature below critical level of storage temperature leads to imbalance of metabolism and causes injury to tissue.
Cholecystokinin (CCK)	A hormone that stimulates enzyme release from the pancreas and bile release from the gallbladder.
Cholesterol	A waxy lipid. It has a structure containing multiple chemical rings.
Chromatography	Chromatography can be defined as a primarily as a process separation which is used for the separation of molecular mixture.
Chronic	Long-standing, developing over time; slow to develop or resolve. When referring to disease this indicates that the disease progress, one developed, is slow and tends to remain; a good example is coronary heart disease.
Chronic toxicity	Refers to an effect that requires some time to develop toxicity.
Chylomicrons	Dietary fat surrounded by a shell of cholesterol, phospholipids and protein. These are made in the intestine after fat absorption and travel through the lymphatic system to the bloodstream.

Terminology	*Definition/Explanation/Principles*
Cis isomer	An isomer form seen in compounds with double bonds, such as fat where the hydrogen on both ends of the double bond lie on the same side of that bond.
Cleaning of wheat	Process of removal of impurities from wheat.
Climacteric fruits	Fruits which ripens after harvesting.
Clinching	Loosely covering of lids before exhausting in canning technology.
Cling film wrapping	It is the pretreatment in which the fresh fruits are wrapped in very thin film of polyethylene having a glossy edible adhesive, reduced water vapour transpiration rate, high gas permeability and do not require vents which helps to reduce the physiological loss in weight and increases shelf life.
Clostridium botulinum	A bacterium that can cause a fatal type of food poisoning.
Cloud point	The temperature at which the first permanent cloud, due to the crystallization of high-melting materials, appears in the body of the oil when cooled under standardized conditions.
Cloud test	The time (h) required for the appearance of visible cloud in the salad oil held at 0 °C.
Cloying	A taste sensation that stimulates beyond the point of satiation; frequently used to describe overly sweet products.
Coagulation	Random aggregation reactions with denaturation and aggregation reactions where protein-protein interactions predominate over protein-solvent interactions are defined as coagulation.
Coarse dispersion	Dispersion having particle size greater than 0.5 μm is coarse dispersion.
Coccus	Comma shape bacteria.
Cold chain	It is the provision of lowest safe temperature to the commodities immediately after its harvest till it reaches the consumer.
Cold test	The temperature at which the oil becomes solid.
Colostrum	The first fluid secreted by the breast during late pregnancy and the first day after birth. This thick fluid is rich in immune factors and protein.
Colostrums	It is first milk secreted by mammary/milking glands of all living females after birth of young one. It is rich in immunoglobulins, anti-microbial peptides (lactoferrin and lactoperoxidase) and other bioactive molecules, including growth factors. Human colostrums have much higher concentrations of epidermal growth factor.

Terminology	Definition/Explanation/Principles
Colour blindness	A condition in which individuals lack the ability to discriminate among different wavelengths of light.
Colour constancy	The tendency for a familiar object to be perceived as a constant colour even though the light that it reflects changes the sensation.
Coma	A state of deep unconsciousness below normal sleep.
Compatibility	In flavor terminology the ability of one substance to enhance the flavor characteristic of another.
Compatibility	It is nothing but package neither should change its colour, flavor, foreign test to food material nor should it absorb it from material.
Compensation	The result of interaction of the components in a mixture of stimuli, each component of which is perceived as less intense than it would be alone.
Competence	Uptake of free DNA fragments and incorporation into its genome by a prokaryotic cell.
Competence	Ability of recipient cell to take up DNA through changes in cell wall.
Complete proteins	Proteins that contain ample amounts of all ten essential amino acids.
Complete proteins	Proteins that contain ample amounts of all ten essential amino acids (PVT TEAM HALL).
Compressible fluid	When the density of fluid is sensitive to change in temperature and pressure, the fluid is said to be compressible fluid.
Concentrate	This is a fruit juice, which has been concentrated by removal of water either by heat or by vacuum drying.
Conditioning	Process of adjustment and throughout distribution of moisture in grain
Conditioning	Process of soaking grains in water to dehydrate them and for even distribution of moisture in grain.
Conduction	It is the transfer of heat between adjacent molecules without appreciable displacement of the particles of the body.
Conformity	Adopting the attitudes and behaviors of other people as a result of real or imagined pressure from others.
Congeners	Congeners are the flavor constituents in beverage spirits that are responsible for its flavor and aroma and that results from the fermentation, distillation, and maturation processes.
Constraints	Physical limitations on the set of decision variable.

Terminology	*Definition/Explanation/Principles*
Contribution	The difference between sales revenue and variable cost is known as contribution.
Convection	Convection is the transfer of heat from one point to another within a fluid, gas or liquid by the mixing of one portion of the fluid with another. The motion of the fluid may be entirely the result of differences in density resulting from the temperature differences. This is known as natural convection. The motion may also be produced by mechanical means as in forced convection.
Convergence	The tendency of a test sample, regardless of quality, to be perceived as similar to prior sample(s); sometimes called the halo effect.
Conversion	Conversion describes the enzymatic starch hydrolysis processes, liquification, and saccharification.
Convolutions	The irregular "hills and valleys" that make up the surface of the cortex in humans and other complex animals.
Convulsion	It is a condition marked by involuntary contraction of the muscles
Cooking	Cooking is the gelatinization by heat treatment and alpha-amylase liquification, and raw material starch.
Cooling	Cooling sensations occur when certain chemicals contact the nasal or oral tissues and stimulate a specific saporous receptor.
Cooperation	Form of association of people to work together in order to active a particular end.
Cordial	It is a sparking, clear sweetened fruit juice from which all pulp and other suspended materials have been completely eliminated. Limejuice cordial is a typical example of cordial.
Cornea	The transparent, domelike, outer covering of the lens and iris that allows light to enter the eye.
Cortical bone	Dense, compact bone that comprises the outer surface and shafts of bone.
Cortisol	A hormone made by the adrenal gland that, among other functions, stimulates the production of glucose from amino acids.
Covalent bond	A union of two atoms formed by the sharing of electrons.
Critical control point	Identify critical control points in the process at which the potential hazard can be controlled or eliminated.

Terminology	*Definition/Explanation/Principles*
Critical point in grain drying	It is the point on the drying curve at which the constant rate period ends and falling rate period starts.
Critical speed of mill	Speed at which the small sphere inside the mill just begins to centrifuge.
Crop insurance	Crop insurance scheme was launched to protect the farmers from heavy loses of crop due to rain fall, floods, drought etc.
Crude fiber	What remains of dietary fiber after acid and alkaline treatment? This consists of primary cellulose and lignin.
Crush	Type of fruit beverage which contains at least 25 per cent fruit juice or pulp and 55 per cent TSS.
Crusting	Formation of dry crust on surface of dough due to evaporation of water.
Cryptosmia	Impairment of olfaction by obstruction of the nasal passages.
Crystallized fruits/vegetables	Candied fruits/vegetables covered or coated with crystals of sugar or finely powdered sugar are called crystallized fruits/ vegetables.
Custard	Sweet mixture of egg and milk baked or cooked over hot water
Daily Reference Values (DRV)	Standards of intake for certain parts of a diet, such as carbohydrate, fat, saturated fat, cholesterol, sodium, potassium and dietary fiber, set by FDA for which no U. S. RDA exist. These values are intended to be used for comparing intake of these factors to desirable levels of intake.
Dampness	The paper boards observed moisture during rainy season when humid atmosphere is there the absorption of moisture package become loose its strength is a dampness.
Data	Records of the observations or measurements in an empirical study.
Deaeration	Freshly extracted and screened juices contain appreciable quantity of oxygen which should be removed before packing this process is known as deaeration.
Debittering	The process of removing bitter substances from oils or fats.
Decimal reduction time	Time over which the spore concentration is reduced by tenfold.
Defatted meal	Meal obtained after extraction of oil, utilization, defatted meal.
Degerming	The process of separating germ or embryo from the seed or kernel.
Degree of milling	The extent to which undesirable parts like husk, bran etc. are separated from paddy.

Terminology	*Definition/Explanation/Principles*
Degreening	It is the process in which treatment with ethylene under controlled conditions hastens the loss of chlorophyll.
Delinting of cotton seed	Cottonseed contains residual fibers after ginning called as lint. The process of removal of these lints in battery machines is known as delinting of cottonseed.
Denaturation	A process in which a protein unravels and loses its native conformation, thereby becoming biologically inactive. Denaturation occurs under extreme conditions of pH, salt concentration, and temperature.
Dental caries	Erosions in the surface of a tooth caused by acids made by bacteria as they metabolize sugars.
Deodorization	A process in which the heated oil is placed in a vaccumized tower and allowed to cascade over steam moving in countercurrent direction, to remove steam-volatile substances contributing to undesirable flavor and odor to the oil.
Depression	An extreme mood of despondency, hopelessness, and lowered self-esteem.
Desolventizers	Used for removal of traces of solvent from oil and miscella.
Dextrin	Partial breakdown products of starch that contain few too many glucose molecules. These appear while starch is being digested into many units of maltose.
Dextrose equivalent	It is total reducing power of glucose syrup expressed on a dry solids basis as dextrose. It is measure of degree of hydrolysis.
Diabetes mellitus	A disease characterized by high blood sugar levels (hyperglycemia), resulting from either an insufficient insulin released by the pancreas or a general inability for insulin to act on certain body cells, such as adipose cells.
Dietary fiber	Substances in food that are not digested by the processes present in the stomach and small intestine.
Dietetics	It is a science that deals with the adequacy of diet during normal life cycle and modification required during diseased condition.
Digestion	The process of breaking down food into molecules small enough for the body to absorb.
Direct calorimetry	A method to determine energy use by the body by measuring heat that emanates from the body, usually using an insulated chamber.

Terminology	*Definition/Explanation/Principles*
Disaccharides	Class of sugars formed by the chemical bonding of two monosaccharides.
Discrimination	In conditioning, a process whereby the organism learns to respond differently to stimuli that are differentially reinforced. In social psychology, the expression of prejudice in hostile behavior toward members of a particular group.
Distillation	The separation of components of a solution on the basis of their volatilities is known as distillation.
Distortion	The process by which information in memory is altered, making it inaccurate.
Dockage	The impurities found in grains such as chaff, stalks, grain dust, other seeds, immature grains are collectively called as dockage.
Docosahexaenoic acid (DHA)	An omega 3 fatty acid with 22 carbons and six carbon–carbon double bonds. DHA is also present in fish oils and also may be synthesized from alpha-linolenic acid [22:6].
Documentation	Establish effective record-Keeping procedures that document and provide a historical record of the facility's food safety performance.
Dominant gene	A member of a gene pair that controls the expression of a physical trait regardless of the nature of the other member of the gene pair.
Doubler	A doubler is a pot still used to redistill whiskey and low wines from a beer still. The low wines are fed into the doubler where they are redistilled by way of steam enclosed in a scroll at the bottom of the still. The bottoms, the organic components remaining at the bottom of the still, are returned to the still to extract the alcohol.
Dough improvers	The chemical substances used to improve viscosity, loaf volume, cellular texture, freshness, extensibility etc. of the bread.
Down syndrome	A form of organic retardation resulting from an abnormality in which the child has one more chromosome than normal.
Drying oil	Oil which when applied as a thin film readily absorbs oxygen from the air and dries to form a relatively tough elastic substance.
Dysomia	Difficulty in ability to smell.
Edema	Edema is the presence of an abnormally large amount of fluid in the tissue spaces of the body

Terminology	*Definition/Explanation/Principles*
Edible films and coating	It can be defined as thin layers of edible material applied on foods by wrapping immersing, brushing or spraying in order to offer selective baser against transmission of gases vapours and solutes while also offering mechanical protection.
Eicosanoids	Hormone like compounds synthesized from polyunsaturated fatty acids. Within this class of compounds are prostaglandins, thromboxanes and leukotrienes.
Eicosapentaenoic acid (EPA)	An omega-3 fatty acid with 20 carbon atoms and five double bonds; present I fish oils and may be synthesized from alpha-linolenic acid [20:5].
Electrocardiogram	A plot of electrical activity of the heart over the cardiac cycle; measured via multiple skin electrodes.
Electrophoresis	It is the migration of colloidal particles through a solution under the influence of an electric field.
Emotion	A state made up of a characteristic subjective experience, a characteristic pattern of physiological arousal, and a characteristic pattern of overt expression.
Emulsification/ Emulsion	Operation in which two normally immiscible liquids are intimately mixed or when an immiscible liquid is dispersed as small droplets in another immiscible liquid by mechanical agitation, an emulsion is formed.
Emulsion	Emulsion is a mixture of two immiscible liquids. Milk cream, mayonnaise, salad dressings, ice cream mix, and cake batters are all O/W emulsions. Butter and margarine are W/O emulsions.
Emulsion capacity	The volume (ml) of liquid emulsified by 1 g of flour or protein.
Emulsion capacity (EC)	It is the volume of oil (milliliters) that can be emulsified per gram of protein before phase inversion occurs.
Emulsion stability (ES)	It is ratio between volumes of final emulsion x 100 divided by volume of initial emulsion.
Enrichment	Addition of specific amounts of selected nutrients in accordance with a standard identity as defined by the U.S. Food and Drug Administration.
Enzymes	Proteins that catalyze specific biological reactions with extraordinary catalytic power.
Equilibrium moisture content	It is moisture content of solid in equilibrium with surrounding conditions when there vapour pressure are equal.

Terminology	*Definition/Explanation/Principles*
Erythroprotein	A protein secreted by the kidneys that enhances red blood cell synthesis and stimulates red blood cell release from bone marrow.
Essence recovery	An essence recovery system which employs a "stripping column" to separate the volatile essence from the less volatile water.
Essential amino acids	Amino acids not efficiently synthesized by humans that must therefore be included in the diet. There are ten essential amino acids [PVT TIM HALL].
Essential fatty acids	The fatty acids, which are not synthesized in the human body that must therefore be included in the diet. They are linoleic and alpha-linolenic acid.
Essential mineral	It is an inorganic substance which is needed by at least two species of animals for good health
Essential nutrient	A chemical element required for a plant to grow from a seed and complete the life cycle. A nutrient substance that an animal cannot make itself from raw materials but that must be obtained in food in prefabricated form.
Essential oil	The volatile material, derived by a physical process, usually distillation, from odorous plant material of a single botanical form and spices with which it agrees in name and odor.
Ester value	The difference between the saponification value and the acid value.
Esterification	With regards to fats, the process of attaching fatty acids to a glycerol molecule, creating an ester bond. Removing a fatty acid is called deesterification; reattaching a fatty acid is called reesterification.
Ethylene	The only gaseous plant hormone, responsible for fruit ripening, growth inhibition, leaf abscission and aging.
Exoenzyme	These are the enzymes produced by microbial cells outside the cell structure or excreted in fermentation media during metabolism.
Expectancy	One aspect of the cognitive view of conditioning, which holds that the organism comes to expect an enforces.
Expel out	Separation of liquids from solids by the application of compressive forces.
Experiment	A study in which a scientist treats an object of study in a specific way and then observes the effects of the treatment.

Terminology	*Definition/Explanation/Principles*
Falling rate period	Period of drying in which the role of drying decreases with time due to slow removal of internal moisture.
False fruit	Fruit is said to be false fruit when it develops collectively along with progressive development of ovary and associated parts.
Famine	A time of massive starvation often associated with crop failures, war and political strife.
Fat	A biological compound consisting of three fatty acids linked to one glycerol molecule.
Fat analog	A compound that provides food with many of the characteristics of fat, but altered digestibility and altered nutritional value.
Fat barrier	An ingredient system that provide a barrier for products that use fat as a heat exchange medium.
Fat extender	A fat replacement system, containing a proportion of fat combined with other ingredients, designed to optimize the functionality of fat, thus allowing a decrease in the usual amount of fat in the product.
Fat mimetics	A fat replacer that can mimic one or more of the organoleptic and physical functions of fat, and usually requires a high water content to achieve its functionality.
Fat replacer	An ingredient that can be used to replace fat, yielding fewer calories than fat and may or may not provide nutritional value
Fat substitute	A synthetic compound, usually having a similar chemical structure to conventional fats and oils, designed to replace some or all of the functions of fat without any energy contribution, and is usually indigestible or unabsorbable.
Fatty acid	A long carbon chain carboxylic acid. Fatty acids vary in length and in the number and location of double bonds; three fatty acids linked to a glycerol molecule form fat.
Feints	Feints are the third fraction of the distillation cycle derived from the distillation of low wines in a pot still. This scotch term is also used to describe the undesirable constituents of the wash that are removed during the distillation of grain whiskey in a continuous patent still (Coffey). These are mostly aldehydes and fuel oils.
Fermentable sugars	Fermentable sugars like glucose [50-99-7], maltose [69-79-4], and maltotriose [1109-28-0], can be fermented by distiller's yeast.

Terminology	*Definition/Explanation/Principles*
Fermentation	A catabolic process that makes a limited amount of ATP from glucose without an electron transport chain and that produces a characteristic end product, such as ethyl alcohol or lactic acid.
Fermentation kinetics	Fermentation kinetics is concerned with the rate of cell synthesis and/or fermentation product formation and the effect of environment on these rates.
Fermented fruit beverage	It is a fruit juice, which has undergone alcoholic fermentation by yeast. The product contains varying amounts of alcohol.
Ferritin	A protein compounds that serves as the storage form of iron in the blood and tissues.
Fertility	It refers to ability of plant not able to set and mature fruits but develop a viable seed.
Fetal alcohol syndrome	A group of physical and mental abnormalities in the infant that result from the mother consuming alcohol during pregnancy.
Fiber	A lignified cell type that reinforces the xylem of angiosperms and functions in mechanical support; a slender, tapered sclerenchyma cell that usually occurs in bundles.
Fire point	The temperature at which a substance, which ignited burns freely when the ignition agent is withdrawn.
First class fruits	Fresh fruits which do not satisfy the minimum quality requirement of the extra class and are superior to lower class.
Flaking of soybean	Heat treatment prior to extraction of oil, increases efficiency of oil extraction.
Flash point	The temperature under standardized conditions at which a liquid begins to evolve inflammable vapours.
Flavour	The sensation produced by a material taken into the mouth, perceived principally by the sense of taste and smell, but also by the common chemical sense produced by pain, tactile, and temperature receptors in the mouth.
Flavour adjunct	A substance used in or with a flavor but not essentially a part of it. These include solvents, antioxidants, enzymes, adjusting agents, emulsifiers, and acidulants.
Flavour enhancer	A substance added to supplement, modify, or enhance the original taste and/or aroma of a food without imparting a characteristic taste or odor of its own.
Flavour extract	The dilute extract of flavor prepared from spices and aromatic plant parts usually in alcohol, fortified with WONF, if needed.

Terminology	*Definition/Explanation/Principles*
Flavour or Flavorant	A substance, added to food, whose significant function is to affect odor, imparting a characteristic flavor to that food.
Flavour reversion	This problem is unique to soybean oil and other linolenate containing oils. The off flavor has been described as beany and grassy and usually develops at low peroxide values.
Flavouring ingredients	Any single chemical entity or natural mixture added to food, drugs or other products taken in the mouth, the clearly predominant purpose and effect of which is to provide all or part of the particular flavor of the final product.
Flavours	Those mixtures of ingredients whose exact composition is usually known only to their suppliers, sold in bulk to food and beverages manufactures. They are to be labeled as flavours per CFR 21 part 101 and may contain adjuncts that are nonflavour ingredients.
Flipper	Can appears normal, but when struck against a tabletop one or both ends become convex and springs or flips out, but can be pushed back to normal condition by little pressure such can is termed as flipper.
Flocculation	It refers to random aggregation reactions in the absence of denaturation.
Flour bleaching agent	The chemical substances used to oxidize carotenoid pigments to make maida creamy white.
Flow point	The temperature at which upon heating the solidified fat becomes soft and flows downward through an orifice.
Fluorescence	Fluorescent compounds may develop from the interaction of carbonyl compounds (produced by lipid oxidation) and certain cellular constituents possessing free amino groups.
Fo value	Number of minutes to specific temperature required to destroy a specific number of organisms having a specific 'Z' value.
Foam	Foam is dispersion of gas bubbles in liquid phase or semisolid phase.
Foam spray drying	In foam spray drying a compressed gas is injected into the fluid prior to the spray nozzle. The injected air causes the solids associated with a droplet to agglomerate in a manner, which increases its buoyancy as compared with conventional spray drying.
Foam stability	The ability of foaming agents (flour or protein) to retain the foam volume upon standing.

Terminology	Definition/Explanation/Principles
Foaming capacity	The volume (ml) of foam formed when a known quantity of flour or protein is mixed with distilled water and stirred for 5 min.
Fold	Strength of concentrated flavouring materials. The concentration is expressed as a multiple of a standard, *e.g.* citrus oil is compared to cold pressed oil. In the case of vanilla, folded flavours are compared to a standard extract with minimum bean content.
Folding carton	Folding carton made from sheets of paperboard which have been cut and scored for bending into desired shapes, they are delivered in a collapsed state for erection at the packaging point.
Food	Any edible material, which is consumed by the individual and his/her family. Whatever we eat and that nourishes the body is called food.
Food Chemistry	Food chemistry is major aspect of food science, which deals with the composition and properties of food and chemical changes it undergoes during handling, processing and storage.
Food intolerance	An adverse reaction to food that does not involve an immune response.
Food retailing	Food retailing is a sale of food items in small lots to the consumers.
Food Science	The Science deals with the food, which is edible by living matter
Food Science and Technology	It is a science deals with food and their technology
Food sensitivity	A mild reaction to a substance in a food that might be expressed as slight itching or redness of the skin.
Food web	The elaborate, interconnected feeding relationship in an ecosystem.
Food-borne illness	Sickness caused by ingestion of foods containing toxic substances produced by microorganisms.
Foods for body building	Foods rich in protein, minerals, vitamins and water.
Foods for energy	Foods rich in carbohydrates, fats and protein.
Foods for regulating body process	Foods rich in minerals, vitamins, water and fiber.

Terminology	*Definition/Explanation/Principles*
Foots	The saponification fatty acids in crude oil which settle down to bottom on agitation of oil with alkali.
Foots	The saponified fatty acids in crude oil which settle down to known to bottom on agitation of oil with alkali.
Fore milk	The first breast milk delivered in the nursing session.
Foreshots	Foreshots is the first fraction of the scotch distillation cycle derived from the distillation of low wines in a pot still.
Forgetting	The inability to recall a particular piece of information accurately.
Fortification	Addition of nutrients in amounts significant enough to render the food a good to superior source of the added nutrients. This may include addition of nutrients not normally associated with the food or addition to levels above that present in the unprocessed food.
Fortified	A term generally meaning that vitamins, minerals or both have been added to a food product in excess of what was originally found in the product.
Free fatty acids	The fatty acids in the oil that are not esterified with glycerol and those liberated from triglycerides when subjected to hydrolytic rancidity.
Free radical	Short-lived form of compounds that exist with an unpaired electron in their outer electron shell. This causes it to have an electron-seeking nature, which can be very destructive to electron-dense areas of a cell, such as DNA and cell membranes.
Free water	The water not bound to the components in a food. This is available for microbial use.
Freeze drying	In this process food is first frozen at -18 C on trays in the lower chamber of freeze drier and the frozen material dried under high vacuum in the upper chamber. Direct sublimation of ice taken place without passing through the intermediate liquid stage.
Freezing point	The freezing point of a material is the temperature at which it changes from a liquid to a solid. A liquid freezes when its vapour pressure is equal to the vapour pressure of its solid. The freezing point of water is 0°C.
Fructose	A monosaccharide with six carbons that form a five-membered ring with oxygen in the ring; found in fruits and honey.

Terminology	*Definition/Explanation/Principles*
Fruit	A mature ovary of a flower that protects dormant seeds and aids in their dispersal.
Fruit	The development of ovary and jaunt tissues following blossoming.
Fruit cordial	This is fruit squash from which all suspended material is completely eliminated and is perfectly clear, *e.g.*, limejuice cordial.
Fruit drink	Liquefying the whole fruit makes this and at least 10 per cent of the volume of undiluted drink must be whole fruit. It may be diluted before being served.
Fruit juice	This is a natural juice pressed out of a fruit, and is unaltered in its composition during preparation and preservation.
Fruit juice concentrate	This is fruit juice, which has been concentrated by the removal of water either by heat or by freezing.
Fruit punches	Mixing the desired fruit juices at the time when it is served makes these.
Fruit squash	This consists essentially of strained juice containing moderate quantities of fruit pulp to which sugar is added for sweetening *e.g.*, orange squash, lemon squash, mango squash, etc.
Functional Foods	Foods with a specific additional benefit that goes beyond the nutritional benefits of the nutrients they contain. Or Food that encompasses potentially healthful products, including any modified food or food ingredient that may provide a health benefit beyond that of the nutritional nutrients it contains.
Fuel oil	Fusel oil is an inclusive term for heavier, pungent tasting alcohols produced during fermentation. Fusel oils are composed of mixture of n-propyl, isobutyl, and isoamyl alcohols.
Galactosemia	A disease characterized by the buildup of the monosaccharide galactose in the bloodstream resulting from the inability of the liver to metabolize it. If present at birth and left untreated, this disease causes severe growth and mental retardation in the infant.
Gelatinization	The tendency of starch molecules to swell and form gel when starch solution is heated.
Gelation	When denaturation molecules aggregate to form an ordered protein network the process is referred to as gelation.

Terminology	*Definition/Explanation/Principles*
Gelation	Gelation is the irreversible change in viscosity of egg yolk after thawing.
Gene cloning	Transfer of sequence of DNA (gene) from one organism to other organism.
Gene stability	Major of resistance to change with time of the sequence of gene within DNA molecule or of nucleotide sequence with gene.
Generally recognized as safe (GRAS)	A group of food additives that in 1958 were considered safe, therefore allowing manufactures to use them thereafter when needed in food products. The FDA bears responsibility for providing they are not safe and can be remove unsafe products.
Generation time	The time required by micro-organism to double its population.
Genetics	The branch of biology concerned with study of heredity and variation.
Germination	It is the process in which soaked grains are allowed to sprout and develop vegetative growth or roots.
Gills	Sexual spore *i.e.* basidiospores are formed, borne on the underside of fruiting bodies, mostly seen in *Agaricus* sp. of mushroom.
Glazed fruits/ vegetables	Covering of candied fruits/vegetables with a thin transparent coating of sugar, which imparts them glossy appearance, is known as glazing fruits/vegetables.
Glucagon	A hormone made by the pancreas that stimulates the breakdown of glycogen in the liver into glucose; this raises the blood glucose level. Glucagon also performs other functions.
Gluconeogesis	The production of new glucose molecules by metabolic pathways in the cell. The source of the carbon atoms for these new glucose molecules is usually amino acids.
Glucose	A six-carbon atom carbohydrate found in blood and in table sugar bound to fructose; also known as dextrose, it is one of the simple sugars.
Glycemic index	A ratio used to measure the relative ability of a carboh drate to raise blood glucose levels as opposed to the ability of white bread (or glucose) to raise blood glucose levels.
Glycerol	An alcohol containing three hydroxyl groups (-OH); used to help from triglyceride molecules.

Terminology	*Definition/Explanation/Principles*
Glycogen	An extensively branched glucose storage polysaccharide found in the liver and muscle of animals; the animal equivalent of starch.
Glycolysis	The splitting of glucose into pyruvate. Glycolysis is the one metabolic pathway that occurs in all living cells, serving as the starting point for fermentation or aerobic respiration.
Goiter	An enlargement of the thyroid gland often caused by a lack of iodine in the diet.
Goitregens	Substances in food that interferes with thyroid hormone metabolism and so may cause goiter if consumed in large amounts.
Grading	Sorting on the basis of quality criteria.
Grain whiskey	Grain whiskey s an alcoholic distillate from fermented wort derived from malted and unmalted barley and corn, in varying proportions, and distilled in a continuous patent still (Coffey).
Growth	The progressive development or increase in size of a living thing, such as the growth of a child
Gustation	A taste sense, the receptors of which lie in the mucous membrane covering the tongue, and the stimuli for which consist of certain soluble chemicals, *e.g.* salts, acid, and sugar.
Haemoglobin	The oxygen-carrying pigment of red blood cells
HANES	Health and Examination Surveys: A survey conducted between 1971-1972 on the United States Population aged 1 to 74 years to evaluate their nutritional status. The sample size of the population was about 20,000.
Hard to cook legume	Rate of water absorption and cooking is low, retards cooking.
Hard water	Water containing chlorides and sulphates of calcium and magnesium.
Harris-Benedict equation	An equation that predicts resting metabolic rate based on a person's weight, height, and age.
Haugh unit	Measurement of height of thick white in relation to the weight of egg.
Heads	Heads is distillate containing a high percentage of low boiling components such as aldehydes.
Health	A state of complete physical, mental and social well-being and not merely the absence of disease or infirmity (WHO, 1948)

Terminology	*Definition/Explanation/Principles*
Health	It is defined as state of complete physical, mental and social well being and not merely the absence of disease and infirmity.
Heartburn	A pain emanating from the esophagus, caused by stomach acid backing up into the esophagus and irritating the esophageal tissue.
Heat	Heat is energy transferred from one system to another solely by reason of a temperature.
Heat transfer	The transfer of heat is the principal unit operation in many food processes.
Heat transfer coefficient	Heat transfer coefficients for drop wise condensation of vapor: The heat transfer coefficients for drop wise condensation are very high in the range of 10,000 to 70,000 Btu/hr. (sq.ft.) (°F).
Hehner number (value)	The percentage of water and fatty acids in 1 g of fat.
Hemoglobin	The iron-containing protein in the red blood cell that carries oxygen to the cells and carbon dioxide away from the cells. It is also responsible for the red colour of blood.
Hemolysis	It is the process whereby the red cell bursts, releasing the haemoglobin from it
Hemorrhage	It is a condition in which blood escapes from the vessels. Some time it also called as bleeding
Herd immunity	Resistance of a group to a pathogen as a result of immunity of a large proportion of the group to that pathogen.
Heterotrophic nutrition	A mode of obtaining organic food molecules by eating other organisms or their by-products.
High wines	High wines are an all-inclusive term for beverage spirit distillates that have undergone complete distillation.
High-density-lipoprotein (HDL)	The lipoprotein synthesized by the liver and intestine that picks up cholesterol from drying cells and other sources and transfers it to the other lipoproteins in the bloodstream. A low HDL level increases the risk for heart disease.
Hind milk	The milk secreted at the end of a nursing session; it is higher in fat than fore milk.
HIV (Human immunodeficiency virus)	The infectious agent that causes AIDS; HIV is an RNA retrovirus.
HLB value	Hydrophile-lipophile balance is ratio of weight of percentage of hydrophilic and hydrophobic groups in an emulsifier.

Terminology	Definition/Explanation/Principles
HMG-CoA reductase	An enzyme in the cytosol that catalyzes the conversion of hydroxymethylglutaryl-CoA (HMG-CoA) to form mevalonate. The action of HMG-CoA reductase is the committed step in cholesterol biosynthesis.
HMG-CoA reductase	An enzyme in the cytosol that catalyse the conversion of hydroxymethyl-CoA to form mevalonate. The action of HMG-CoA reductase is the committed step in cholesterol biosynthesis.
Holding cost	The cost associated with holding inventories in a stock is known as holding cost.
Hops	Hops are dried female flowers of hope plants which are grow in Ciregon and Washington.
Hormone	A compound secreted into the bloodstream that acts to control the function of target organ cells. Hormones can be either protein like or fatlike, such as insulin or estrogen.
Hue	Hue is the aspect of colour that we describe by words such as green, blue, yellow and red.
Hydrogenation	The chemical addition of hydrogen to a material or The addition of hydrogen atoms to the double bonds of polyunsaturated and monounsaturated fatty acids to reduce the extent of unsaturation. This process turns liquid vegetable oils into solid fats.
Hydrogenation of oil	It is the process in which liquid fat or oil is converted into solid or semi-solid fat by addition of hydrogen.
Hydrophilic	Attracts water (water loving).
Hydrophilic-Lipophilic balance (HLB)	It is a ratio between hydrophilic emulsifier and lipophic emulsifier. As rule emulsifiers with HLB values in the range 3-6 promote W/O emulsions; values between 8 and 18 promote O/W emulsions.
Hydrophobic	Repels water (water fearing).
Hyperglycemia	High blood glucose levels, above 140 milligrams per 100 milliliters of blood.
Hyperosmia	Unusually keen olfactory sensitivity.
Hypertension	A condition in which blood pressure remains persistently elevated, especially when the heart is between beats.
Hypobaric storage	The storage in which pressure in the storage atmosphere is less than atmospheric pressure.
Hypogeusia	Diminished sense of taste.

Terminology	*Definition/Explanation/Principles*
Hypoglycemia	Low blood glucose levels, below 4 to 50 milligrams per 100 milliliters of blood.
Hyposmia	Diminished sense of smell.
Hypothesis	A proposition or assertion about the possible relationship between variables.
Imbibed water	May not be different from water held as hydrate some substances pickup water and swell, when they come in contact with water.
Immobilized enzymes	An immobilized enzyme or cell is chemically or physically restricted in movement so that it can be physically reclaimed from the reaction medium.
Immune system	A complex of organs and cells that function to protect the body against disease agents such as viruses, bacteria and pollutants.
Imperfect flower	Species that have only stamens or stigma within a flower.
Incidental food additives	Additives that gain access to food products indirectly from environmental contamination of food ingredients or during the manufacturing of food process.
Incomplete proteins	Food protein that lacks ample amount of one or more of the essential amino acids needed to support human protein needs.
Index of nutrition quality	It is percentage of nutrient need provides by food divided by percentage of calorie need provided by from food.
Infusion mashing	Infusion mashing is the process of simultaneously cooking and converting small grains (rye, barley and wheat).
Inherent toxicants	These are metabolites produced via biosynthesis of food under normal growth conditions and by organisms.
Inland fishery	Process of raising and harvesting fresh water fish, rivers, lakes are the places of habitat of fresh water fish. *e.g.* rohu, katla and mrigal etc.
Insipid	Tasteless, fat, vapid.
Insoluble fiber	Fibers that, for the most part, do not dissolve in water nor are digested by bacteria in the large intestine. These include cellulose, some hemicelluloses and lignin.
Instant dhal	Less time required for cooking as compared to ordinary dhal/ ready to eat dhal.

Terminology	*Definition/Explanation/Principles*
Insulin	A hormone produced by the beta cells of the pancreas. Insulin increases the synthesis of glycogen in the liver and the movement of glucose from the bloodstream into muscle and adipose cells.
Integral protein	A protein of biological membranes that penetrates into or spans the membrane.
Intelligence	The capacity to understand the world and the resourcefulness to cope with its problems.
Intelligence quotient (IQ)	The relationship between mental age and chronological age, expressed as MA/CA x 100.
Intermediate density lipoprotein (ILD)	The product formed after a very low-density lipoprotein (VLDL) has most of its triacylglyceride removed.
International unit (IU)	A crude measure of vitamin activity, often based on the growth rate of animals. Today these units have been replaced by more precise microgram quantities.
Intoxication	The state in which an excess of a sedative poisons the body, leading to uncoordinated actions, slowed reaction times and slurred speech.
Invert sugar	The hydrolyzed mixture of sucrose, containing glucose and fructose is known as invert sugar.
Iodine value	The number of iodine absorbed by 100 g of oil. It is a measure of the unsaturation of fat.
Iodine value	This test is a measure of the unsaturated linkages in a fat and is expressed in terms of percentage of iodine absorbed.
Iron	A metallic element occurring in the haemoglobin of red blood cells, stored in tissues in the form of ferritin and is an essential part of important respiratory enzymes
Isoelectric point pH	It is defined as the pH at which a solute (compound) has no net electric charge or carry equal positive and negative charge (*i.e.* zwitterions).
Isolate	A relatively pure chemical produced from natural raw materials by physical means, *e.g.* distillation, extraction, crystallization, etc. and therefore natural; or by chemical means, *i.e.* via hydrolysis, bisulfate addition products, and regeneration, etc. and therefore artificial by 1993 U.S. labeling regulations.
Isomer	Different chemical structures for compounds that share the same chemical formula.

Terminology	*Definition/Explanation/Principles*
Isotope	An alternate form of a chemical element. It differs from other atoms of the same element in the number of neutrons in its nucleus.
Isozymes/ Isoenzymes	Multiple forms of the same enzyme arising from genetically determined differences in primary amino acid sequences.
Jam	It is a product prepared by boiling fruit pulp with sufficient quantity of sugar to reasonably thick consistency, firm with enough to hold fruit tissues in position. It should contain not less than 68.5 per cent T.S.S.
Jaundice	A yellow staining of the skin and sclera resulting from a buildup of bile pigments in the bloodstream. Liver or gallbladder disease is often the cause.
Jelly	It is a product prepared by mixing strained extract of fruit with sugar and boiling the mixture to a stage at which it will set a clear gel.
JunK DNA	The genetic material of animal and plants consists of mainly two types coding and non coding. The non coding is also known as no functional or junk DNA.
Kernel Hardness	Kernel hardness measurements are mainly useful in differentiating between soft or hard grains in plant breeding programme. Extensively hard reflects in to increase in power for grinding while extensively softness reflects into bolting and increases the requirements for sieving space.
Ketone bodies	Products of acetyl-CoA (Fat) metabolism containing three to four carbon atoms; acetoacetic acid, beta-hydroxybutric acid, and acetone. These contain a ketone group, hence, the name.
Ketosis	The condition of having high levels of ketones in the bloodstream.
Key column	The optimum column, the one with largest negative index number.
Kreis test	This is one of the first tests used commercially to evaluate oxidation of fats. It measure a red colour believed to result from reaction of epihydrin aldehyde or other oxidation products with phloroglucinol.
Kries test	A test in which the oxidized fats react with phloroglucinol in acid solution to give a red colour.
Kwashiorkor	A disease seen primarily in young children who have an existing disease and who consume a marginal amount of calories and considerably insufficient protein in the face of high needs. The child will suffer from infections and exhibit edema, poor growth, weakness, and an increased susceptibility to further illness.

Terminology	*Definition/Explanation/Principles*
Kwashiorkor	Means a sickness in a child develops when another baby is born.
Lamination	Combining two or more films into single film by means of solvent or heat to compliment all the properties in a single film is called lamination.
Lard	It is an animal fat from hogs.
LDL (low density lipoprotein) cholesterol	The fraction of cholesterol carried in the blood as a part of low-density lipoproteins.
Lean body mass	The part of the human body that is free of all but essential body fat. About 2 per cent of body weight as fat is essential to retain. The rest of the fat in the body represents storage and so is not part of lean body mass. Lean body mass includes muscle, bone, organs, connective tissue, skin and other body parts.
Leathery crust	Crust of bread is crisp and breaks easily when pulled but crust becomes tough and exhibits the quality of leatheriness when pulled.
Lecithin	A group of phospholipids containing two fatty acids, a phosphate group, and a choline molecule.
Lectins	Antinutritional factor present in plant foods some time it is called hemaglutinins.
Liarrage	The place where animals are brought 24 hrs before slaughtering.
Life span of spore	The life span of spore is defined as the length of time for which spores exposed to given temperature, remain viable.
Lignin	An insoluble fiber made up of a multiringed alcohol (no carbohydrate) structure.
Limit dextrin	Limit dextrin are oligosaccharides containing one or more 1, 6-α-linkages.
Limiting amino acid	The essential amino acid in the lowest concentration in a food in comparison with the body's need.
Lipogenesis	The building of fatty acids using derivatives of acetyl-CoA molecules.
Lipogenic	Means creating lipid. The liver is the major lipogenic organ in the human body.
Lipolysis	The breakdown of lipids.
Lipolysis	Hydrolysis of lipids to produce glycerol and free fatty acids.

Terminology	*Definition/Explanation/Principles*
Lipolysis	Hydrolysis of ester bonds in lipids may occur by enzyme action or by heat and moisture, resulting in the liberation of free fatty acids. The release of short chain fatty acids by hydrolysis is responsible for the development of a rancid flavour (hydrolytic rancidity) in raw milk. Lipolysis is a major reaction occurring during deep fat frying due to the large amounts of water introduced from the food and the relatively high temperature at which the oil is maintained.
Lipoprotein	A compound found in the bloodstream containing a core of lipids with a shell of protein, phospholipid and cholesterol.
Low wines	Low wines is the term for the initial product obtained by separating (in a pot or Coffey still) the beverage spirits and congeners from the wash. Low wines are subjected to at least one more pot still distillation to attain a greater degree of refinement in the malt whiskey.
Lozenges	These are the sugar dough which has been flavoured, cut to shape and subsequently dried to remove the most of the added water.
Macrosmatic	Abnormally keen olfactory sense.
Malnutrition	Impairment of health resulting from a deficiency or imbalance of nutrients. It can be under nutrition or over nutrition.
Malnutrition	The chronic deficiency of one or more nutrient in daily diet.
Malt whiskey	Malt whiskey is an alcoholic distillate made from fermented wort derived from malted barley only and distilled in pot stills. It is the second fraction (heart of the run) of the distillation process.
Malting	Malting is a controlled germination process, which activates the enzymes of the resting grain, resulting in conversion of cereal starch to fermentable sugars, partial hydrolysis of cereal proteins and other macromolecules.
Malting	It is processing of grain where grains are steeped, germinated and kilned to increase maltose content of grain.
Marasmus	Results from a person consuming insufficient protein and kcalories; usually seen in infancy. It is the equivalent of protein-energy malnutrition in adults. The person will have or no fat stores and show muscle wasting.
Margarines	Mixtures of fats blended to provide desired properties for use as table spreads or for bakery use.
Marketing channels	It is the chain of intermediaries through whom the various food grains pass from producer to consumers.

Terminology	*Definition/Explanation/Principles*
Marmalade	It is a fruit jelly in which slices of fruit or of the peel are suspended. The term marmalade is generally associated with product made from citrus fruit like oranges. Lime/Lemon shredded peel is included as the suspended materials.
Mayonnaise	An emulsion of oil-in-water that readily breaks if crystals begin to form in oil.
Meat	Meat may be defined as contractive tissue from all animals used for food.
Megaloblast	A large, immature red blood cell that results from an inability for cell division during red blood cell development.
Melanoids	The brown to black amorphous, unsaturated heterogeneous polymers are called melanoids.
Melting point	The temperature at which a fat becomes completely clears.
Memory	The ability to store information so that it can be used at a later time.
Mental age	A measure of an individual's performance on an intelligence test expressed in terms of months and years.
Merosmia	A condition analogous to colour blindness, in which certain odors are not perceived.
Micro-aerophillic	Microbe requiring small amount of air.
Microsmatic	Having a poorly developed sense of smell.
Milk	Milk is normal secretion of mammary gland is defined as the lacteal secretion. Practically free from colostrums, obtained by complete milking of one or more than one cows which contains not less than 8 per cent MSNF and not less than 3–35 per cent milk fat.
Milk fats	Milk fats mostly contain palmitic, oleic and stearic fatty acids and its fat is unique among animal fats in that it contains C_4 to C_{12} chain fatty acids.
Milkfeed	Mixture of bran, germ and shorts.
Millard reaction	The browning reactions occur in the presence of an amino bearing compound, usually a protein, a reducing sugars and some water. It changes colour, flavor, odour and texture of the product. These types of reaction occur by different four ways. 1. Sugars + nitrogenous compounds + heat 2. Sugar + organic acids + heat 3. Nitrogenous compounds + organic acids + heat 4. Organic acids themselves + heat

Terminology	*Definition/Explanation/Principles*
Milling	Process of separation of bran and germ from endosperm and reduction of endosperm into fine flour.
Milling loss	Loss of moisture by evaporation during milling.
Minimum support price	This is the price fixed by the Government to protect the producer-Farmers against excessive fall in the price during bumper production year.
Miso	Fermented food product prepared from soybean and rice using microorganisms for fermentation process.
Mixing	Two or more components are interspersed in space with one another by means of flow.
Molar extinction coefficient	In the specific absorption coefficient for a concentration of one mole and path length 1 cm.
Monod equation	$\mu = \mu$ max – S/ks + S
Motivation	The desires, needs and interests that arouse an organism and direct it toward a specific goal.
Moulds	Moulds are multicellular, filamentous fungi belonging to the division Thallophyta but are devoid of chlorophyll. *Penicillium* sp. Blue moulds *Aspergillus* sp. Black moulds *Mucor* sp. Gray moulds
Mulching	It is system of orchard soil management in which orchard soil is covered with curry material to reduce moisture loss due to evaporation.
Multiple personality	A form of dissociative disorder in which large segments of the personality are split off from conscious awareness, so that the person seems to fluctuate between two or more distinct personalities.
Mutarotation	The change in specific optical rotation representing inters conversions of α and β forms of D glucose to an equilibrium mixture.
Mycotoxin	Toxin produced by molds *e.g.* Aflatoxin.
Nature identical flavor material	A flavor ingredient obtained by synthesis, or isolated from natural products through chemical processes, chemically identical to the substance present in a natural product and intended for human consumption either processed or not; *e.g.* citral obtained by chemical synthesis or from oil of lemongrass through a bisulfate addition compound.

Terminology	*Definition/Explanation/Principles*
NCHS	National Centre for Health Statistics, United States-NCHS collected and analyzed statistics on health status, human needs and resources. NCHS also published data in the form of tables on weight and height for children and adolescents upto 18 years old (sex separated)
NDF (neutral detergent fiber)	The representative of the fibrous cell wall constitutes, containing lignin, cellulose, hemicelluloses and some fiber-bound proteins.
NDP (net dietary protein) cal, per cent	The percentage of total calories proved by the protein.
Nectar	Nectar is obtained by blending the thin pulp of the fruit with sugar and citric acid. The finished product has 15-20 °Brix and a mild acid taste.
Nectar	Type of fruit beverage which contains at least 20 per cent fruit juice/pulp 15 per cent TSS and 0.3 per cent acid.
Neurotoxin	Toxin affects central nervous system..
Neurotransmitter	A chemical substance involved in the transmission of neural impulses from one neuron to another. Neurotransmitters are released when an action potential reaches the end of the axon (axon terminal). Upon release, neurotransmitters cross the synapse and attach to receptors on either the dendrite or cell body of the adjacent neuron.
Neutralization of oil	Process for removal of free fatty acids from oil.
Newtonian fluid	Fluid obeys Newton's viscosity law.
Newtonian fluid	For the fluids the ratio of shear stress to shear rate is constant and equal to the viscosity of the fluid. Such fluids are called Newtonian fluids.
Nonbreak oil	Edible oil, which does not deposit solid material when heated to at least 300 °C.
Non-climacteric fruits	Fruits those are showing respiratory pattern slow drift down wards after detachment from parent plant.
Nosocomical infection	Hospital acquired infection.
Nougats	These are the confections prepared by using sugar, glucose syrup and whipping agents which increase the desired volume of air.
NPR	Weight gain in test protein group (g) + weight loss in non-protein group (g) ÷ Protein consumed (g).

Terminology	*Definition/Explanation/Principles*
NPSH	The NPSH is the amount by which the pressure at the suction point of pump (*i.e.* sum of velocity and pressure heads) is in excess of the vapour pressure of the liquid.
NPU	N intake – fecal N – urine N x 100 ÷ N intake + (metabolic N + endogenous N) x 100 ÷ N intake.
Nuclear magnetic resonance (NMR)	Brain-imaging technique in which the head is exposed to magnetic fields of varying strengths and a computer-assisted three-dimensional image of the brain is produced.
Nutraceuticals	Nutricuticals are products and their derivatives that occur in Nature and are constituents of plants and animal, including humans. These constituents confer a health benefit above and beyond basic nutrition or basic fortification.
Nutrient	The combination of processes by which a subject receives and utilizes materials (foods) necessary for the maintenance of body components.
Nutrients	Nutrients are the constituents in food that has specific function in body.
Nutrition	It is a science of food and its interactions with an organism to promote and maintain health.
Nutrition assessment	Includes the measurement and description of the nutritional status of an individual or population in relation to economic, socio-demographic and physiological variables
Nutritional status	The condition of the body of individual (s) resulting from intake, absorption and utilization of food or nutrients over a period of time.
Nutritionist	A person who advise about nutrition and/or works in the field of food and nutrition. In many states in the United States a person does not need formal training to use this title. Some states reserve this title for Registered Dietitians.
Nyctalopia	It is night blindness, failure or imperfection of vision at night or in a dim light
Obesity	A condition characterized by excess body fat, usually defined as 20 per cent above desirable weight.
Obesity	When the energy intake exceeds over expenditure, the excess is deposited as a fat over a period of time obesity occurs.
Objective evaluation	Objective evaluation involving instruments. These instruments may be categorized into two types, viz., imitative measures and non-imitative measurements.

Terminology	Definition/Explanation/Principles
Odor and odorant	That which is smelled. Odor may refer to the odorant or to the sensation resulting from the stimulation of olfactory receptors in the nasal cavity by gaseous material.
Odoriphore or Osmophore	Odor-producing group.
Olericulture	It deals with cultivation of vegetable crops.
Oligosaccharides	Carbohydrates containing three to ten monosaccharide units.
Omega-3 (w-3) fatty acid	A fatty acid with its first double bond first appearing at the third carbon atom from the methyl end ($-CH_3$).
Omega-6 (w-6) fatty acid	A fatty acid with its first double bond first appearing at the sixth carbon atom from the methyl end ($-CH_3$).
Opsonization	Promotion of phagocytosis by a specific antibody in combination with complement.
Optimum feasible solution	Any basic feasible solution which optimizes objective function of general linear programming problem is called as optimum feasible solution.
Osmics	The science of smell.
Osmosis	The movements of micro-organism in response to salt or sugar solution.
Osmyl	An odorant.
Osteoporosis	A bone disease that develops primarily after menopause in women and is characterized by a decrease in bone density.
Oven spring	It is phenomenon of increases in loaf volume of dough during first 15 minutes of baking.
Overweight	Weight in excess of the average for given sex, age and height in relation to the NCHS reference tables
Oxidation	The loss of electron from a substance involved in a redox reaction.
Oxirane test	This method is based on the addition of hydrogen halides to the oxirane group. Epoxide content is determined by titrating the sample with HBr in acetic acid, in the presence of crystal violet to a bluish green end point.
Oxygen absorption	The amount of oxygen absorbed by the sample as determined by the time to produce a specific pressure decline in a closed chamber or the time to absorb a pre-established quantity of oxygen under specific oxidizing conditions is taken as a measure of stability. This test has been particularly useful in studies of antioxidant activity.

Terminology	*Definition/Explanation/Principles*
Oxygen bomb method	The samples are dispersed in filter pulp in the glass liner of a sealed bomb at 50 psig oxygen pressures. The bomb is placed in boiling water and the time required to reduce the pressure to 2 psig is noted.
Packaging	It is an external means of preserving the food during storage, transportation, and marketing.
Parageusia	Gustatory disturbance resulting in erroneous identification of taste stimuli.
Parasite	A microbe which depends on living matter for its growth and survival.
Parboiling	It is a hydrothermal treatment given to paddy followed by drying before milling.
Parboiling of legumes	Partial boiling of legumes to reduces leaching losses.
Parosmia	A disturbance to the sense of smell resulting in smelling the wrong odors, usually perceived as repulsive.
Partition chromatography	If the moving phase is liquid and the stationary phase is a liquid film on same support, the resulting chromatography is called partition chromatography.
Passive absorption	Absorption that uses no energy. It requires permeability for the substance through the wall of the intestine and a concentration gradient higher in the lumen of the intestine than in the absorptive cell.
Pasteurization	Pasteurization means making free the food from human pathogens and most vegetative microorganisms by heat treatment upto 100 °C for shorter period.
Pathogencity	The ability of a parasite to infect or to cause the disease in host.
Pelagic fish	Fishes that found on ocean surface.
Pellagra	A disease characterized by inflammation of the skin, diarrhoea and eventual mental incapacity resulting from the lack of the vitamin niacin in the diet.
Pellagra	Disease associated deficiency of nicotinic acid, characterized by dermatitis, diarrhoea, and dementia, usually seen in area where sorghum is staple food.
PEM	Protein energy malnutrition is defined as protein energy malnutrition occurring due to deficiency of protein and energy in the supplied food.
Pepsin	A protein-digesting enzyme produced by the stomach.

Terminology	*Definition/Explanation/Principles*
PER	The gain in body weight per gram of protein or Mean gain in body weight of rat (g) ÷ Protein consumed (g).
Percentile	A value on a scale of one hundred that indicate the percent of a distribution that is equal to or below it
Perception	The process, by which the organism selects, organizes and interprets sensations.
Pernicious anemia	The anemia that results from a lack of vitamin B-12 absorption. It is pernicious (deadly) because of the associated nerve degeneration that can result in eventual paralysis and death.
Peroxide value	It is expressed as milliequivalents of oxygen present in one kilogram of fat.
Peroxide value	Peroxides are the main initial products of autoxidation. They can be measured by techniques based on their ability to liberate iodine from potassium iodide or to oxidize ferrous to ferric ions. Their content is usually expressed in terms of milliequivalents of oxygen per kilogram of fat.
PGI_2 (prostaglandin type I_2)	A type of prostaglandin (hormone-like substance) derived from arachidonic acid.
Phosphatides	A group of lipids which on hydrolysis give fatty acids, phosphoric acid and nitrogenous base.
Photo-oxidation	The oxidation of fatty acids by attack of singlet oxygen molecules produced from the usual triplet from by the action of light energy and sanitizers.
Photophos-phorelation	The process of generating ATP from ADP and phosphate by means of a proton-motive force generated by the thylakoid membrane of the chloroplast during the light reactions of photosynthesis.
Photorespiration	The metabolic pathway that consumes oxygen, evolves carbon dioxide, generates no ATP, and decreases photosynthetic output; generally occurs on hot, dry, bright days, when stomata close and the oxygen concentration in the leaf exceeds that of carbon dioxide.
Phytate phosphorus	Phosphorus binds with phytic acid, no free phosphorus is available.
Phytic acid	Inositol hexaphosphoric acid.
Pickles	The process of preservation of food in common salt or in vinegar is called pickling and the product is called pickles. Spices and edible oil are added for taste.

Terminology	*Definition/Explanation/Principles*
Pitting of legumes	Scratching of seed coat for oil penetration in to seed and it is given as a premilling treatment to the legumes.
PK_a	PK_a is defined as the pH at which given ionic compound is half dissociated.
Plant growth regulators	These are enzymes or chemicals which stimulate the vegetative growth of plant.
Plant layout	The plan or act of planning an optimum arrangement of industrial facilities is called as plant layout.
Polenske value	The number of milliliters of 0.1 N alkalis required neutralizing the volatile water-insoluble fatty acids present in 5 g of fat.
Polymerization	Polymerization or aggregation reactions generally involve the formation of large complexes.
Polymorphism	Certain triglycerides exist in several different crystal systems each of which has a characteristics melting point, X-ray diffraction pattern and infrared spectrum. This phenomenon is known as polymorphism.
Polymorphism	Polymorphic forms are solid phases of the same chemical composition that differ among themselves in crystalline structure but yield identical liquid phases upon melting (a, b′ and b polymorphic structure).
Pork	Uncured meat of pig or swine.
Postharvest technology	It is defined as techno economical activity applicable to all agricultural commodities or produced originated from farm, forest livestock and aqua culture for their conservation, handling, storage, marketing and value addition to make them useful as food, fed, fiber, fuel and industrial raw material.
Pour point	The lowest temperature at which a liquid will flow when a test container is inverted.
Poverty	A situation in which the level of living of an individual family or group is below the standard of living of the community either in terms of subsistence or in contrast to normal standards of income required for at least modest participation in community life
Powder	When juice is converted into free flowing highly hydroscopic powder to which natural flavour in powder from is incorporated to compensate for any loss of flavour in concentration.

Terminology	Definition/Explanation/Principles
Power number	The ratio of external to internal forces per unit volume of liquid is defined as the power number.
Pre cooling	Removal of field heat from the freshly harvested commodities (Fruits and vegetables) using cooling media.
Pre packaging	Operation of packaging of fresh produce in thin plastic films.
Prebiotics	Prebiotics are a collective term for non-digestible but fermentable dietary carbohydrates that may selectively stimulate growth of certain bacterial group resident in the colon, such as biofidobacteria, lactobacilli and eubacteria, considered to be beneficial for human host.
Precipitation	It includes all aggregation reactions leading to a total or partial loss of solubility.
Preservatives	Compounds that extend the shelf life of foods by inhibiting microbial growth or minimizing the destructive effect of oxygen and metals.
Preserves	It is a product made from properly matured fruit by cooking it whole or in the form of large pieces in heavy sugar syrup until it becomes tender and transparent. In this case cooking is continued until a concentration of T.S.S. is reached to 68.5 °Brix
Primary metabolite	A metabolite produced by organism during the growth phase.
Primary packaging	It is direct contact with the contained product.
Principle of canning	Destruction of spoilage organisms within the sealed container by means of heat.
Prions	Smallest infective forms of life.
Probes	Different instruments used for instrumentation and their control (sensitive)
Probiotics	Probiotics are living microorganisms that following ingestion from part of the colonic flora at least temporarily and are used with a view to improving the health and well being of the host, *e.g.* Lactobacillus strain GG.
Proof	The alcoholic concentration of beverage spirits is expressed in terms of proof in Canada, the United Kingdom, and the United States. U.S. regulations defines this standards as follows: proof spirit shall be held to be that alcoholic liquor which contains one-half its volume of alcohol of a specific gravity of 0.7939 at 15.6 °C *i.e.* the figure for proof is always twice the percent alcohol content by volume. For example,

Terminology	*Definition/Explanation/Principles*
	100° proof means 50 per cent alcohol by volume. In the United Kingdom as well as Canada, proof spirit is such that at 10.6°C alcohol weighs exactly twelve-thirteenths of the weight of an equal bulk of distilled water. A proof of 87.7° indicates an alcohol concentration of 50 per cent. A conversion factor of 1.142 can be used to change British proof to U.S. proof.
Proof gallon	Proof gallon is a U.S. gallon of proof spirits or the alcoholic equivalent thereof, *i.e.*, a U.S. gallon 3785 cm^3 (231 cubic in.) containing 50 per cent of ethyl alcohol by volume. Thus a gallon of liquor at 120° proof is 1.2 proof gallons; a gallon at 86° proofs is 0.86 proof gallons. A British and Canadian proof gallon is an empirical gallon of 4546 cm^3 (277.4 cubic in.) at 100° proof (57.1 per cent of ethyl alcohol by volume). An empirical gallon is equivalent to 1.2 U.S. gallons. To convert British proof gallons to U.S. proof gallons, multiply by 1.37. Since excise taxes are paid on the basis of proof gallons, this term is synonymous with tax gallons.
Proofing	Process of exposing dough to 31 -32 °C with 88 per cent relative humidity in closed proofing chamber to increase loaf volume of dough by stimulating process of fermentation.
Pro-oxidants	Chemical substances and basic natural conditions or environmental states, which actively promote the oxidation of fats.
Protein	A three-dimensional biological polymer constructed from a set of 20 different monomers called amino acids.
Protein association	Protein association reactions generally refer to changes occurring at the subunits or molecular level.
Protein calorie malnutrition (PCM)	Clinical and biochemical disorders caused by various degrees of deficiency and additional physical insults and stress. Inadequate dietary intake of good quality protein and calories
Protein efficiency ratio (PER)	A measure of protein quality determined by the ability of a protein to support the growth of a young animal.
Protein isolate	Percent protein content is more than 90 per cent.
Protein-energy malnutrition (PEM)	This results when a person regularly consumes insufficient amounts of kcalories and protein. The deficiency eventually results in body wasting and an increased susceptibility to infections.
Prothrombin	A blood protein needed for blood clotting that requires vitamin K for its synthesis.

Terminology	*Definition/Explanation/Principles*
Proton pump	An active transport mechanism in cell membranes that consumes ATP to force hydrogen ions out of a cell and in the process generates a membrane potential.
Psychometric chart	A psychometric chart is a graph of the properties of air-water vapour mixtures as a function of temperature. Lines of constant wet bulb in psychometric chart are a plot of equation. If the 'wet bulb' temperature Tw and the dry bulb temperature Ta are known it is possible to establish the humidity Ha.
PUFA	Poly unsaturated fatty acids, more in vegetable oil.
Pungency	Certain compounds found in several spices and vegetables cause characteristic hot, sharp, and stinging sensations that are known collectively as pungency.
Radiation	A hot body gives off heat in the form of radiant energy, which is emitted in all directions. When this energy strikes another body, part is reflected and part may be transmitted unchanged through the body, depending on its degree of opacity. The remainder is absorbed and quantitatively transformed into heat except where photochemical or special reactions are induced.
Radiation dose	The quantity of radiation energy absorbed by a food as it passes through in processing.
Radioactivity	It is property of an atom whose nucleus or center is physically unstable and spontaneously releases radiation energy.
Rancidity	The development of off-flavour in stored oil/fat due to lipolysis or autoxidation.
Rancidity	Development of off flavor in food items due to break down of unsaturated fatty acids to saturated one.
Reactor	Reactor is a well designed vessel used for different bioconversion reaction.
Recommended dietary intake (RDI)	Recommendations from the original tenth edition RDA Committee that were published in 1987 in the American journal of Clinical Nutrition after the National Academy of Sciences refused to publish the original tenth edition of the RDA.
Recommended Nutrient Intake (RNI)	The Canadian version of RDA.
Reducing agent	In one sense a compound capable of donating electrons (also hydrogen ions) to another compound.
Reduction	To gain an electron or hydrogen atom.

Terminology	Definition/Explanation/Principles
Reference Daily Intake (RDI)	Standards of expressing nutrient content on nutrition labels. RDI figures are based on average 1989 RDA values set for a nutrient that span a particular age range, such as children over 4 years through adults. RDI soon should replace U.S. RDA.
Refractive index	A measure of the bending or refraction of a beam of light on entering a dense medium. It is the ratio of the sine of the angle of incidence of the ray of light to the site of the angle of refraction.
Refrigeration	If the primary purpose is to discharge heat to a certain high temperature region, the system is called a heat pump. If the purpose is to absorb heat from a certain low temperature region, the system is called a refrigerator.
Regulated market	In these markets business is done accordance with rules and regulations.
Reichert-Meissel value	The R-M value of a fat is the number of milliliters of 0.1 N potassium hydroxide required to neutralize the stem-volatile fatty acids in 5 g of fat.
REM sleep	A periodic state during sleep identified by rapid eye movements and correlated with vivid dreaming.
R-enamel cans	Lacquered cans used for acid are known as R-enamel cans.
Rendering	Steam heating to oil seeds in vacuum container.
Rennin	An enzyme formed in the kidney in response to low blood pressure; it acts on a blood protein to produce angiotensin I.
Respiration	Oxidative breakdown of the complex materials of cell into simples' molecules with the concurrent production of energy required by the cell for the completion of chemical reactions.
Respiration	Fundamental process of conversion of potential energy to kinetic energy.
Respiration rate	It is postharvest phenomenon where grains take in O_2 and give out CO_2.
Respiratory quotient	It is an arithmetic ratio of mole of CO_2 evolved to the moles of O_2 absorbed.
Restoration	Addition to restore the original nutrient content.
Retention time	Time required coming out of eluent.
Retina	The innermost layer of the vertebrate eye containing photoreceptor cells (rods and cones) and neurons; transmits images formed by the lens to the brain via the optic nerve.

Terminology	Definition/Explanation/Principles
Retinal	The light-absorbing pigment in rods and cones of the vertebrate eye.
Retrogradation	The insolubilizating effect, initiated when the long and somewhat unwieldy molecules begin to crystallize is retrogradation when it occurs in starch.
Reversion	The development of a characteristic beany, buttery, fishy, grassy, hay like, or painty flavor in oil before the onset of autoxidation.
Reynolds number	The ratio inertial force to viscous force.
Rheology	Rheology is concerned with stress-strain relationships of materials that show, flavor, intermediate between those of solids and liquids.
Rheology	It is viso-elastic property that studies deformation of dough including elasticity and flow.
Rhodopsin	A protein involved in vision; it is made in the eye and incorporates a protein called opsin and a form of vitamin A; especially important in night vision.
Rickets	A disease characterized by softening of the bones because of poor calcium deposition. This deficiency disease arises from lack of vitamin D activity in the body.
Rickets	The bone abnormalities due to deficiency of Ca, P and vit. D.
Rigor mortis	Stiffening of muscles due to loss of ATP and fall in pH during muscle postmortem.
Ripening	It is defined as the sequences of changes in colour, flavor, and texture which lead to state at which the fruit is acceptable to eat.
Roasting of groundnut	Dry heat treatment to groundnut for desirable flavor, light roasting.
Ropyness of bread	It is bread fault where sticky, gummy material is developed in center of loaf 1 to 3 days after baking due to contamination of microorganism *B. mesentericus*.
Royal jelly	It is used by bees to include development of larvae into the queen phenotype and is currently sold as a general health tonic
Saccharin	An alternate sweetener that yields no energy to the body; it is 500 times sweeter than sucrose.
Salt	Generally refers to a mixture of sodium and chloride in a 40:60 ratio.

Terminology	*Definition/Explanation/Principles*
Salty taste	Only salts are salty; however, not all salts are salty. Some are sweet, bitter, or tasteless. Monovalent cations, especially sodium, can pass directly through ion channels in the tongue, leading to an action potential leading to the salty percept. Sodium chloride in foods may be analyzed using a specific ion electrode. To measure other salts, ion chromatography or atomic absorption emission spectroscopy are generally used.
Sapid	Having the power of affecting the taste receptor.
Sauce	Sauces are mainly two kinds; thin and thick sauces. The thin sauce contains vinegar, extract of spices and herbs. Thick sauce is more viscous and does not flow easily. It has at least 3 per cent acetic acid for improving the keeping quality. Tomato pulp without skin and seeds are used as basic raw material for tomato sauce. Apples, peaches, plums, apricots, mangoes, cauliflower, carrots are also used for preparation of sauce. Onion, garlic, spices, and herbs are used as flavouring agents while vinegar; salt and sugar are added for taste. The mixtures of above all material cooked to get consistency of jam *i.e.* upto 68 to 69 °Brix.
Savory	Appetizing; having an agreeable flavor.
Scalding	It is process of treating birds with hot water and agitating for short period of time. It facilitates removal of feathers by expanding or relaxing muscles that surrounds feathers.
Scale up	Scale up is the study of the problem associated with transferring data obtained in laboratory and pilot equipment to industrial production.
Schaal oven test	The sample is stored at about 65 °C and periodically tested until oxidative rancidity is detected.
Scientist P Fizer	In the year 1923 obtained 1st successful plant of citric acid fermentation with the help of sugar and M. O. *Aspergillus niger*.
Score	Which are used to represent their achievement are average scores of the two or more examinations
SCP	Single cell protein *i.e.* protein derived from microbial cell for use as food or a feed supplement.
Screening	The programme which consist of highly selective procedure used for detection and isolation of industrial important microorganisms.

Terminology	Definition/Explanation/Principles
Scurvy	The deficiency disease that results after a few weeks of consuming a diet free of vitamin C; pinpoint hemorrhages on the skin are an early sign.
Scutellum	It is cementing layer which separates endosperm from germ.
Secondary response	Antibody made on second (subsequent) exposure to antigen mostly of class Ig G.
Self-fertile	It refers to ability of plant to mature viable seeds without aid of pollens from some other plant flower.
Semi perishable foods	Food products which contain water in the range of 10-30 per cent are called semi perishable food.
Semolina	Coarsely ground white endosperm chemically same to white flour.
Senescence	It is define as period during which anabolic (synthetic) biochemical processes give way to catabolic (degradative) processes leading to aging and finally the dead of tissue.
Senescence	It is defined as period during which anabolic biochemical processes gives way to catabolic processes leading to aging and finally the death of tissue.
Sensory analysis	The science of measuring and evaluating the properties of food products by one or more human senses.
Septicemia	Infection of the bloodstream by microorganisms.
Sequestrants	These are chelating agents or sequestering compounds. They react with trace elements such as iron and copper present in foods and remove them from solution. The trace elements are active catalysts of oxidation and discolouration of food products. Sequestrants such as ethylene diamine tetra acetic acid (EDTA), poly phosphates and citric react with trace elements and inactivate them.
Sequestrants	Chemical substances used to chelate or complex free metals in food system.
Serbet (Juice)	It is clear sugar syrup, which has been artificially flavoured.
Settling	It is a method of oil refining. It involves storing heated fats quiescently in tanks with conical bottoms. Water and materials associated with water settle into the cone from where they are drawn off.
Settling down period	Period at which solid surface conditions comes into equilibrium with drying air.
Sherbats	This is cooling drink of sweetened diluted fruit juice.

Terminology	*Definition/Explanation/Principles*
Sherry	It is term given to number of related type of dessert wine originally developed in the area around Serez in the South of Spain.
Shortenings	These are compounded from mixtures of fats prepared by hydrogenation and are called lard compounds or lard substitutes.
Simultaneous reaction	Nutrients converted to products in a variable proportion without accumulation of intermediate
Single whiskey	Single whiskey is the whiskey, either grain or malt, produced by one particular distillery. Blended Scotch whiskey is not a single whiskey.
Size reduction	The breakdown of solid materials through the application of mechanical forces.
Smoke point	The temperature at which the column of solidified oil or fats starts rising in the capillary tube.
Smoking	Decreases the available moisture on surface of the meat, preventing microbial growth and spoilage. It enhances colour and falvour of meat.
Soaking of legume	Absorption of water/medium through hull or seed coat.
Sodium-potassium pump	A special transport protein in the plasma membrane of animal cells that transport sodium out of and potassium into the cell against their concentration gradients.
Softening point	The temperature at which the column of solidified oil or fat starts rising in the capillary tube.
Soil	It can be defined as superficial earth crust which functions as store lower time providing necessary physical support to plant.
Solubility	The tendency of molecule to dissolve in a liquid.
Solute	A substance that is dissolved in a solution.
Solvent	The dissolving agent of a solution. Water is the most versatile solvent known.
Sour mash	Sour mash is made with a lactic culture and not less than 20 per cent stillage added back to the fermentor and fermented for at least 72 h.
Sour taste	Sourness indicates acidity, through not all acids are sour. The detection of acids facilitates maintaining the body fluid compositional balance. The pH is characteristic of the carbon dioxide levels in blood and cerebrospinal fluid.

Terminology	*Definition/Explanation/Principles*
Specific gravity	The density of a substance is defined as mass per unit volume. The density of a substance is characteristic property and has a definite value at a given temperature and pressure. The density of one substance in relation to the density of another material (*e.g.* water) is known as specific gravity.
Spirits	Spirits are distilled spirits including all singular whiskey, gin, brandy, rum, cordials, and others made by a distillation process for nonindustrial use.
Spore	Resistant, resting stage of bacteria.
Spray drying	Spray drying is usually applied to fluids high in moisture content. The fluid is divided into droplets by its passage through a spray nozzle at high pressure or its passage through a centrifugal disk at high speed.
Sprouting of legumes	Soaking of legumes in water and then kept for vegetative growth.
Squash	It is essentially strained juice containing moderate quantities of fruit pulp to which cane sugar is added for sweetness.
Starter culture	Culture of desirable microbes used to initiate fermentation.
Stature	Height of the body in a standing position
Stepwise reaction in kinetic pattern	Nutrients completely converted to intermediate before conversion to product.
Sterilization	Sterilization means the destruction of all viable microorganisms by heat treatment.
Steroids	A group of hormones and relate compounds that are derivatives of cholesterol.
Stevens power law, $S = I^n$	The increase in perceived intensity, S, is equal to the concentration, I, to the nth power.
Stillage	Stillage is dealcoholized fermented mash.
Strecker degradation	It involves the interaction of α-dicarbonyl compounds and α-amino acids. Volatile products, such as aldehydes, pyrazines and sugar fragmentation products from the strecker reaction may contribute to aroma and flavour.
Stress	A pattern of disruptive psychological and physiological functioning that occurs when an environmental event is appraised as a threat to important goals and one's ability to cope.
Stunning	Of making animal unconscious before slaughtering

Terminology	*Definition/Explanation/Principles*
Subjective evaluation	Human sense organs for taste, smell, sight, touch and hearing are the ways for subjective evaluation of food.
Superoxide dismutase	An enzyme that can neutralize a superoxide free radical.
Sweating	Moisture condensation on surface of endosperm during wheat conditioning.
Sweet taste	Two types of receptors systems correspond with sweet taste; one responds to certain carbohydrates and the other to high potency sweeteners. The structural requirements for a compound to active sweet receptors have not been fully defined. High pressure liquid chromatography is a key tool used in analyzing sweet components.
Syneresis	The phenomenon of spontaneous exudation of fluid from gel.
Syrup	It is clear sugar syrup, which has been artificially flavoured. It contains 25-30 per cent juice, 59 per cent sugar and 1 per cent citric acid. Syrup used as a sherbet by diluting 1 cup of syrup with 3 cups of water.
Tails	Tails is a residual alcoholic distillate.
Tallow	It is a lipid obtained from beef by the process of rendering.
Tannins	The phenolics with MW 500 to 3000 that form complexes with proteins.
Taste	Chemoreceptive events in the mouth lead to taste perception. Taste is typically described by five modalities coupled with chemesthesis: salt, sour, sweet, and bitter and umami.
TBA (thibarbituric acid) test	Thiobarbituric acid is a more sensitive reagent than phloroglucinol and gives a yellow coloration with various saturated and unsaturated aldehydes and a red color with a more restricted number.
TD (true digestibility)	N intake – (fecal N – metabolic N) x 100 ÷ N intake.
Tempering	Process of equalizing the moisture of grain by temporary holding the paddy/any grains between drying passes.
Tempering of chocolate	It is the most important process in chocolate manufacturing which decides the quality. It consists of cooling the chocolate with continuous mixing to produce cocoa butter seed crystals and uniform distribution it is a technique of uniform distribution.

Terminology	*Definition/Explanation/Principles*
Ten State Nutrition Survey (TSNS)	A survey, which was funded by the Department of Health, Education and Welfare (United States) to study the problem of malnutrition in the United States.
Tetany	A syndrome marked by sharp contraction of muscles with failure to relax afterward; usually caused by abnormal calcium metabolism.
Texture	Closely related to the rheology of food materials the texture of food. It is a structural quality of a material.
Texturizers	The chemical substances used to improve texture of food products.
Thermal reactions	Thermal reactions occurring with carbon-carbon bond cleavage yield as primary products volatile acids, aldehydes, ketones, diketones, furans, alcohols, aromatics, carbon monoxide and carbon dioxide.
Thermodynamics	Thermodynamics is the science of energy transfer in relation to the physical properties of substances.
Thickeners	The chemical substances used to improve thickness, viscosity and water holding capacity of liquid or semi-liquid foods.
Thiobarbituric acid (TBA) test	This is one of the most widely used tests for evaluating the extent of lipid oxidation. Oxidation products of unsaturated systems produce a colour reaction with TBA. It is believed that the chromagen results from condensation of two molecules of TBA with one molecule of malonaldehyde.
Thiocyanogen value	This value is the result of determining the percentage of oleic, linoleic and linolenic acids, provided that other unsaturated acids are not present and the iodine number and the amounts of saturated acids are known.
Titer test	When molten fatty acids are cooled and begin to solidify, the latent heat of fusion is liberated and a sudden rise in temperature can be observed.
Tocopherols	The chemical name for some forms of vitamin E.
Tomato juice	It is a juice, which contains T.S.S. not less than 5.6 per cent at 20 °C (4-6 per cent salt is added to counteract the astringent taste of the juice and 1 per cent sugar is added to improve the taste).
Tomato ketchup	It is a product made by concentrating tomato juice or pulp (without seed and pieces of skin). Spices, salt, vinegar, onion, garlic are added to the extract that the ketchup contains not less than 12 per cent tomato solids and 28 per cent total solids.

Terminology	*Definition/Explanation/Principles*
Tomato paste	It is a concentrated tomato juice (without skin and seeds) containing not less than 25 per cent tomato solids. It is further concentrated to 33 per cent tomato solids.
Tomato puree	It is a concentrated tomato pulp without skin or seeds containing not less than 8.37 per cent tomato solids. It is further concentrated to not less than 12 per cent tomato solids. It is called heavy purees.
Total and volatile carbonyl compounds	Methods for the determination of total carbonyl compounds are usually based on measurement of the hydrazones arising from reaction of aldehydes and ketones (oxidation products) with 2,4-dinitrophenylhydrazine.
Totipotency	Capacity of cell to regenerate phenotype of the complete and differentiated organism from which it is derived.
Toxin	Any harmful chemical substance released by microbes.
Transduction	Bacteria phases function as intermediate in transfer of bacterial genetic information from one bacteria to another.
Transpiration	It is important physiological activity of fruit and vegetables in which water loss from lenticels or pores to surrounding atmosphere takes place relates to physiological loss in weight.
Tricep Skin fold Thickness	Measurement, with a calibrated caliper, of thickness of a fold of skin at the upper arm. It is a measurement of subcutaneous fat. Since subcutaneous fat make up about 50 per cent of the adipose tissue stores, tricep skin fold thickness can be an important measurement to estimate body fat reserves of an individual.
Turbidity point	The temperature at which a solution of oil or fat starts showing turbidity when it is cooled slowly.
Turbulent flow	The flow in which the fluid instead of flowing in orderly manner moves erratically in the form of cross and eddies is called turbulent flow.
Ultimate pH	The lowest pH of muscles after slaughter of animal.
Ultra filtration	The process of separating components of a solution largely on the basis of molecular size, utilizing a membrane as a molecular sieve.
Ultra filtration membranes	Membrane is preventing the passage of larger solute molecules in solution by means of filtration through micro pores in a membrane structure.

Terminology	*Definition/Explanation/Principles*
Umami taste	Umami is the taste of a few amino acids (*e.g.*, glutamate, aspartate, and related compounds) and was classically not included as a taste modality. Sometimes described as savory, brothy, or meaty, it is the dominant taste of such foods as chicken broth, meat, and ageing cheese. Umami perception results from activation of two receptor systems, with one that overlaps with the receptor systems for artificial sweeteners. There are many HPLC systems that integrate sample preparation along with data analysis specific to amino acids analysis.
Unbalanced transportation problem	The transportation problem in which the requirement and capacity are not balanced.
Under nutrition	Inadequate intake of one or more nutrients including calories. Chronic under nutrition (chronic malnutrition) refers to a long-term inadequate food intake and is reflected by low height-for-age levels. Acute under nutrition refers to a short-term severe inadequate food intake and is reflected by low weight for height levels.
Under nutrition	State of insufficient supply of essential nutrients.
Under weight	Weight below the average for a given sex, age and height in relation to the NCHS reference tables.
Unit operations	Aerobic fermentation involves unit operation the mixing of three heterogeneous phase microorganisms, medium and air mass transfer of oxygen from the air to the organisms and heat transfer from the fermentation medium.
Unit processes	Fermentation processes can be classified by the reaction mechanisms involved in converting the raw materials into products; these include reductions, simple and complex oxidations, substrate conversions, transformations, hydrolysis, polymerization, complex biosynthesis and formation of cells.
Unsaponifiable matter	This term includes all those constituents of fats, which are not saponified by alcoholic caustic potash. They include sterols, carotenoids, long-chain alcohols and hydrocarbons.
Unsaturated fatty acids	Hydrocarbon chain of fatty acid contains double bonds between carbon atoms.
Uric acid	An insoluble precipitate of nitrogenous waste excreted by land snails, insects, birds and some reptiles.

Terminology	Definition/Explanation/Principles
UV Spectro-photometry	Measurement of absorbance at 2 nm (conjugated dines) and 268 nm (conjugated trienes) is sometimes used to monitor oxidation.
Variability	In a frequency distribution, the separation, dispersion, or spread of the scores on the x-axis.
Variable cost	The cost which varies with the amount produced, raw material, labour.
Vegetable oil	Vegetable oils mostly contain large amounts of oleic and linoleic acids and less than 20 per cent saturated fatty acids. The most important group is cottonseed, corn, peanut, sunflower, safflower, olive, palm, and sesame oils.
Verification	Routinely check the system for accuracy to verify that it is functioning properly and consistently.
Very low calorie diet (VLCD)	Known also as protein sparing, modified fast (PSMF), this diet allows a person 400 to 700 kcalories per day, often in liquid form. Of this, 30 to 120 grams are carbohydrate; the rest is high biological value protein.
Very low density lipoprotein (VLDL)	The lipoprotein that initially leaves the liver. It carries both the cholesterol and lipid newly synthesized by the liver.
Virulence	It is the degree or capacity or intensity of pathogencity of an organism *i.e.* to cause the disease in host.
Viscosity	The resistance to flow of a fluid. In the cgs (centimeter-gram-second) system, the absolute unit viscosity is the poise.
Vitamin	An organic molecule required in the diet in very small amounts; vitamins serve primarily as coenzymes or parts of coenzymes.
Vitamins	Organic compounds occurring in small quantities in the different natural foods and necessary for the growth and maintenance of good health in human being.
Viticulture	It can be defined as the science and art of growing grapes.
Volatibility	The tendency of a molecule to pass in to the vapour state.
Volatile oil	That portion of a botanical that co-distills with water during steam distillation and is generally flavourful.
Water activity (a_w)	It is a ratio between the partial pressure of water in a sample (P) to the vapor pressure of pure water at the same temperature (Po): $a_w = P/Po$. F/Fo = The fugacity of the solvent/The fugacity of the pure solvent. Percent equilibrium relative humidity surrounding the product.

Terminology	*Definition/Explanation/Principles*
Water holding capacity	It is a term frequently employed to describe the ability of a matrix of molecules usually macromolecules to entrap large amounts of water in a manner such that exudation is prevented.
Water potential	The physical property predicting the direction, in which water will flow, governed by solute concentration and applied pressure.
Water structure breakers ions	K^+, Rb^+, Cs^+, NH_4^+, Cl^-, Br^-, I^-, NO_3^-, BrO_3^-, IO_3^-, and ClO_4^-
Water structure formers ions	Li^+, Na^+, H_3O^+, Ca^{2+}, Ba^{2+}, Mg^{2+}, Al^{3+}, F^-, and OH^-
Wave length	In measuring light and sound waves, the distance from the crest of one wave to the crest of the next.
Waxes	Esters of long-chain monohydroxy alcohols and higher-chain fatty acids.
Waxy rice	Rice with low amylose content
Weber's law	Principle stating that the amount by which a stimulus must be increased or decreased to be perceived as different is always a constant proportion of the initial stimulus intensity.
Wet milling of legume	Related to pre-milling treatment, soaking in water.
Wheat shorts	Mixture of bran and germ produced from milled wheat.
Whey	It is by product of cheese and casein and contains approx. 20 per cent of the original milk proteins. These proteins include α-lactalbumin, β-lacto globulin, lactoferrin, lactoperoxidase, immunoglobulins, glycomacropeptide and a variety of growth factors.
Whitening of rice	The process of removal of bran from brown rice called as whitening of rice.
Winterizing	A process where oil is held at 5 °C until crystallization is well advanced, and then filtered in chilled room to remove solids. The winterized oil is used as salad oil.
Work	A force is a means of transmitting an effect from one body to another.
Wort	It is the clear liquid obtained through the filtration of insoluble materials during mashing process of brewing.
Xantham	It is polysaccharides produced by Xanthomonas compestris.

Terminology	*Definition/Explanation/Principles*
Xerophthalmia	A cause of blindness that results from infection of the eye secondary to vitamin A deficiency. The specific cause is a lack of mucus production by the eye, which then leaves it more vulnerable to surface dirt and bacterial infections.
Xylitol	An alcohol derivative of the five-carbon monosaccharide, xylose.
Yeast	A unicellular fungus that lives in liquid or moist habitats, primarily reproducing asexually by simple cell division or by budding of a parent cell.
Yeast	Single cell micro organism which multiplies by process of budding. They are single cell protein and rich source of enzymes like lipase, amylase, protease and zymase.
Yogurt	It contains live lactic acid bacteria (probiotic).
Yolk index	It is used to evaluate egg quality height of the yolk in relation to width of yolk. Yolk index is high in fresh egg.
Yo-Yo dieting	The practice of losing weight and then regaining it, only to lose it and regain it again.
Z value	Number of degree required for a specific thermal death time curve to pass through one log cycle.
Z-Score	A score's distance from the mean of the group expressed in units of the standard deviation; used to identify unusual values.
Zwitter ion	Amino acid having net charge zero *i.e.* equal no positive and negative charges.
Zymogen	An inactive form of an enzyme.

APPENDIX–II
Abbreviations Used in Food Science and Technology

Short Form	*Long Form/Full Form*
λ	Latent heat of vaporization, kcal/kg
1-MCP	1-methyl cyclopropene
AACC	American Association of Cereal Chemists
AAO	Ascorbic Acid Oxidase
ACC	1-aminocyclopropane-1-1carboxylic acid
ACNFP	Advisory Committee on Novel Foods and Processes
ACS	American Chemical Society
ADA	Azodicarbonmide
ADF	Acid detergent fiber
ADH	Antidiuretic hormone
ADI	Acceptable Daily Intake
ADPI	American Dry Products Institute
AECA	Aroma Extract Concentration Analysis
AFLP	Amplified Fragment Length Polymorphism
AGMARK	Agricultural Produce (Grading and Marketing) Act, 1937
AGP	Alpha 1-acid glycoprotein
AI	Adequate Intake
AICSIP	All India Co-ordinated Sorghum Improvement Project
AIDS	Acquired Immunodeficiency Syndrome
ALP	Alkaline phosphatase
ALV	Available Lysine Value
AMC	Automatic Machinery and Electronics Inc.
AMCP	Anhydrous Monocalcium Phosphate
AMD	Age-related Macular Degeneration
AML	Amylose Leaching
AMS	Agricultural Marketing Service
ANOVA	Analysis of Variance
AOAC	Association of Official Analytical Chemists

Short Form	*Long Form/Full Form*
AOM	Active Oxygen Method or Atmospheric oxygen metabolizing
APEDA	Agricultural and Processed Food Export Development Authority
APHIS	Animal & Plant Health Inspection Service
APMC	Agricultural produce market committees
APP	Acute Phase Proteins
ARI	Acute Respiratory Infection
ASAE	American society of agricultural engineers
ASTM	American society of testing materials
ATA	Alimentary Toxic Aleukia
ATP	Adenosine triphosphate
a_w	Water activity
BAM	Bacteriological Analytical Manual
BAPN	β-Amino Propionitrile
BATF	Bureau of Alcohol, Tobacco and Firearms
BBT	Bright Beer Tank
BHA	Butylated Hydroxy Anisole
BHI	Body Mass Index
BHT	Butylated Hydroxy Toluene
BIS	Bureau of Indian Standard
BMC	Bone Mineral Content
BMD	Bone Mineral Density
BOAA	β-N-Oxalylamino-L-alanine
BOD	Biological Oxygen Demand
BOP	Broken Orange Pekoe
BOPE	Broken Orange Pekoe Fannings
BRAP	Bilateral research activities program
BSA	Bovine Serum Albumin
BSE	Bovine Spongiform Encephalopathy
BU	Brabender Units
BV	Biological Value
BVA	Brabender Visco Amylogram
BVO	Brominated vegetable oils

Short Form	Long Form/Full Form
CA	Controlled atmosphere
CAC	Codex Alimentarius Committee or Citric Acid Cycle
CACP	Commission for agricultural costs and prices
CAF	Calcium Activated Factor
CANP	Calcium Activated Neutral Proteinase
CAP	Controlled Atmosphere Packaging
CAPP	Calcium Acid Pyrophosphate
CAS	Controlled Atmosphere Storage
CBER	Center for Biologics Evaluation and Research
CC	Column Chromatography
CCI	Cotton corporation of India
CCK	Cholecystokinin
CCMP	Cooked cured-meat pigment
CCP	Critical Control Point
CDC	Centers for Disease Control and Prevention, Atlanta, USA
CDER	Center for Drug Evaluation and Research
CDT	Come-down time
CE	Capillary Electrophoresis
CEPCI	The cashew nuts export promotion council of India
CER	Carbon dioxide Evolution Rate
CFA	Corrugated Fiber Board
CFAM	Cyclic Fatty Acid Monomers
CFC	Chloro Fluoro Carbon
CFTRI	Central Food Technology Research Institute Mysore
CFU	Colony forming units
CGC	Canadian grain commission
CHARM	Combined Hedonic and Response Measurements
CHD	Coronary Heart Disease
CIP	Clean-in-Place
CLA	Conjugated Linoleic Acid
CM	Carboxymethyl
CM	Chloroform-Methanol

Short Form	*Long Form/Full Form*
CMC	Carboxy Methyl Cellulose
COD	Chemical Oxygen Demand
COSAMB	Council of state agricultural marketing boards
CPg	Specific heat of grain, Kacl/kg °C
CPS	Counts Per Second
CPw	Specific heat of water, Kacl/kg °C
CRD	Complete Randomized Design
CROs	Contract Research Organizations
CRP	C-Reactive Protein
CRS	Chinese Restaurant Syndrome
CS	Chemical Score
CSA	Cross-sectional open area
CSIR	Council of Scientific and Industrial Research
CTC	Crush Tear Curl or Charge Transfer Certificate
CUT	Come-up time
CVA	Cerebrovascular accident
CVD	Cardio Vascular Disease
CW	Continuous Wave
CWC	Central ware housing corporation
d	Level of absolute precision
D.B.	Dry bulb temperature
d.b.	Dry weight basis
DAGs	Diacylglycerols
DAP	Diamino Pimelic Acid
DAP	Diammonium Phosphate
DAS	Diacetone-L-Sorbose
DATE	Diacetyl Tartaric Acid Ester
DATEM	Diacetyl Tartaric Esters of Monoacylglycerols
DBS	Dried blood spot
DCPD	Di-calcium Phosphate Dihydrate
DDB	Dialkyl Dihexadecyl Malonate
DDG	Distillers Dry Grains

Short Form	*Long Form/Full Form*
DDM	Dialkyl Dihexadecyl Malonate
DDS	Distillers Dry Solubles
DE	Dextrose Equivalent or Degree of Esterification
DE	Diatomaceous earth
DEAE	Diethylaminoethyl
DEFF	Design Effect fortified food
DEFT	Direct Epifluorescent Filter Technique
DEPC	Diethyl Pyrocarbonate
DFD	Dark, firm and dry
DHA	Docosahexaenoic acid
DHA	Dehydroalanine
DHAA	Dehydroascorbic acid
DHS	Demographic and Health Survey
DISCUS	Distilled Spirits Council of the United States
DLVO	Derijaguin and Landau and Verwey and Over beek
DM	Degree of Methylation
DMA	Dynamic mechanical analysis
DMA	Dimethylamine
DMAPP	Dimethyl allyl Pyrophosphate
DMDC	Dimethyl Dicarbonate
DMI	Directorate of marketing and inspection
DMTA	Dynamic mechanical thermal analysis
DNA	Deoxyribonucleic Acid
DO	Dissolved Oxygen
DOPA	Dihydroxy Phenylalanine
DP	Degree of Polymerization or Deep Press or Diastatic Power
DRR	Dough Rate of Reaction
DRV	Daily reference values
DS	Degree of Substitution
DSC	Differential Scanning Calorimetry
DSM	Dutch States Mines
DTT	Dithiothreitol

Short Form	Long Form/Full Form
DUS	Distingness University and Stability
Dv	Diffusivity, m^2/hr
EA	Emulsion Activity
EAR	Estimated Average Requirement
ECA	Essential Commodity Act.
ECD	Electron capture detector
ECOST	European corporation in the field of scientific and technical research
EDB	Ethylene dibromide
EDCT	Ethylene dichloride Carbon tetrachloride
EDTA	Ethylene Diamine Tetra Acetic Acid
EDTA	Ethylenediaminetetraacetic acid
EFEMA	European Food Emulsifier Manufacture's Association
EGF	Epidermal Growth Factor
Eh	Redox potential
EHEDG	European Hygienic Equipment Design Group
EIA	Enzyme Immuno Assay
EIC	Export inspection council
ELISA	Enzyme Linked-Immuno Sorbent Assay
ELMC	Equilibrium moisture content
EM	Electromagnetic
EMA	Equilibrium modified atmosphere
EMC	Equilibrium Moisture Content
EME	Electro-magnetic energy
EMIT	Enzyme Multiplied Immuno Assay Technique
EPA	Environmental Protection Agency
EPA	Eicosapentaenoic acid
EPA	Esterified Phenolic Acid
EPC	Epicatechin
EPG	Esterified Propoxylated Glycerol
EPI	Expanded Programme on Immunization
EPR	Electron Paramagnetic Resonance
ERH	Equilibrium Relative Humidity

Short Form	*Long Form/Full Form*
ES	Emulsion Stability
ESADDI	Estimated safe and adequate daily dietary intake
ESI	Electrospray Ionization
ESLR	Extended Shelf life Refrigerated
ESP	Epithio Specifier Protein
EST	Expressed Sequence Tag
ETO	Ethylene Oxide
EVA	Ethylene vinyl acetate
EVOH	Ethylene vinyl alcohol copolymer
FACS	Fluorescence Activated Cell Sorting
FAD	Flavin adenine dinucleotide
FAD	Food and Drug Administration
FAME	Fatty Acid Methyl Ester
FAO	Food and Agriculture Organization
FAS	Fetal alcohol syndrome
FASEB	Federation of American Societies for Experimental Biology
FCC	Food Chemical Codex
FCI	Food corporation of India
FDA	Food and Drug Administration
FEMA	Flavour Fragrance Materials Association
FEP	Free erythrocyte protoporphyrins
FFA	Free Fatty Acid
FFDCA	Federal Food Drug and Cosmetic Act
FIA	Fluorescence Immuno Assay
FID	Flame Ionization Detector
FIFRA	Federal Insecticide, Fungicide and Rodenticide Act.
FM	Freezing-melting
FMA	Fragrance Materials Association
FNB	Food and Nutrition Board Council
FOS	Fructooligosaccahrides
FOS	Fructosyl Oligosaccharides
FOSHU	Foods for Specific Health Use

Short Form	*Long Form/Full Form*
FPA	Free Phenolic Acid
FPC	Fish Protein Concentrate
FPO	Food Product Order
FPP	Famesyl Pyrophosphate
FRM	Fat-reduced Meat
FSIS	Food Safety and Inspection Service
FSO	Food Safety Objective
FTIR	Fourier transformation infrared spectroscopy
G	Mass flow rate of base dry air, kg/hr m^2
GAP	Good Agricultural Practices
GC	Gas Chromatography
GC-MS	Gas Chromatography-Mass Spectroscopy
GCOH	Gas Chromatography Olfactometry of Headspace Samples
G-CSF	Granulocyte Colony-Stimulating Factor
GDL	Glucono-δ-lactone
GGPP	Geranyl Geranyl Pyrophosphate
GHP	Good Hygiene Practice
GL	Glycolipids
GLC	Gas Liquid Chromatography
GLP	Good Laboratory Practice
GMO	Genetically Modified Organisms
GMP	Good Manufacturing Practice
GMPR	Good manufacturing practice regulations
GOT	Glutamate-Oxalacetate Transaminase
GPP	Geranyl Pyrophosphate
GPT	Glutamate-Pyruvate Transaminase
GPV	Gross Protein Value
GRAS	Generally Recognized as Safe
GRE	Glucocorticoid Response Element
GSC	Granular stationary phase
GSH	Glutathione Peroxidase
GTF	Glucose Tolerance Factor

Short Form	Long Form/Full Form
Gy	Gray
H	Humidity, kg/kg
HAA	Heterocylic Aromatic Amines
HACCP	Hazard Analysis and Critical Control Points
HADY	High Active Dry Yeast
HART	Hybrid Arrested Translation
HAZOP	Hazard and operability study
Hb	Hemoglobin
HCN	Hydrogen Cyanide
HDA	Homochiral Derivatizing Agents
HDI	Human Development Index
HDL	High Density Lipoprotein
HDP	Heat pump dehumidifier
HEMF	4-hydroxy-2-(or 5-) ethyl-5-(or 2-) methyl-3 (2H) furnone
HFCS	High Fructose Corn Syrup
HH	Household
HHS	Health and Human Services
HIC	Hydrophilic interaction column
HIV	Human Innuno-deficiency Virus
HKI	Helen Keller International
HLB	Hydrophilic Lipophilic Balance
HM	High-methoxyl
HMF	Hydroxy Methyl Furaldehyde
HMF	Hydroxymethyl Furfural
HMFP	High moisture fruit products
HMG CoA	Hydroxy Beta-methyl Glutaryl CoA
HMP	Hexose Monophosphate Pathway
HMW	High molecular weight
H_o	Humidity of atmospheric air, kg/hr m^2
HOCl	Hypochlorous acid
HPC	Hydroxypropyl cellulose
HPCE	High Performance Capillary Electrophoresis

Short Form	*Long Form/Full Form*
HPLC	High Pressure/Performance Liquid Chromatography
HPLC	High-Pressure Liquid Chromatography
HPMC	Hydroxypropyl methyl cellulose
HRGC	High-Resolution Gas Chromatography
HSH	Hydrogenated starch hydrolysates
HTC	Hard-to-Cook
HTS	High Throughput Screening
HTST	High Temperature Short Time
IADY	Instant Active Dry Yeast
IARI	Indian Agricultural Research Institute
IBC	Iodine Binding Capacity
IBPA	Insoluble Bound Phenolic Acid
IC_{50}	Inhibitory Concentration
ICAR	Indian Council of Agricultural Research
ICC	International Association of Cereal Science and Technology
ICCIDD	International Council for the Control of Iodine Deficiency Disorders
ICUMSA	International Commission for Uniform Methods of Sugar Analysis
ID	Iron Deficiency
IDA	Iron Deficiency Anemia
IDD	Iodine Deficiency Disorders
IDF	International Dairy Federation
IDF	Insoluble Dietary Fiber
IDL	Intermediate density lipoprotein
IEF	Isoelectric focusing
IF	Intrinsic Factor
IFS	International Foundation for science
IFTM	Instron Food Testing Machine
IGF	Insulin like Growth Factor
ILPS	International Lecithin and Phospholipids Society
ILSI	International Life Science Institute
IMF	Intermediate moisture foods
IMMPaCt	International Micronutrient Malnutrition Prevention and Control Program, CDC

Short Form	*Long Form/Full Form*
IMP	Industrial Membrane Process
IMS	Immuno Magnetic Separation
INACG	International Nutritional Anemia Consultative Group
INCAP	Institute of Nutrition of Central America and Panama
IOM	Institute of Medicine
IPNSs	Integrated Plant Nutrition Systems
IPRs	Intellectual Property Rights
IQF	Individually Quick Frozen
IR	Infra Red
IRDA	Insurance regularity & development authority
ISFE	International society of food engineering
ISI	Indian Standards Institution
ISM	Industrial Scientific and Medical Applications
ISO	International Organization for Standards
ISTA	International Seed Testing Association
IU	International unit
IV	Iodine Value
IVACG	International Vitamin A Consultative Group
IVPD	In Vitro Protein Digestibility
JCI	Jute corporation of India
K	Constant
K	Drying constant, 1/hr
KAP	Knowledge, Attitudes, and Practices
kD_a	Kilo Dalton
KI	Potassium Iodide
KV	Kilo Volt
LAB	Lactic acid bacteria
LBG	Locust Bean Gum
LBW	Low birth weight
LCD	Liquid Crystal Display
LCE	Low-Cost Extrusion cooking
LCFA	Long-Chain Fatty Acids
LCP	Liquid cyclone process

Short Form	Long Form/Full Form
LCR	Ligase Chain Reaction
LDL	Low Density Lipoproteins
LEC	Low-cost Extrusion Cookers
LF	Lactoferrin
LH	Low Hydroxy group
LIFDCs	Low-income food deficit countries
LISA	Low input sustainable agriculture
LM	Low-methoxyl
LMP	Low Methoxy Pectin
LMW	Low molecular weight
LNG	Liquefied natural gas
LOAEL	Lowest Observed Advance Effect Level
LOEL	Lowest Observed Effect Level
LOX	Lipoxygenase
LP	Lactoperoxidase
LP	Lacto Peroxidase or lipid peroxidase
LPC	Leaf Protein Concentrates
LSE	Lant Stanol Esters
LTI	Lysine Tri-isocyanate
LUVs	Large unilamellar vesicles
LYC	Lycopene cyclase
M	Molarity
MA	Modified atmosphere
MAGs	Monoacylglycerols
MAHPD	Modified atmosphere heat pump dehumidifier
MALDITOF	Matrix-assisted Laser-desorption Ionization Time-of-Flight
MAP	Modified atmosphere packaging
MAPit	Micronutrient Action Plan instructional tool
MAS	Monoacetone-L-Sorbose
Mbo	Oxymyoglobin
MC	Methyl cellulose
MCC	Microcrystalline cellulose

Short Form	*Long Form/Full Form*
MCH-PCR	Magnetic Capture Hybridization Polymer Chain Reactions
MCP	1-methyl cyclopropene
MCP	Mono-Calcium Phosphate
MCTs	Medium-chain triacylglycerols
MDA	Malondialdehyde
MDSC	Modulated differential scanning calorimetry
MetMb	Metmyoglobin
MF	Modified filtration or Microfiltration
MFGM	Milk Fat Globule Membrane
MHP	Modified humidity packaging
MI	Micronutrient Initiative
MICS	Multiple Indicator Cluster Survey
MIG	Mercury-in-glass
MLVs	Multilamellar vesicles
MM	Micronutrient Malnutrition
MPC	Milk Protein Concentrates
MPEDA	Marine products export development authority
MR	Moisture ratio
MRI	Magnetic resonance imaging
MRL	Maximum pesticide Residue Limits
MRP	Maximum Retail Price
MS	Methyl Sulphonate
MS	Moles of substitution or Mass Spectrometry
MSG	Mono Sodium Glutamate
MSR	Multistage Recycle Designs
MTD	Maximum tolerated Dose
MUFA	Monounsaturated Fatty Acid
MVS	Machine vision system
MWCO	Molecular weight cutoff
MWM	Molecular Weight Marker
N	Normality
NA	Not Analyzed

Short Form	Long Form/Full Form
NACMCF	National advisory committee on the microbiology criteria for foods
NACMF	National agricultural cooperative marketing federation
NAD	Nicotine Amide Dinucleotide
NAFED	National Agricultural Finance and Economical Development
NAIP	National Agricultural Innovative Project
NAS	National Academy of Sciences
NASBA	Nucleic Acid Sequence-based Amplification
NATP	National Agricultural Technology Project
NBI	Nitrogen Balance Index
NCA	N-carboxy-alpha-Amino Acid Anhydride
NCCF	National consumers cooperative federation
NCDC	National cooperative development corporation
NCHS	National Center for Health Statistics
NCTGF	National cooperative tobacco growers federation
ND	Not Detected
NDDB	National dairy development board
NDF	Neutral Detergent Fiber
NDGA	Nordihydro Guaiaretic Acid
NDMA	N-nitrosodimethylamine
NDPE	Net Dietary Protein Energy Ratio
NDPV	Net Dietary Protein Value
NDR	Neutral Detergent Residue
NDUS	Novelty Distingness, Uniformity and Stability
NF	Nanofiltration
NFC	Not From Concentrate
NFDM	Non fat dry milk
NFE	Nitrogen free extract
NHB	National Horticulture Board
NIF	Nasal Impact Frequency
NIH	National Institute of Health
NIN	National Institute of Nutrition
NIS	Nitrogen Solubility Index

Short Form	*Long Form/Full Form*
NISCAIR	National Institute of Science Communication and Information Resources
NLEA	Nutrition Labeling and Education Act.
NMN	Nicotinamide mononucleotide
NMR	Nuclear Magnetic Resonance
NNMB	National Nutrition Monitoring Bureau
NOAEL	No Observed Advance Effect Level
NOEL	No Observed Effect Level
NPAGE	Non-Denaturing Polyacrylamide Gel-Electrophoresis
NPN	Non-Protein Nitrogen
NPR	Net Protein Ratio
NPU	Net Protein Utilization
NPV	Net Protein Value
NR	Not Recorded or Never-ripe
NRCS	National Research Centre for Sorghum
NSI	Nitrogen Solubility Index
NSP	Non-Starch Polysaccharides
NSP	Non storage protein
NVMCE	Nonvolatile Methylene Chloride Extract
OIE	Office International Epizootics
OMFs	Oscillating magnetic fields
ONC	Ocean nutrition Canada
OP	Orange Pekoe
ORP	Oxidation-reduction potential
OSHA	Occupational Safety and Health Administration
OUR	Oxygen Uptake Rate
P	Phosphorus
PA	Phytic Acid or Polyamides
PABA	Para-Amino benzoic Acid
PACMS	Primary agricultural cooperative marketing societies
PAHs	Polycyclic Aromatic Hydrocarbons
PAL	Phenylalanine ammonialyase

Short Form	*Long Form/Full Form*
PAO	Palm acid oil
PATH	Program for Technology in Health
PBB	Polybrominated Biphenyls
PC	Polycarbonate
PCA	Perchloric Acid
PCB	Polychlorinated Biphenyls
PCCMP	Powdered cooked cured-meat pigment
PCM	Protein Calorie Malnutrition
PCMB	Para-Chloro-Mercuric Benzoate
PDA	Photodiode Array Detection or Potato Dextrose Agar
PDAs	Personal Digital Assistants
PDCAAS	Protein Digestibility-Corrected Amino Acid Score
PDCB	Partially Defatted Chopped Beef
PDI	Protein Dispersibility Index
PDS	Phytoene Desaturase
PE	Polyethylene
PEF	Pulsed electric field
PEM	Protein energy malnutrition
PER	Protein Efficiency Ratio
PET	Polyethylene terephthalate
PF	Pekoe Fanning's
PFA	Prevention of Food Adulteration Act.
PFAD	Palm fatty acid distillate
PFDA	Pure Food and Drug Act
PG	Propyl Gallate
PGE	Polyglycerol Esters
PHT	Post Harvest Technology
PI	Protein Isolate
PIC	Paired Ion Chromatography
PIT	Phase Inversion Temperature
PKC	Palm kernel cake
PKOL	Palm kernel olein

Short Form	*Long Form/Full Form*
PL	Phospholipids
PME	Pectin methyl esterase
PMO	Pasteurized milk ordinance
POD	Peroxidase
ppm	Parts Per Million
PPO	Polyphenol oxidase
PPO	Propylene Oxide
PPPP	Phytoene Pyrophosphate
PPS	Probability Proportionate to Size
PR	Protein Rating
PRE	Protein Retention Efficiency
PS	Protein Score
PS	Polystyrene
PSE	Pale, soft and exudative
PSU	Primary Sampling Unit
PSY	Phytoenesynthase
PT	Press-twist
PTKs	Protein Tyrosine Kinases
PU	Pasteurization Unit
PUFA	Poly Unsaturated Fatty Acids
PV	Peroxide value
PVC	Polyvinyl chloride
PVDC	Polyvinylidene chloride
PVOH	Polyvinyl alcohol
PVP	Polyvinyl Pyrrolidone
PW	Propanol-Water
QA	Quality Assurance
QAE	Quaternary aminoethyl
QC	Quality Control
QPM	Quality Protein Maize
QUATS	Quaternary ammonium compounds
R	Solvent-to-Seed Flour Ratio

Short Form	*Long Form/Full Form*
R	Regression Coefficient
RA	Risk Analysis
RAPD	Randomly Amplified Polymorphic DNA
RAR	Retinoic Acid Receptor
RBP	Retinol Binding Protein
RCA	Rolling Circle Amplification
RD	Registered dietitian
RDA	Recommended Dietary Allowance
RDI	Recommended dietary intake
RDI	Reference Daily Intake
Rf	Response Factor
RFLP	Restricted Fragment Length Polymorphism
RFLP-PCR	Restricted Fragment Length Polymorphism-Polymerase Chain Reaction
RH	Relative humidity
RI	Refractive Index
RIA	Radio Immuno Assay
RIAU	Research institutions and agricultural universities
RNA	Ribonucleic Acid
RNI	Recommended nutrient intakes
RO	Reverse Osmosis
ROS	Reactive Oxygen Species
RQ	Respiratory Quotient
RSV	Respiratory Syncytial Virus
RTD	Resistance temperature device
RTD	Resistance Thermometer Device
RTE	Ready-to-eat
SALP	Sodium Aluminum Phosphate
SAM	S-adenosyl-L-methionine
SAMB	State agricultural marketing boards
SAPP	Sodium Acid Pyrophosphate
SAS	Statistical Analytical System
SAS	Sodium Aluminum Sulfate

Short Form	*Long Form/Full Form*
SCC	Somatic cell count
SCCMO	Special commodity cooperative marketing organizations
SCF	Supercritical Fluid
SCFA	Short-Chain Fatty Acids
SCMF	State cooperative marketing federation
SCP	Single Cell Proteins
SD	Standard Deviation
SDA	Strand Displacement Amplification
SDAM	State directorate of agricultural marketing
SDF	Soluble Dietary Fiber
SDS-PAGE	Sodium Dodecylsulphate Polyacrylamide Gel-Electrophoresis
SEM	Scanning Electron Micrograph/Microscopy
SEPC	Silk export promotion council
SF	Swelling Factor
SF	Serum Ferritin
SFDA	Small Farm Development Agency
SFEs	Sucrose Fatty Acid Esters
SFPE	Sucrose Fatty Acid Polyester
SGA	Small for gestational age
SGE	Starch Gel Electrophoresis
SH	Sulfhydryl
SHMP	Sodium Hexametaphosphate
SIP	Sterilization-in-Place
SMB	Simulated Moving Bed
SMER	Specific moisture extraction rate
SMEs	Sucrose Monomers
SMFs	Static magnetic fields
SMUF	Simulated milk ultrafilterate
SNIF	Surface of Nasal Impact Frequency
SO	Soybean Oil
SOD	Super Oxide Dismutase
SOPP	Sodium o-phenylphenate

Short Form	*Long Form/Full Form*
SOPs	Standard operating procedures
SP	Sulphoproyl
SP	Storage protein
SPC	Standard plate count
SPS	Sanitary and Phytosanitary
SSL	Sodium Stearoyl Lactylate
SSLL	Sodium Stearoyl Lactoyl Lactate
SSP	Shelf-stable Products
STC	State trading corporation
STP	Standard Pressure and Temperature
SUVs	Small unilamellar vesicles
SWCS	State ware housing corporations
SWM	Standards of Weight and Measures
TAC	Trialkoxycitrate
TAGs	Triacylglycerols
TATCA	Trialkoxy Tricarballylate
TBA	Thio Barbituric Acid
TBHQ	Tetra Butylated Hydro Quinone
TBT	Technical Barriers to Trade
TCA	Trichloro Acetic Acid
TCA	Trichloroanisol/Trichloroacetic acid/Tricarboxylic acid
TCD	Thermal conductivity detector
TD	True Digestibility
TDF	Total Dietary Fiber
TDI	Tolerable Daily Intake
TEF	Toxic Equivalency Factor
TFA	Trifluoro Acetic Acid
TFA	Transfatty acids
TFC	Thin Film Composite
TFR	Tempeh Fish and Rice Mixture
TfR	Transerrin Receptor
THBP	2,4,5-Trihydroxy ButyroPhenone

Short Form	*Long Form/Full Form*
TLC	Thin Layer Chromatography
TMA	Transcription-Mediated Amplification
TMA	Thermo-mechanical analysis
TMA	Trimethylamine
TMAO	Trimethylamine Oxide
TMCT	Thermal mechanical compression test
TNF	Tumor Necrosis Factor
TOFMS	Time of Flight Mass Spectrometry
TOS	Total Organic Solids
TQM	Total Quality Management
TRIFED	Tribal cooperative marketing federation
T-RNA	Tetrahymena Relative Nutritive Value
TS	Total Solids
TSAI	Transition State Analog Inhibitors
TSH	Thyroid-stimulating Hormone
TSH	Thyroid Stimulating Hormone (or thyrotropin)
TSNS	Ten State Nutrition Survey
TSP	Texturized Soya Proteins
TTB	Tax and Trade Bureau
TTI	Time-temperature integrator/Time-temperature indicator
TTT	Time-temperature-tolerance
TVP	Textured Vegetable Proteins
TWT	Traveling Wave Tube
UF	Ultra filtration
UHT	Ultra High Temperature
UI	Urinary iodine
ULO	Ultra-low oxygen
ULV	Unilamellar Vesicles
UNICEF	United Nations International Children Fund
UNICEF	United Nations Children's Fund
UNU	United Nations Union
UP	Utilizable Proteins

Short Form	*Long Form/Full Form*
USDA	United State Department of Agriculture
USFDA	United State Food and Drug Administration
USI	Universal Salt Iodization
USS	United States Standards
UTLIEF	Ultra Thin Layer Isoelectric Focusing
UV	Ultra Violet
V/V	Volume-to-Volume
VAD	Vitamin A Deficiency
VFA	Volatile Fatty Acids
VFD	Variable Frequency Drive
VFMPT	Vacuum freezing multiple-phase transformation
VLCD	Very low calorie diet
VLDL	Very Low Density Lipoproteins
VP	Vacuum packaging
VSP	Vacuum Skin Packaging
VVM	Vessel Volume Ratio
W.B.	Wet bulb temperature
W/V	Weight-to-Volume
W/W	Weight-to-Weight
WARC	World Administrative Radio Conference
WBP	Water Binding Potential
WFI	Water-for-Injection
WHC	Water Holding Capacity
WHO	World Health Organization
WLF	Williams-Landel-Ferry
WOF	Warm-over flavour
WRA	Women of Reproductive Age
WSC	World Summit for Children
WSSN	World Standards Services Network
WTO	World Trade Organization
XME	Xenobiotic Metabolising Enzymes
XOD	Xanthine Oxidase
XRE	Xnobiotic Response Element
Z	Z-score
ZDS	ξ-Carotene Desaturase

APPENDIX–III
Conversion Factors (Weights and Measures)

Length	1 m	10^{10} A	39.37 in	3.2808 ft
Area	1 m^2	10^4 cm^2	1550.0031 in^2	10.7639 ft^2
Volume	1 m^3	1000 L	10^6 cm^3	264.17 gal (US)
Mass	1 kg	2.2046 lbm	35.274 oz	1.1023 x 10^{-3} ton (short)
Density	1 g/cm^3	1000 kg/m^3	62.428 lbm/ft^3	8.3454 lbm/gal (US)
Velocity	1 m/s	3.6 km/h	2.2369 min/h	196.85 ft/min
Force	1 kg-m/s^2 (N)	0.2248 lbf	10^5 g.cm/s^2 (dyn)	7.2330 lbm.ft/s^2 (pdf)
Pressure	1 atm	101.325 kg/ms^2 (pa)	14.696 lbf/in^2 (psi)	760 mm Hg
Energy	1 kg.m^2/s^2 (J)	9.4782 x 10^{-4} Btu	0.2388 cal	0.7376 ft.lbf
Power	1 J/S (W)	1 kg/m^2/s^3	3.4121 Btu/h	1.341 x 10^{-3} hp
Thermal conductivity	1 W/m.k	0.5778 Btu/h.A.°F	2.388 x 10^{-3} cal/s.cm.°C	6.9335 Btu.in/h.ft^2.°F
Specific heat	1 KJ/kg.K	0.2388 cal/g.°C	0.2388 Btf/lbm.°F	185.86 ft.lbf/lbm.°F
Viscosity (dynamic)	1 pa.S	1 kg/m.s	1000 cp	0.6720 lbm/ft.S
Temperature	°C=5/9 (F-32)	°F = 9/5 (°C) + 32	K = °C + 273.15	°R = °F + 45967

APPENDIX–IV
Equipments and Machinery Used in Fruit and Vegetable Processing Industry

Washing Equipments

1. Over-head sprayer of water
2. Continuous woven wear belt
3. Rotating rod-cylinder with helical screw
4. Cement or galvanized iron tank

Sorting and Grading Equipments

1. Continuous broad belt made of woven metal
2. Roller grader
3. Screen grader
4. Rope or cable grader

Coring and Peeling Equipments

1. Core remover
2. Cutting knife
3. Peeling knife
4. Pitting knife
5. Corer or seed remover
6. Potato peeler

Juice/Pulp Extraction Equipments

1. Halving and burring machine
2. Continuous screw expeller press
3. Plunger-type press
4. Roller-type press
5. Grater
6. Crusher
7. Basket press
8. Rack and cloth press
9. Hydraulic press

Straining/Screening Equipments

1. Pulper with brush paddles
2. Filter press

Deaeration and Pasteurization

1. Deaerator
2. Flash-pasteurizer

Bottling Equipments

1. Bottle washing machine
2. Bottle filling machine
3. Crown corking machine

Canning Equipment

1. Cans
   (*a*) Acid-resistant cans
   (*b*) Sulphur resistant cans
2. Vacuum can tester
3. Can reformer
4. Can opener
5. Refractometer
6. Retorts

Equipments for Making Jams, Jellies and Marmalade

1. Steam jacketed kettles
2. Jelmeter/viscosimeter
3. Jelly thermometer
4. Cookers

Driers

1. Tray drier
2. Puff drier
3. Foam mat drier
4. Drum drier
5. Spray drier
6. Vacuum drier
7. Freeze drier

Other Equipments and Accessories

1. Stainless steel sieves
2. Electronic balances
3. Avery weighing scale
4. Salometer
5. Preparation tables
6. Aprons, rubber gloves
7. Laboratory glassware

APPENDIX–V
Addresses for Suppliers of Equipments/Machinery

Food and Beverage Processing/Preservation Equipments/ Machinery

M/s. Adam Fabriwerk Pvt. Ltd., 203, Rajguru Apts., New Nagardas Road, Andheri (E) Mumbai-400069, Maharashtra

M/s. A. E. C. Instruments, 6, Akashdeep Welfare Society, Near Salvi Chawl, Shankarwadi, Jogeshwari (E), Mumbai – 400 060, Maharashtra.

M/s. Akshay Industries, 101, Rajshree Indl. Estate II, Chitalsar Manpada, Ghodbunder Road, Thane-400607, Maharashtra

M/s. Alok Technical & Marketing Service Pvt. Ltd., P.O. Box-8, Chandra Bldg., Kalkaji Temple, New Delhi-110019

M/s. Alven Foodpro System (P) Ltd., 24, Goldfield Plaza, 45, Sassoon Road, Pune-411 00 1, Maharashtra

M/s. Apple Bakery Machinery Pvt. Ltd. 20/1, Hansraj Damodarwadi, 1st Flr. Off. Kennedy Bridge Opera House, Mumbai – 400004, Maharashtra.

M/s. Asian Chemical Works (Bombay) Pvt. Ltd., Asian House, 29, Ramakrishna Mandir Road, Andheri-Kurla Road, Andheri (E), Mumbai-400059, Maharashtra

M/s. B. Sen Barry and Co., 65/11,New Rohtak Road, Karol Bagh, New Delhi-110005

M/s. Bajaj Maschinen Pvt. Ltd., 0-14, Lajpat Nagar-II, New Delhi-110024

M/s. Baker Enterprises, 23, Bhera Enclave, Near Peera Garhi, New Delhi-110087

M/s. Cantech Machines, 13, Vora Bhavan, Maheshwari Udyan, King's Circle, Matunga (c. Rly), Mumbai-400019, Maharashtra

M/s. OSI Industries, C-10, Devatha Plaza, Residency Road, Bangalore-560025, Karnataka

M/s. Eastend Engineering Co., 173/1, Gopal Lal Thakur Road, Kolkata-700016, West Bengal

M/s. Environmental Products (India) Pvt. Ltd. 160/3, Rajani House, Opp. Don Bosco School, L. T. Road, Borivali (W), Mumbai – 400 091, Maharashtra.

M/s. Fab Flavours & Fragrances Pvt. Ltd. S. P. Mukerjee Marg, New Delhi-110006.

M/s. Filtron Engineers Ltd., 117-A, Vithalwadi Road, Pune-411030, Maharashtra

M/s. FMC Food Tech, 7, Shivaji Housing Society, Pune-411053, Maharashtra

M/s. Frigoscandia Winner Food Process System Ltd., Shreesh Chambers, 3rd Floor, 25/1, Yeshwant Niwas Road, Indore-452003, M.P.

M/s. Ganesh Benzoplast Ltd. Ganesh House, A1/A2, Gurudutt CHS Ltd. Near Jankalyan Bank, J. B. Nagar, Andheri (E), Mumbai-400059, Maharashtra.

M/s. Gardners Corporation, 6, Doctors Lane, Near Gole Market, Post Box 299, New Delhi- 110001

M/s. Guru Nanak Engg. & Foundary Works (Regd.), 166, Indl. Focal Point, Mehta Road, Amritsar-143001, Punjab

M/s. Hindusthan National Glass & Industries Ltd. 2, Red Cross Place, Kolkata-700 001 (W.B.).

M/s. HMT Ltd. (Food Processing Mlc Div.), H-2, MIDC, Chikalthana Indl. Area, P.B. No. 720, Aurangabad-431210, Msharashtra

M/s. Ion Exchange (India) Ltd. Ion House, Dr. E. Moses Road, Mahalaxmi, Mumbai-400011, Maharashtra.

M/s. Jay Chem Marketing, 101, Labh Sarita, Opp. Manek Nagar, M.G. Road, Kandivali (W), Mumbai-400067, Maharashtra.

M/s. K.S. Seetharamiah & Sons Pvt. Ltd., 29/1, Jaraganahalli, 10th Km. Karakapura Road, Bangalore-560078, Karnataka

M/s. Malladi Specialities Ltd. 9, GST Rd, St. Thomas Mount, Chennai-600016.

M/s. Mather and Platt (lndia) Ltd., 805-806, Ansal Bhawan, 16, Kasturba Gandhi Marg, New Delhi-110001

M/s. M.M.M. Buxabhoy and Co., 141-A, Sarang Stree, 1st Floor Near Crawford Market, Mumbai-400003, Maharashtra.

M/s. Micron Industries Pvt. Ltd. R-710, TTC Indutrial Area, MIDC Rabale, Navi Mumbai, Maharashtra.

M/s. Mojj Engg. Systems (P) Ltd., 81-B/15, M.LD.C., Opp. Morris Electronics, Bhosari, Pune-411026, Maharashtra

M/s. Mukul Brothers Engg. Works, P.B. No. 325, Kishan Flour Mill Compound, Tirthankar Mahavir Mart (Rly. Road), Meerut-250002, Ll.P

M/s. Pharmalab Engg. India Ltd., Star Metal Compound, L.B.S. Marg, Vikhroli (W), Mumbai-400083, Maharashtra

M/s. Quasar Engineers, Plot No. 53, Sector 'A' Indl. Area, Sanwer Road, Indore-452003, M.P.

M/s. Satellite Plastic Industries, 2A, Court Chambers, 35 New Marine Lines, Mumbai-400020, Maharashtra.

M/s. Shiva Engineers, Patel Avenue, Plot No. 165, Flat No. 1, Right Bhusari Colony, Paud Road, Pune-411038, Maharashtra.

M/s. Sri Rajalakshmi Commercial Kitchen Equipments Pvt. Ltd. 57, (1st Floor) Silver Jubilee Park Road, Bangalore-560002, Karnataka.

M/s. Sri Rajalakshmi Industrial Agency, 57(30/1), Silver Jubilee Park Road, Rajalakshmi Corner House, Post Box No. 6690, Bangalore- 560002, Karnataka

M/s. Stern Ingredients India Pvt. Ltd. 211 Nimbus Center, Oberoi Complex Off. Link Road, Andheri West, Mumbai – 400 053, Maharashtra.

M/s. Raylons Metal Works, Ramakrishna Mandir Road, Kondivita Village, opp. Marol Bazar, J.B. Nagar, Andheri (E), Mumbai-400059, Maharashtra

M/s. Riddhi Pharma Machinery Ltd. Om-Shivam Buiding 2, Tarun Bharat Sahar Road, Andheri (E), Mumbai-400009, Maharashtra.

M/s. Rita Bottling Machines Ltd., 1B & 1C, Suvarna Darshan, 47, 2nd Main Road, Gandhi Nagar, Adyarn, Chennai-600020, Tamil Nadu

M/s. Uma Brothers, C-110, Bhaveshwar Plaza, 189, L.B.S.Marg, Ghatkopar (W), Mumbai-400086, Maharashtra.

M/s. Ultra International Ltd. 304, AVG Bhavan, M-3 Connaught Circus, New Delhi – 110001.

M/s. V.S. Enterprises, Gala No. 7, Laxmi Industrial Estate Penkar Pada Road, Near Dahisar Check Naka, Mira, Thane-401104, Maharashtra.

M/s. Veenu Hitech, Plant Manufacturing Pvt. Ltd., F-6 St. Soldier Tower, G-Block, Vikas Puri, New Delhi-110 018

Canning Equipments/Machinery

M/s. Asian Consolidated Industries Ltd., "Asian House", 0-193, Okhla Indl. Area, Phase-I, New Delhi-110020

M/s. Ganga Singh Engg. Works P, Ltd., No.1, Vishal Indl. Estate, Village Road, Bhandup (W) Mumbai- 400078, Maharashtra

M/s. Quality Equipment Co., 89, Mahavira Street, Haiderpur Indl. Area Delhi-110052

M/s. Recon Machine Tools Pvt. Ltd., 37, Sarvodaya Indl. Estate, Mahakali Caves Road, Andheri (E), Mumbai-400093, Maharashtra

M/s. Sangram Engg. Ltd., B-5, Super Con, Opp. LT.L, Aundh, Pune-411007, Maharashtra.

M/s. Thakar Equipment Co., 66, Okhla Indl. Estate, New Delhi-l10020

Drying/Dehydration Equipments/Machinery

M/s. Admir Enterprises, Plot No. IIE, 4, Shivaji Nagar Govandi, Mumbai-400043.

M/s. Aifso Industrial Equipment Co., B/13, Veena Beena Apts, P. Thakrey Marg, Sewri (W), Mumbai 400015, Maharashtra

M/s. Aratic India Engg. Pvt. Ltd., 20, Rajpur Road, Delhi-110054

M/s. Bombay Industrial Engineers. 430, Hind Rajasthan Chambers, D.S. Phalke Road, Dadar (C Rly.), Mumbai-400014. Maharashtra

M/s. Bry-Air India Pvt. Ltd., 419-420, Udyog Vihar, Phase-Ill, Gurgaon, Haryana

M/s. Burman Plant & Machinery Co. Pvt. Ltd., 36, Sarkar Lane, Kolkata-700007, West Bengal

M/s. CM. Equipments & Instruments (India) Pvt. Ltd., B-194, 5th Main Road. P.B. No. 5847, Peenya 2nd Stage, Bangalore-560058. Karnataka

M/s. Eec Cee & Co., 1, Anant Indl. Estate, Opp, Cemet Fruit & Chemicals, Rakhial Ahmedabad-380023, Gujarat

M/s. Orbit International Technologies (P) Ltd., 404, Taramandal Complex, Secretariat Road, Hyderabad -500004, A.P.

M/s. Ratan Equipments 69, Lake View Road, Kamakati St., West Mambalam, Chennai-600003, Tamil Nadu

M/s. Raylon Engg. Works, 31 A, Ghanshyam Indl. Estate., Near Veera Desai Road, Andheri (W), Mumbai-400058, Maharashtra

M/s. Wintech Taparia Ltd., 25/1, Yeshwant Niwas Road, Shreesh Chambers, III Floor, Indore-452003, M.P.

Product Specific Equipments/Machinery

Jam Making

M/s. Foram Foods Pvt. Ltd., 397, Swami Vivekanand road, Vile Parle (W), Mumbai-400056, Maharashtra

M/s. Inventure India BV, 24, Gold Field Plaza, 45, Sasso on road, Pune-411001, Maharashtra

Pickle and Chutney Making

M/s. Geeta Food Engineering, Plot C-7/1 TIC Area, Pawana M.ID.C., Thanc-Belapur Road, Behind Savita Chemical Ltd., Navi, Mumbai-400705, Maharashtra

M/s. Techno Equipments, Saraswati Sadan, Girgaum Court, Girgaum, Mumbai-400004, Maharashtra

Tomato Processing

M/s. Goma Engineering Pvt. Ltd., Majiwada, Behind Universal Pertol Pump, Thane-40060 1, Maharashtra

M/s. K.S. Seetharamiah & Sons. Pvt. Ltd., 29/1, Jaraganahalli, 10th Km. Kanakapura Road, Bangalore-560078, Karnataka

M/s. Penwalt India Ltd., 507, Kakad Chambers, 32, Dr. Annie Besant Road, Worli, Mumbai-400018, Maharashtra

Mushroom Processing

M/s. Jwala Engg. Co., 12, Surve Indl. Estate, Sonawala Cross Road No.1, Goregaon (E), Mumbai-400063, Maharashtra

M/s. Wintech Taparia Ltd., 25/1, Yeshwant Niwas Road, IIIrd Floor, Shreesh Chambers, Indore-452003, M.P.

Wrapping/Filling/Packaging Equipments/Machinery

M/s. Aarkay Wrapping Machines Pvt. Ltd., 1, Hormurz, 131, August Kranti Marg, Mumbai-400036, Maharashtra

M/s. Abhay & Abhay Pvt. Ltd., B-84-1, Okhla Indl. Area, Phase-Il, New Delhi-ll0020

M/s. APT Packaging, 184/5007, Pantnagar Ghatkopar (E), Mumbai-400075, Maharashtra

M/s. Bombay Engineering Industry, R. No.6 (Estns.), Sevanthibhai Bhavan, Chimatpada, Marol Naka, Andheri (E), Mumbai-400059, Maharashtra

M/s. Compack Systems, E-211, F.F. Complex, Okhla III, New Delhi-ll0020

M/s. Debes Industries, 11, Govt. Place East, Kolkata-700069, West Bengal

M/s. E.C. Packaging Pvt. Ltd., 14/7 Mile Stone, Mathura Road Faridabad- 121003, Haryana

M/s. Europack Machines (I) Pvt. Ltd., Akash Business Centre, CST Road, Kurla (W), Mumbai-400070, Maharashtra

M/s. Multipack Systems Pvt. Ltd., 2nd Floor, Patrict Complex, Ellora Park, Vadodara-390007, Gujarat

M/s. Packaging Machines of India, 8-3-229/5/3/7, Jai Bharani Nagar Yusufguda Check Post, Hyderabad, A.P.

M/s. Reliance Packaging, 155-A, Ekta Enclave, Peeragarhi, Main Rohtak Road, New Delhi–110041

Refrigeration/Cold Storage Equipments/Machinery

M/s. Airtech Engineers, B-19, Okhala Indl. Area, Phase-II, New Delhi-110020

M/s. Anand Refrigeration Co. Pvt. Ltd., D-Lajpat Nagar-I, New Delhi-110024

M/s. Blue Star Ltd., Kasturi Bldg., Mahan T. Advani Chowk, Jamshedji Tata Road, Mumbai-400020, Maharashtra

M/s. Carrier Refrigeration (P) Ltd., 700/701 A, Lado Sarai, Aurobindo Marg, Mehrauli, New Delhi-11 0030

M/s. Greenfield Agencies, Shop No.6, Anuradha, Opp. Golf Club, Rest House, Tidke Colony, Nasik-422002, Maharashtra

M/s. Industrial Refrigeration Pvt. Ltd., 901, Maker Chambers-V, Nariman Point, Mumbai-400021, Maharashtra

M/s. Kooling System, 35, Mannar Reddy Street, T. Nagar, Chennai- 600017, Tamil Nadu

M/s. Western Refrigeration Ltd., 4, Ready Money Terrace, Dr. A.B. Road, Worli, Mumbai-400018, Maharashtra

Packaging Materials

M/s. Alu Foil Products Pvt. Ltd., 208, Mukti Chambers, 2nd Floor, 4, Clive Row, Kolkata-700001, West Bengal

M/s. Arora Box & Carton Pvt. Ltd., HR 12, Gali No.10, Anand Parbat Indl. Area, New Delhi-II 0005

M/s. Jain. Flexipack Pvt. Ltd., M.A. C-9/9, Sector-5, Rohini, New Delhi-110085

M/s. Jayco Trading Co., 506, Kakad Market Bldg., 306, Kalbadevi Road, Mumbai-400002, Maharashtra

M/s. Oriental Containers Ltd., P.O. Box No. 6584, 1076, Dr. E. Moses Road, Worli, Mumbai-400018, Maharashtra

M/s. Packwell Industries 3746, Netaji Subhash Marg, New Delhi-110 002

M/s. Panchmahal Polypack (P) Ltd., 36, Mithila Society, Opp. Shreyas School, Ambawadi, Ahmedabad-380015, Gujarat

M/s. Swan Packaging, 53/54, Unique Indl. Estate, Dr. RP. Road, Mulund (W), Mumbai-400080, Maharashtra

M/s. Vindhyachal Process Corporation, 116, Malviya Nagar, Bhopal-462003, M.P.

Books

M/s. Daya Publishing House, 4760-61/23, Ansari Road, Darya Ganj, New Delhi-110002, Phone: +91-011-23244987, E-mail: dayabooks@vsnl.com; Website: www.dayabooks.com

APPENDIX–VI
Project Reports (Standard Format) of Processing Industries

1. Primary Information

(*a*) Name of Unit: ....

(*b*) Units' address: ....

(*c*) Constitution: ....

(*d*) Name of proprietor: ....

(*e*) Birth Date: .... Age: ....

(*f*) Qualification: ....

(*g*) Nature of business: ....

(*h*) Registration No. of business: ....

(*i*) Loan for: ....

(*j*) Loan amount: ....

2. Introduction

In this section the information on following points should be included:

(*a*) Experience of the proprietor in this field

(*b*) Facilities and arrangement for product sale

(*c*) Information regarding machinery requirement and their availability as well as importance

(*d*) Availability of the skilled and unskilled labour

(*e*) Information regarding self-sale counter

(*f*) Importance of this business to the society

3. Scope of Business

Give the location of the processing unit and availability of raw material in that area. Provide the information regarding marketing of the processed product.

4. Manufacturing Process

Give the details of manufacturing process/technology employed.

5. Land

Land for proprietor owns unit or it is in partnership give details. If in partnership then make agreement and submit with this proposal.

6. Building

Give the details about building construction (shed, RCC or load bearing etc.).

7. Water Supply

Water supply from bore well or Municipal Corporation etc.

8. Electric Power Supply

MSEB connection deposit, main switchboard, wiring, fitting, tubes, fans, labour charges etc.

9. Printing

For labeling, advertisement, specification on the product etc.

10. Furniture

Expenditure occurred on tables, chairs, cupboards, stools etc.

11. Quantitative Details

(Rs. in Lacs)

Sl.No.	*Description*	*1st Year*	*2nd Year*	*3rd Year*	*4th Year*	*5th Year*
1.	Installed capacity 300 days & 20 hrs					
2.	Utilization of capacity (per cent)					
3.	Raw material consumption at installed capacity (Rs./Ton)					
4.	Sale at installed capacity (Rate Rs./Ton)					
5.	Raw material consumption at utilized capacity					
6.	Sales at utilized capacity					

12. Calculation of Working Capital

Sk.No.	Particular	. Holding Period (days)	Amount
1.	Raw material		
2.	Consumable packing		
3.	Finished goods		
4.	Other Exp.		
5.	Debtors		
Gross current assets:			
	Less credits		
	Net working capital		
	Own margin	35 or 25 per cent	
	Bank Finance	56 or 75 per cent	

13. Cost of Project and Sources of Finance

Sl.No.	Item	Cost Already Incurred	Cost to be Incurred	Total
A.	**Cost of project**			
1.	Land			
2.	Site development			
3.	Civil & Technical work			
4.	Plant & Machinery			
5.	Total working capital			
6.	Preliminary & preoperative Exp. (Bank fee, mortgage fee, Prof. Fee etc.)			
	Total			
B.	**Source of finance**			
1.	Equity capital			
2.	Subsidy/Quasi equity			
3.	Others			
4.	Unsecured Loan			
5.	Term Loan			
	Total			

14. Project Profit and Loss Account

(Rs. in Lacs)

Sl.No.	Particular	1St Year	2nd Year	3rd Year	4th Year	5th Year
1.	Gross sales					
2.	Job work charges					
A.	*Net sales*					
B.	**Cost of production**					
1.	Raw material (as per qty. details)					
2.	Power & Fuel					
3.	Direct labour & Wages					
4.	Consumables stores					
5.	Repair & Maintenance					
6.	Other Mfg. Exp.					
7.	Depreciation					
8.	Add opening stock in process & finished goods					
9.	Deduct closing stock in process & finished goods					
C.	Cost of sales					
D..	Gross profit (A- C)					
E.	Interest					
F.	General & administrative over					
G.	Profit before Tax					
H.	Provision for Tax					
I.	Net Profit					
J.	Depression added back					
K.	Net cash accruals					
L.	**Repayment obligations**					
M.	Debt service ratio (Average DSCR 3.87)					

15. Projected Balance Sheet

(Rs. in Lacs)

Sl.No.	Particular	1st Year	2nd Year	3rd Year	4th Year	5th Year
A.	**Assets**					
1.	Fixed assets					
2.	Investment/L. C. Deposits					
3.	Current assets					
	a. Raw material, packing Exp.					
	b. Finished goods					
	c. Debtors					
	d. Other current assets					
	Total					
B.	**Liabilities**					
1.	**Equity capital**					
2.	Profit & loss A/C					
3.	Unsecured Loans					
4.	Quasi equity					
5.	Subsidy (Fr. MFPI)					
6.	Secured Loans					
7.	Working limits Fr. Bank					
	a. Term Loan					
	b. Creditors					
	c. Other current Liabilities & Provision					
	Total					

16. Projected Cash Flow and Fund Flow Statement

(Rs. in Lacs)

Sl.No.	Particular	1st Year	2nd Year	3rd Year	4th Year	5th Year
A.	**Sources**					
1.	Net cash accruals					
2.	Subsidy					
	Total					
B.	**Application**					
1.	**Repayment of Term Loan**					
2.	Repayment of unsecured & quasi equity					
3.	Increase in investment					
4.	Increase in W. C.					
	Total					

17. Projected Breakeven

(Rs. in Lacs)

1.	Fixed cost (Average)
2.	Average selling price
3.	Average variable cost (per ton)
	a. Materials
	b. Labours
	c. Other overheads
4.	Contribution/Unit
5.	Breakeven qty
6.	Breakeven value
7.	Percentage on installed capacity

18. Repayment Schedule of Term Loan and Interst on Term Loan

Period	*Principle Amount*	*Interest*	*Total*
Interest in ... per cent, period in quarters			
1st Year			
I Quarter			
II Quarter			
III Quarter			
IV Quarter			
Total			
2nd Year			
I Quarter			
II Quarter			
III Quarter			
IV Quarter			
Total			
3rd Year			
I Quarter			
II Quarter			
III Quarter			
IV Quarter			
Total			
4th Year			
I Quarter			
II Quarter			
III Quarter			
IV Quarter			
Total			
5th Year			
I Quarter			
II Quarter			
III Quarter			
IV Quarter			

19. Calculation of Depreciation as per IT Rules and WDV Method

Period	*Building*	*Plant and Machinery*	*Total*
Initial			
1st Year			
Total			
2nd Year			
Total			
3rd Year			
Total			
4th Year			
Total			
5th Year			
Total			

20. List of Plant and Machinery

Sl.No.	*Name of Plant or Machinery*	*Price (Rs.)*	*Quantity (Nos.)*	*Total Amount (Rs.)*
1.				
2.				
3.				
4.				
5.				
6.				
7.				
8.				
9.				
10.				
	Total			

21. Total Interest Payable Yearwise

Period	*Term*	*Working*	*Other*	*Total*
@p. a.	11 per cent	15 per cent	18 per cent	
1st year				
2nd year				
3rd Year				
4th Year				
5th Year				
Total				

22. Salary and Wages Details

Sl.No.	*Particular*	*Nos.*	*Salary P.M.*	*Total*
1.	Operator			
2.	Skilled workers			
3.	Unskilled workers			
	Total			

Proprietor/Partner

(Name & Address of the Unit)

Index